**Everyone knew it was crazy to try to extract
oil and natural gas buried in shale rock deep**

The Frackers also tells the story of the angry opposition unleashed by this
revolution, and explores just how dangerous fracking really is.

ABOUT THE AUTHOR

Gregory Zuckerman is a special writer at the *Wall Street Journal* and the bestselling author of *The Greatest Trade Ever*. He is a two-time winner of the Gerald Loeb Award and a winner of the New York Press Club Journalism Award. He lives in New Jersey with his wife and two sons.

Further praise for *The Greatest Trade Ever*:

'A must-read for anyone fascinated by financial madness'
Mail on Sunday

'A forensic, read-in-one-sitting book' *Sunday Times*

'Extraordinary, excellent' *Observer*

'Compelling' *Economist*

'The most successful entrepreneur on Wall Street – certainly of the past decade and perhaps even of the postwar era – is a hedge fund manager named John Paulson . . . Gregory Zuckerman's account of Paulson's triumph offers a fascinating perspective on the predator thesis. Rarely in human history has someone made so much money in so short a time'
Malcolm Gladwell, author of *The Tipping Point, Blink* and *Outliers*

'Gritty . . . Paulson's achievement stands as a landmark in an era of folly. Zuckerman skillfully captures the mounting tension of Paulson's wait to be proved right and the greed-driven delusion of everyone in the market who believed he was wrong' *Mail on Sunday*

'Zuckerman takes us to Wall Street's heart of darkness, where mushroomed a $1 trillion subprime mortgage market that only the few, the brave, the smart dared short. This is at once a great page-turner and a great illuminator of the market's crash'
John Helyar, co-author of *Barbarians at the Gate*

'Greg Zuckerman was the first to tell the world about John Paulson's sensational trade . . . He's written the definitive account of a strange and wonderful subplot of the financial crisis'
Michael Lewis, author of *Liar's Poker* and *The Big Short*

GREGORY ZUCKERMAN

THE
FRACKERS

The Outrageous Inside Story of the
NEW ENERGY REVOLUTION

PORTFOLIO
PENGUIN

PORTFOLIO PENGUIN

Published by the Penguin Group
Penguin Books Ltd, 80 Strand, London WC2R 0RL, England
Penguin Group (USA) Inc., 375 Hudson Street, New York, New York 10014, USA
Penguin Group (Canada), 90 Eglinton Avenue East, Suite 700, Toronto, Ontario, Canada M4P 2Y3
(a division of Pearson Penguin Canada Inc.)
Penguin Ireland, 25 St Stephen's Green, Dublin 2, Ireland (a division of Penguin Books Ltd)
Penguin Group (Australia), 707 Collins Street, Melbourne, Victoria 3008, Australia
(a division of Pearson Australia Group Pty Ltd)
Penguin Books India Pvt Ltd, 11 Community Centre, Panchsheel Park, New Delhi – 110 017, India
Penguin Group (NZ), 67 Apollo Drive, Rosedale, Auckland 0632, New Zealand
(a division of Pearson New Zealand Ltd)
Penguin Books (South Africa) (Pty) Ltd, Block D, Rosebank Office Park,
181 Jan Smuts Avenue, Parktown North, Gauteng 2193, South Africa

Penguin Books Ltd, Registered Offices: 80 Strand, London WC2R 0RL, England

www.penguin.com

First published in the United States of America by Portfolio Penguin,
a member of Penguin Group (USA) LLC 2013
First published in Great Britain by Portfolio Penguin 2013
001

Printed in Great Britain by Clays Ltd, St Ives plc

ISBN: 978–0–670–92367–0

www.greenpenguin.co.uk

To Michelle

For your support, humor, and love

CONTENTS

CAST OF CHARACTERS

MITCHELL ENERGY

George Mitchell *Founder*
Dan Steward *Senior executive on the shale-drilling team*
Nicholas Steinsberger *Engineer focused on hydraulic fracturing*
Kent Bowker *Exploration geologist on the shale-drilling team*
Jim Henry *Veteran geologist*

ORYX ENERGY

Robert Hauptfuhrer *Chief executive*
Kenneth Bowdon *Geologist*

CHESAPEAKE ENERGY

Aubrey McClendon *Cofounder*
Tom Ward *Cofounder*
Marc Rowland *Chief financial officer*

CONTINENTAL RESOURCES

Harold Hamm *Founder*
Jeff Hume *Senior executive and engineer*
Jack Stark *Senior executive and geologist*

CHENIERE ENERGY

Charif Souki *Founder*
Meg Gentle *Head of strategic planning*

EOG RESOURCES

Mark Papa *Chief executive officer*
Bill Thomas *Senior executive*

OTHERS

Sanford Dvorin *Son of Newark butcher; tried to drill in the Barnett Shale*
Ray Galvin *Senior executive at Chevron determined to find gas in shale*
Michael Johnson *Septuagenarian who discovered the largest field in the Bakken*
Buck Butler *Cowboy in Nixon, Texas, sitting on shale*
Elizabeth Irish *Oregon native who moved to Williston, North Dakota, with her family*

INTRODUCTION

William Butler was up nights, full of worry.

The grizzled eighty-three-year-old rancher in South Texas owed millions of dollars to various lenders, had almost nothing in the bank, and feared his two sons wouldn't be able to manage when he was gone.

Butler had the look of someone just off the set of a John Ford movie. Tall and broad, he tended his cattle in a flannel shirt, blue jeans, and muddy boots. He went by the nickname "Buck," spent seven days a week working with thousands of cows on his ranch, and in his old age relied on a walking stick made from the manhood of a two-thousand-pound Brahma bull.

Buck Butler was no cinema hero, however. A series of local schemers and connivers had taken advantage of him over the years. Butler compounded his problems by plowing all his free cash into nearby land, usually telling his nervous wife, Vera, about the purchases only after they had been completed.

By 2009, Butler faced growing difficulties with his business and was coping with a nervous system disorder. Vera began taking medication to calm her own nerves.

"When you owe over three million dollars you worry," she explains.

Less than four years later, on a warm January afternoon in 2013, I bounced around the front seat of Butler's new Dodge pickup as he told the story of how his difficult days had come to an end and a new life had begun.

Pointing to his rolling acreage like a proud parent, Butler described how one day—just over two years earlier—a representative of Conoco-Phillips had come knocking on his office door to ask if the huge oil company could drill on his property. It turned out that a type of rock called shale was buried more than a mile below its surface. The rock was soaked with oil that suddenly had become accessible. Almost overnight, Butler's land was transformed into some of the most valuable acreage in the world.

Butler parked his truck outside a Mexican restaurant in Nixon, Texas, population twenty-four hundred, and turned to me with piercing blue eyes.

"It's goddamn unbelievable what's happened to me in the last two years," he said, a smile of relief forming on his rugged face. "I have to reach out and pinch myself, it's too good to be true."

I was a business reporter from New York on a visit to South Texas in search of a story. My crisp blue Yankees cap seemed to clash with Butler's scuffed cowboy hat, and his honeyed Texas twang sometimes sounded like an entirely new language. I had spent a career at the *Wall Street Journal* writing about men and women who traded stocks and bonds, not livestock. Before I began my research, "frack" was the kind of word I'd caution my kids to avoid.

At that moment, though, I was sure my Marlboro Man's tale, and the stories of others I had heard in places like Williston, North Dakota, New Milford, Pennsylvania, and Lexington, Oklahoma, were among the most compelling a writer could hope to find. Buck Butler and others at the heart of one of the greatest energy revolutions in history had experienced astonishing and unexpected change thanks to American oil and gas discoveries deemed unthinkable just a few years ago. The nation itself has been transformed, as has the world.

The more work I put into the topic, the clearer it became that a burst of drilling in shale and other long-overlooked rock formations had created the biggest phenomenon to hit the business world since the housing and technology booms. In some ways, the impact of the energy bonanza might be even more dramatic than the previous expansions, especially if shale drilling catches on around the globe. Surging oil and gas production likely will affect governments, companies, and individuals in remarkable ways for decades to come.

Consider the following:

- As recently as 2006, business and government leaders fretted that America was running out of energy. By 2013, however, the United States was producing seven and a half million barrels of crude oil each day, up from five million in 2005. The country enjoyed its largest production increase in history in 2012 and could pump more than eleven million barrels a day by 2020, its highest figure ever and more than even Saudi Arabia currently produces.

  So much oil is flowing that in a few more years, the United States may not need to import any crude, or might only rely on friends such as Canada and Mexico, ending a fifty-year addiction to oil from countries with interests that many years ago diverged from ours. In 2013, Saudi Arabia's billionaire prince Alwaleed bin Talal said the kingdom's oil-dependent economy had become vulnerable to rising U.S. energy production, a shocking turnaround from a few years ago when America seemed hopelessly dependent on Middle Eastern oil.

- America already is the world's largest producer of natural gas, thanks to shale drilling, and the country sits on two of the world's largest gas fields. Gas production has soared 20 percent in five years, and the United States now should have enough gas to last generations. Soon, the nation will begin *exporting* gas, an unimaginable possibility just a few years ago when energy supplies looked set to run out and the construction of gas importing facilities was considered a matter of national urgency.

- Rising production from dense rock has sent natural gas prices tumbling 75 percent since 2008. Because gas is used to heat and cool homes, produce electricity, grow food, power some vehicles, and make plastic, steel, and other products, the American gusher has been a boon to consumers and businesses, many of whom are still suffering from the worst economic downturn since the Great Depression. Meanwhile, growing U.S. oil production has allowed the country to enforce a boycott of crude from Iran at relatively little cost, and it could help keep a lid on global prices for years to come.

- The energy boom could generate more than two million new jobs by 2020, offsetting the jobs lost in the housing market's collapse. Hiring is on the rise in Texas, Oklahoma, and Louisiana, as well as in Ohio, Wyoming, West Virginia, and Pennsylvania, a shot in the arm for many long-struggling regions. North Dakota enjoys an unemployment rate of about 3 percent, a Walmart in the heart of the state's oil region pays employees twenty-two dollars an hour, and some local McDonald's outlets have resorted to offering bonuses of $300 and thirty-two-inch flat-screen televisions to lure new employees.[1]

- Electricity and natural gas prices are so much cheaper in the United States than in most other countries that they could help usher in a new era of American economic dominance. A "reshoring" trend already is under way, as steel, chemical, fertilizer, plastics, tire, and other companies move production back to the country or expand existing factories, while foreign firms build new plants in the United States. The shift is helping to bring back some jobs once believed to have been lost forever to China and other low-cost economies. Some even see a manufacturing rebound in the making as "made in the USA" again becomes de rigueur.

- All the newfound oil and gas, along with expected energy exports, could boost the value of the U.S. dollar and reduce the nation's trade deficit. The explosion of oil also has defense specialists debating whether the United States may be able to avoid certain future military actions aimed at securing energy supplies, allowing the country to trim its bloated defense budget and perhaps cede some security responsibilities to other countries, like China, that remain dependent on Middle East oil production.

- China, Russia, Argentina, and Mexico are among the countries with their own deep pockets of shale that may be tapped in the years ahead, while government officials in the United Kingdom and elsewhere have urged their countries to embrace shale drilling. In fact, global gas production could rise by 50 percent by 2035, some analysts say, helping consumers and businesses around the world.

But troubling questions have been raised about the environmental consequences of some of the production methods responsible for soaring oil and gas supplies, including hydraulic fracturing, or "fracking." Some worry about their impact on air and water quality, while others are concerned that fracking may contribute to climate change or lead to tremors and other disturbances. Hollywood starlets, rock stars, and media moguls, including Yoko Ono, Sean Lennon, Alec Baldwin, Paul McCartney, and Scarlett Johansson, have become activists on the hot-button issue, which figures to dominate headlines for years to come. The Rolling Stones even wrote a song, "Doom and Gloom," that disparages fracking.

Many of the environmental threats can be addressed or are overstated. But progress has been too slow, there have been damaging mishaps, and some say there's too little regulation. While soaring natural gas production actually might help alleviate global warming by reducing demand for dirtier coal in places like China, the resurgence of fossil fuels threatens to sap interest in still cleaner alternative energy sources. The full impact of the new drilling may take years to be fully understood, and some companies continue to resist sharing full details of how they fracture rock to get all that oil and gas, adding to the unease.

The shifts that have taken place in the United States, and those on the way around the globe, are in many ways not nearly as astonishing as the story of how a small group of individuals made it all happen, against all odds. These modern-day wildcatters ignored the skepticism and derision of experts, major oil companies, and even colleagues to drill in rock they believed was packed with oil and gas miles beneath the earth's surface. These men have altered the economic, environmental, and geopolitical course of the world while scoring some of the swiftest windfalls in history.

George Mitchell, who discovered a novel way to extract gas from shale formations, pocketed more than $2 billion. His impact eventually might even approach that of Henry Ford and Alexander Graham Bell. Aubrey McClendon and Tom Ward turned $50,000 into one of the nation's largest natural gas producers, one that would control the mineral rights to fifteen million acres, around three times the area of New Jersey. Mark Papa built a $43 billion oil power from the discards of the disgraced Enron Corporation.

Another pioneer, Harold Hamm, amassed a fortune of more than $12 billion, making him one of America's richest individuals. Hamm, who owns more oil in the ground than any American, has more wealth than Rupert Murdoch, Steven Cohen, Sumner Redstone, or even the estate of Steve Jobs. Hamm's ongoing divorce likely will set a record for the costliest in history and could leave his wife with more money than Oprah Winfrey. Even Hamm's right-hand man is worth as much as $400 million, a sign of the outsized profits racked up by innovators of the age.

Some of the architects of the "shale gale" were upended by a revolution they themselves helped create, however. They would see fortunes slip through their fingers and experience ridicule and scorn rather than wealth and admiration. And if the worst fears about the drilling are borne out, those at the forefront of the movement will be remembered for the damage they wrought rather than the blessings they bestowed.

But how did a few unlikely, ambitious, and headstrong wildcatters—some without college degrees or much background in geology or drilling—manage to tap massive energy deposits dismissed by the largest energy powers? ExxonMobil's corporate headquarters are directly above a huge shale formation, but the oil giant disregarded the area, even as George Mitchell worked on coaxing historic amounts of gas from rock in the region.

Why did a new age of energy emerge from the depths of the Great Recession, even as Federal Reserve chairman Alan Greenspan warned of dwindling U.S. supplies, investors Warren Buffett and Henry Kravis bet on a dearth of natural gas, and Vladimir Putin predicted a Russian gas monopoly?

Why did private enterprise revitalize the nation's energy outlook with a focus on fossil fuels, of all things, even as governments funneled $2 trillion toward cleaner, alternative energy? How did obscure energy entrepreneurs develop technologies to produce a surge of energy, even as a chorus of experts, including Peter Thiel, an original investor in Facebook, derided the country for no longer making dramatic technological advances? And why did it all happen in the United States and not in China, Russia, or other countries that boast their own enormous deposits of oil and gas in similar rock?

This book, based on over three hundred hours of interviews with more than fifty of the key players of the era, attempts to answer these questions. It also anticipates how the new age of energy might influence global financial markets, economies, military activities, and international politics. Government experts charged with developing energy policy were caught flat-footed by the dramatic recent shifts, as were environmental specialists and top oil-and-gas executives, suggesting that there is much to learn from those who managed to lead the shale revolution.

Those responsible for the remarkable period skirted danger every step of the way, risking their reputations and livelihoods for the discoveries of a century. Their saga unfolded in barren fields, in cluttered pickup trucks, and in high-pressure boardrooms. It's one that will continue to impact the world for years to come.

PROLOGUE

The phone call came as a jolt.

It was May 2007 and Harold Hamm was enjoying dinner at the Brown Palace Hotel and Spa, an elegant hotel in downtown Denver. To Hamm's right were two longtime colleagues from his energy company, Continental Resources. A pair of Merrill Lynch investment bankers sat to Hamm's left. The group was two-thirds through a grueling ten-day coast-to-coast trip aimed at wooing investors ahead of a crucial initial public offering of shares of Continental.

When rock bands go on road trips, the days are slow and the nights furious; when executives take to the road to sell pieces of their companies, the opposite usually is true. Hamm and the bankers had spent a full day convincing mutual fund honchos that Continental was set to strike it big in North Dakota, a long-overlooked part of the country. Now the group relaxed over dinner and drinks, their jackets off and ties loosened, as they discussed ways to improve their pitch.

As he'd courted the investors, Hamm hadn't dwelled on the personal challenges he had overcome, such as growing up dirt poor in rural Lexington, Oklahoma, the youngest of thirteen children. He'd also avoided mention of his more recent turbulent love life.

Instead, Hamm shared an upbeat message about his company and a promising 15,000-square-mile formation of rocks called the Bakken that was under parts of North Dakota, Montana, and Canada. Hamm believed he was at the vanguard of a revolution that would deliver new oil and gas supplies to a nation running out of energy.

For all his big talk, though, Continental was producing just seven thousand barrels of oil a day from the Bakken's dense rock. It was a mere trickle, representing a fraction of 1 percent of what Exxon pumped on a daily basis. The results offered little proof that much more would flow from a region that had frustrated wildcatters for years, making Hamm's pitch to investors challenging.

Still, the trip appeared to be on track, and the final leg—meetings with investors in Los Angeles—was just days away. The Merrill Lynch bankers told Hamm there seemed to be enough interest in Continental for the company to sell its shares for as much as eighteen dollars each in its prospective offering. With the IPO just a few weeks away, Hamm, a sixty-one-year-old with thinning auburn hair, a full stomach, and a playful grin, was in good spirits as the dinner wound down.

A lot was riding on the offering. Continental already owed $253 million to lenders, and Hamm planned to use some of the IPO's proceeds to pay down a portion of that debt. He knew his company would have to raise hundreds of millions of additional dollars at some point to tap the vast quantities of oil he was convinced were buried in the Bakken's layers of rock. Being public was the only way his company could raise enough cash at reasonable rates to find all that oil.

"We had to be public to be a major player in the Bakken," Hamm recalls. "But at that point we had very little product."

Outside the hotel, gloom was setting in. The real estate market was on its last legs, financial firms were wobbling, and the global economy soon would experience its worst downturn in eighty years. Technology and management gurus said America had lost its creative spirit and the country's economic dominance seemed doomed. Many economists predicted a passing of the baton to India, Brazil, and China. Energy producers were coming up dry, leading to hand-wringing on Wall Street and in Washington, D.C., about how the United States would meet its future energy needs.

As Hamm looked around the table, though, he was convinced he was close to realizing his life's dream, one that would help confer extraordinary blessings on the nation just when it needed them most. Most executives his age were negotiating retirement packages and scouting golf courses. In a sense, though, Hamm's real life's work was just beginning. He was convinced a historic gusher was in the offing that would reenergize the entire country.

Just then, one of the Merrill Lynch bankers, Christopher Mize, heard his cell phone ring. Hamm watched as Mize picked it up.

"Wow . . . that's a surprise . . . okay . . . thanks."

Hamm sensed something was wrong.

"It's not good," Mize told him.

Two competitors in North Dakota had just reported results from their own exploration efforts in the Bakken rock layer. They were big-time disappointments, suggesting that Continental would also come up dry in the area.

"They busted their pick," one of the bankers told Hamm, using industry lingo for a huge strikeout.

We can pull the plug on the IPO or keep it going and try to make it happen, though it will be more of a challenge, the bankers told Hamm. It's up to you, Harold. Think it through carefully.

Hamm barely uttered a good night and went straight to his room, shaken by the news.

The next morning, Hamm told his team they'd continue to pursue the IPO, even if it meant lowering the price to try to entice investors. They might not appreciate what Continental was working on, but Hamm was sure they'd catch on.

Behind Hamm's confidence was an understanding of the dramatic advances a few American companies like his own were achieving in the way they drilled and extracted oil and gas, breakthroughs that Wall Street investors, industry experts, and even the largest oil companies didn't fully appreciate. For decades, prospectors had fractured, or broken up, rock formations by pummeling them with various liquids, creating pathways for natural gas to flow to the surface. The process was called hydraulic fracturing, or fracking.

But Hamm, along with other adventurous wildcatters around the country, had begun to combine improved fracking techniques with cutting-edge methods of drilling sideways deep in the ground. They were targeting long, wide rock layers thick with oil and gas. These were formations of shale and other rock that geologists once only dreamed of accessing. Hamm was receiving daily updates detailing how his crew was boring ten thousand feet into the ground, turning drill bits to go another two miles horizontally, and locating layers of rock brimming with oil. One of Hamm's men boasted he now could hit a target miles below the surface that was no larger than a tiepin.

Hamm was determined to use this new technology to tap a modern-day gusher, one that would change the direction of his company and even his country. He just needed to get investors on board.

Several weeks later, Continental sold its shares at a reduced price of fifteen dollars a share. Wall Street is accustomed to IPOs that soar on takeoff. This one barely fluttered. For months, the stock did little, falling below fifteen dollars in early September 2007. One of Hamm's key staff members sold all his shares—even he worried the company wouldn't amount to much.

Hamm still believed an abundance of oil lay untapped in overlooked areas of America, and that his company was getting closer to finding it. Few agreed with him, however.

Don't they get what's happening? Hamm thought.

There was good reason for deep skepticism about Harold Hamm and his quest to unlock huge amounts of oil in North Dakota. Hamm was a genial dreamer, maybe even past his prime, not unlike his nation.

The oil industry got its start in the United States and the country spent over a century as the world's energy giant. At one time, American oil-and-gas supplies seemed endless. But during the 1970s, promising fields became more difficult to locate. American oil production peaked in 1970 at 9.6 million barrels a day, and imports began to soar as the country scrambled to get enough crude to meet its growing demand. The 1973 Arab oil embargo served notice that the nation had become reliant on others for its energy needs. Those others usually weren't our best friends, either. Oil magnates J. R. Ewing on *Dallas* and Blake Carrington on *Dynasty* may have seduced television audiences in the 1970s and 1980s, but real-life drillers were finding it tough to regain their swagger amid a string of dry holes. A global glut of oil during the 1980s and 1990s resulted in weak prices—good news for consumers but bad news for American wildcatters, who found it difficult to keep up with foreign rivals.

Power brokers in energy capitals such as Dallas, Houston, and Tulsa ignored what Hamm was up to in North Dakota. Instead, it was Aubrey McClendon—an Oklahoma upstart employing his own improved fracking and drilling techniques—who was gaining admiration and even envy. In late spring of 2008, McClendon became a multibillionaire, as shares of his new energy power, Chesapeake Energy, soared. He also helped bring a pro basketball team to Oklahoma City, electrifying the state. Later, when McClendon sat courtside at games with his relative, *Sports Illustrated* swimsuit cover model Kate Upton, the jealousy and buzz grew. Hamm owned his own front-row seat close by. Few noticed him, though.

Talk in the energy patch, in Washington, and on Wall Street in early 2008 was of "peak oil," a popular and vaguely Malthusian notion that the growth of global energy supplies had reached its limit, a dreaded shift sure to lead to rising prices and global economic strains. McClendon's company was among the few still making huge new discoveries by focusing its drilling on shale, a dense rock long ignored by oil giants.

In March 2008, McClendon hosted a dinner at New York's swanky '21' Club for a group of Wall Street billionaires and others hoping to understand the new energy world. Investors George Soros and Stanley Druckenmiller were there. So were some of the leading minds of the global energy business. Over dinner, McClendon's guests agreed that an era of oil and gas scarcity, and rising prices, had begun.

As he watched the conversation unfold from the head of the table, McClendon couldn't hide a confident smile. His company was extracting growing quantities of natural gas, despite the pessimism around the room. McClendon had plans to pump even more of it, giving him a chance to become a modern-day Getty or Rockefeller.

Far from that New York hot spot, however, a revolution was quietly under way, one that star investors, energy experts, and most oil executives were oblivious to. It had started a decade earlier in Texas when a wildcatter named George Mitchell searched for a way to keep his business alive.

Soon Hamm, McClendon, and other once obscure drillers would unleash a dramatic transformation of the nation and the world. That evening in New York, though, few at the table had any idea what was ahead.

"It never occurred to any of us what was about to happen," says Druckenmiller.

THE
BREAKTHROUGH

CHAPTER ONE

The meek shall inherit the earth, but not its mineral rights.

—J. Paul Getty

George Mitchell wasn't out to change history. He just wanted to keep his company going.

It was the summer of 1998 and Mitchell Energy was sending huge amounts of natural gas each day from its Texas fields to a pipeline serving the city of Chicago. For decades, the arrangement had provided the company with steady profits, helping Mitchell become wealthy.

But his gas reserves were depleting, slowly but surely. Mitchell Energy's shares were falling, the industry was on its back, and time was running out for both the seventy-nine-year-old executive and his business. A slow, inevitable decline seemed certain, one that Mitchell was desperate to avoid.

He convened a meeting of his top executives in a large conference room at the company's headquarters outside Houston. They had been searching for new sources of natural gas for nearly two decades, a fruitless quest that left some of the management team frustrated. Mitchell was convinced their best bet was to try to unlock gas deposits from shale, a dense type of rock deep below Mitchell Energy's acreage in North Texas. They had spent years drilling this rock, however, and it still wasn't working.

Larger rivals, such as Exxon, Royal Dutch Shell, and Chevron, already had shuttered operations trying to tap oil and gas from similar rock formations around the country. These huge companies headed off to drill in Africa, Asia, Brazil, and offshore locations. Almost any spot outside the United States appeared more promising than Mitchell's Texas shale.

The United States was running on empty, just like Mitchell Energy. A growing dependence on foreign energy had pressured the country into costly foreign entanglements at least partly aimed at safeguarding sources of oil and gas, such as the invasion of Iraq seven years earlier after that country annexed Kuwait, a Gulf oil power. The United States would have to get used to a deeper reliance on foreign energy and additional military campaigns, it seemed, as Russia and other nations with vast energy resources assumed greater power.

As the 1998 meeting got under way, Mitchell's heir apparent, Bill Stevens, spoke up forcefully. The company was wasting its time, Stevens said. Even if they somehow *could* get a lot more gas to flow from this challenging rock, production costs likely would be too high to make much of it worthwhile. Drilling in the Texas shale wasn't worth the expense of leasing new acreage in the area, Stevens argued to Mitchell and the rest of the company's board of directors.

Mitchell had been getting the same message from his son Todd, an experienced geologist who was on the company's board. Shale drilling might be huge someday, Todd had said, but the company was spending too much on additional wells. Todd repeated his view over lunch, at family occasions, and in private meetings with his father. The company was piling on a frightening amount of leverage, Todd warned, and George Mitchell's personal debts were at worrisome levels.

It seemed clear that George Mitchell was jeopardizing his company's future with his stubborn pursuit of shale drilling.

Mitchell listened patiently to the skepticism at the board meeting. He said little in response, though, ending the meeting abruptly, leaving board members frustrated

Later, Mitchell reached out to his geologists and engineers working on the Texas shale project. He told them he was sure they'd eventually find a way to get substantial amounts of gas from the shale.

"Keep going, it's got to be there," he said, trying to boost their spirits.

Privately, Mitchell had his own growing worries. "I thought perhaps they were right and it wouldn't work," he recalls.

Mitchell couldn't share his concerns with any of his top executives, though, let alone discuss them with his team working in the field. This was Mitchell's last shot. He had to give it a full try.

"I had no choice, really," Mitchell recalls. "We had to get the gas to flow."

Running out of gas was a huge problem for George Mitchell and his company. He and his family of immigrants already had overcome a series of imposing obstacles, however. It gave him some hope he might succeed in the final quest of his life.

M itchell's father, Savvas Paraskevopoulos, was a goatherd in Nestani, a dusty mountain hamlet in southern Greece. In 1901, at the age of twenty, Paraskevopoulos, the fourth of five children, got a glimpse of his meager inheritance in the rural, impoverished country: a quarter acre of land nowhere near any water source. He quickly decided he'd have to do something else with his life.

Impoverished and illiterate, Paraskevopoulos walked fifty miles to a freighter bound for Ellis Island, agreeing to work on the ship in exchange for free passage, becoming one of thousands of Greeks leaving for America. With no skills to speak of and virtually no English, Paraskevopoulos was grabbed by railroad laborers as he disembarked. Soon he was laying tracks with an Arkansas railroad gang working its way toward Texas. It was backbreaking work that helped lay the nation's foundation.

One day, the Irish paymaster of the gang, Mike Mitchell, walked up to Paraskevopoulos, looking annoyed. Mitchell said he was having too much trouble writing Paraskevopoulos's long and ungainly name. He told the immigrant to change it.

"From now on, your name is the same as mine," the paymaster told Paraskevopoulos.

With that, Savvas Paraskevopoulos became Mike Mitchell. He eventually made his way to Houston, joining a cousin who had a shoeshine parlor near a glamorous hotel in town. Soon the young Greek had saved enough money to start his own shoeshine and pressing shop across from the Buccaneer Hotel in Galveston, a city fifty miles away with a growing population of Greek immigrants.

Flipping through a local Greek-language weekly paper one morning, Mitchell saw an article about an attractive young woman, Katina Eleftheriou, on her way to Tampa, Florida, to live with her sister. Taken with her beauty, Mike Mitchell cut the article out and stuffed it in his jacket pocket. Two years later, after saving enough money to afford a train ride to Tampa, he put on a coat and tie and left to track Eleftheriou down.

Mitchell didn't know a soul in Tampa, but he got in touch with someone from the Greek community and managed to locate the young woman, who by then was engaged to a local man. With charm and persistence, Mike Mitchell broke up the relationship and brought Eleftheriou, a fellow Peloponnesian, back to Galveston. They wed in 1905 and raised four children, the third of whom was George Phydias, born in 1919 and named for one of the greatest sculptors of classical Greece.

The Mitchell family lived in a rough immigrant neighborhood that locals nicknamed the "League of Nations," an area where bootlegging and gambling flourished.[1] Money was tight, so George helped his family make ends meet by hunting for duck and dove and by fishing for speckled trout and redfish, which he sold to Gaido's, a historic Galveston fish restaurant, for twenty cents a pound.

Katina Mitchell, who never learned English, was a beacon for the local Greek community, hosting new immigrants at the Mitchell home for weeks at a time until her sudden death from a stroke at forty-four, when George was thirteen. Shortly thereafter, Mike suffered a badly shattered leg in an automobile accident. George's older siblings, Christie and Johnny, were old enough to be on their own, but George and his sister Maria were sent to live with relatives.

George Mitchell didn't speak English until he began grade school, but he managed to become a top student at well-regarded Ball High School in Galveston. He planned to fulfill his mother's dream by attending Rice University and becoming a physician. But a math class taught by a dynamic teacher during his senior year, as well as a summer working in Louisiana's oil fields with his older brother Johnny, helped change his mind.

"Making discoveries by looking at maps seemed exciting," says Mitchell, even though oil was only selling for about $1.20 a barrel and the industry was going through tough times.

Mitchell studied geology at Texas A&M University and was the school's top-ranked tennis player, but he often found himself strapped for money and unable to pay tuition. To raise cash, he began reselling candy to fellow students on campus. "You'd put a stand up and put up a sign saying, 'Drop your money in,' and you'd have twenty or thirty different candy bars," he says. "It was kinda fun."

Well, it was fun until members of the school's football team began grabbing Mitchell's candy bars without paying for them, daring him to do something about it. He realized he had to come up with a fresh moneymaking idea, so he began selling gold-embossed stationery to lovesick freshmen, making $300 a month in his last year in school.[2]

That year, a professor of petroleum engineering gave Mitchell advice that left an impression. "If you want to go work for Humble Oil, fine, then you can drive around in a pretty good Chevrolet," the professor said, referring to a major energy producer at the time that later became part of Exxon. "But if you want to drive around in a Cadillac, you'd better go out on your own someday."

Mitchell earned a degree in petroleum engineering in 1940 and then worked in Louisiana's Cajun country on an Amoco oil rig. Then World War II broke out. Like many of his friends, Mitchell expected to see combat overseas. He was wary, however, after mourning too many friends who had died abroad. Instead, he enlisted with the U.S. Army Corps of Engineers. Mitchell tried to convince his commanding officer that he was indispensable in the Houston area. He worked hard, played tennis with the colonel, and even dated his daughter, all to ensure that he wasn't sent overseas. It worked—Mitchell spent almost five years managing hundreds of men and building projects in Houston and elsewhere.

At the time, Mitchell was more focused on the opposite sex and on avoiding combat than on his future in the energy business. Less than two weeks before the attack on Pearl Harbor, on a train ride back to Houston after a Texas A&M football game, he spotted an attractive young woman, Cynthia Woods, and her twin sister, Pamela, on a double date with two students at the school. Cynthia's date was intoxicated and obnoxious and she was growing miserable.

"The guy was drinking whiskey out of a flask and it made her angry," Mitchell recalls.

Finally, Cynthia asked her sister's date to try to get rid of the jerk so she wouldn't have to deal with him any longer. The date found a poker game in another train car and convinced the rowdy young man to check it out. Mitchell quickly made his move, introducing himself to Cynthia and getting her phone number. Back home, George and Cynthia began dating and later would join Pamela and her date from that evening in a double wedding.

After the war, George consulted on oil and gas drilling projects in Houston, earning a reputation in the area for picking productive wells. It was enough early success to convince him and his brother, Johnny, to try their luck as prospectors, or "wildcatters," modern-day treasure hunters who risk financial ruin in hopes of discovering oil and gas. They would gamble it all to drill their own wells and George would try for that Cadillac, just as his professor had advised.

The brothers didn't choose the most creative name for their company—Oil Drilling, Inc.—but the Mitchells, working with a third partner, opened for business in 1946 in downtown Houston's Esperson Building with high hopes.

An early problem developed: George and his brother couldn't afford the fees charged by local libraries to borrow well logs, the detailed records of geological formations that are crucial to finding oil and gas. To get out of their bind, George convinced a local firm to let him borrow the logs at the end of the day and return them by 8 a.m. the next morning, before they opened for business.

George, the trained geologist, stayed up past 2 a.m. each night poring over the logs, trying to identify promising oil and gas prospects. After George located acreage that seemed attractive, Johnny, more outgoing and upbeat than his younger brother, would approach a group of oil and gas promoters congregated in the building's drugstore, a famous downtown gathering spot and breakfast joint, to try to raise enough cash so the Mitchell brothers could begin drilling some wells.

Soon Johnny became friendly with some Jewish businessmen from Houston and Galveston, including grocery-store owners Abe and Bernard Weingarten, who were looking for ways to invest their spare money. The group agreed to pay the Mitchell brothers fifty dollars a month to find drilling prospects for them. The brothers also received an ownership stake of about one thirty-second of each well they drilled for the investor group. After some early successes, other investors also began giving money to the Mitchell boys to search for oil and gas wells.

"We got a reputation for finding oil, so people found us," Mitchell says. "All the top Jewish families were willing to risk money in oil and gas."

The brothers accumulated an impressive track record, even as they competed with a rush of independent wildcatters flocking to Texas after World War II. The investor group was making money, but they still could be hard on George and Johnny, threatening to withdraw their support if the Mitchell brothers hit too many dry holes in a row.

Once, when George Mitchell told Will Zinn, a Galveston lawyer, that they'd found salt water, not oil, in a key well, Zinn shot back, "Hey, I've got all the salt water I need in Galveston. . . . I don't need to spend all my money" to get any more of it.[3]

"If we did over five dry holes in a row they would leave us," Mitchell recalls.

To avoid that scenario, Mitchell made sure to always retain one especially promising well in reserve, to drill if he ever hit a losing streak, a move that kept the investors from bailing.

Soon cash was coming their way from all over the country. One backer was Galveston gambling magnate and organized crime boss Sam Maceo. Another was Barbara Hutton, the troubled and famous heiress to the Woolworth retailing fortune who also was the daughter of a founder of the brokerage firm E. F. Hutton. (Hutton made headlines for marrying seven men in her lifetime, including actor Cary Grant. She profited from Mitchell's drilling successes but nonetheless died almost penniless, having squandered her wealth on alcohol, drugs, and a string of playboys.)

Another satisfied investor was Robert E. Smith, a well-known independent energy producer and real estate investor in Houston. In his youth, Smith was turned down by an army recruiting officer who noticed he was missing two fingers on his right hand, the result of a hunting accident.

"I've still got my trigger finger and you can pick out the toughest guy in this bunch and I'll bust his butt for you," Smith told the officer, who quickly signed him up.[4]

After the army, Smith was fired twice by major oil companies for punching coworkers. With only a high school education, the young man set out to become a wildcatter, enjoying huge success drilling his own wells in the region. Later, he would help bring the Astros baseball team to Houston.

Smith agreed to take a 25 percent stake in some Mitchell wells, easing financial pressures that had built on the brothers, though adding other kinds of stress they hadn't counted on.

"Smith would call up and raise hell about this or that," Mitchell says, "but he was a good investor."

George and Johnny Mitchell began claiming larger ownership stakes in their wells as they drilled in various spots in Texas, Louisiana, and New Mexico. Their company's name may have referenced oil, but the Mitchells usually hunted for natural gas, the poor stepchild of oil in the energy family. Natural gas is a light hydrocarbon that comes from the compressed remains of plants and animals. It can be converted into liquid form by lowering its temperature, making it a bit more useful, but until the second half of the twentieth century it was generally dismissed as a useless by-product of crude oil, which is more easily stored and transported.

Oil, by contrast, is composed of heavier hydrocarbons that form as liquids. Its liquid form makes oil ideal for a range of uses, such as powering cars, jet engines, and other forms of transportation. That's a key reason why oil became more popular than natural gas, which had its own uses, such as cooking and heating some homes.

At the time, during the early 1950s, natural gas was only selling for about seven cents per thousand cubic feet, a price so low that most large energy companies ignored gas or burned it off as waste if they captured it while drilling for oil. Vast interstate transmission systems hadn't yet been built to easily ship gas to homes, power plants, or businesses, keeping a lid on demand.

The Mitchells found they could score profits by keeping their costs low, though, and they bet that the market for gas might grow as the country's petrochemical industry expanded. Just as important, rivals were busy looking for oil, so the Mitchells faced limited competition.

"The majors didn't care about gas, they just wanted oil," George Mitchell says.

Johnny Mitchell—good-looking, well dressed, and flamboyant—dealt with the company's investors as he drove a maroon 1946 Ford around Houston. He sometimes walked around the city in jungle shorts and a safari helmet, according to a news report at the time. Later he would write a wartime potboiler called *The Secret War of Captain Johnny Mitchell: The Lusty Wartime Reminiscences of One of Texas's Most Colorful Oilmen.* (Much of the book describes Johnny's wartime sexual conquests in prose that can be a bit awkward. A sample passage: "Iceland wasn't barren of good-looking girls; in fact, some were extremely beautiful. They wore wool dresses, which were imported from Britain in exchange for fish.")

"Johnny was a promoter, he was a positive guy who could really sell a deal," recalls T. Boone Pickens, who started his own energy investing and drilling career in Texas in the 1950s.

George Mitchell, tall, balding, and more introspective and intense than his brother, continued to focus on finding and drilling gas wells. Married and already a parent, he lived in a modest home, employees recall. He kept an eye on the company's expenses, as well as his own, though he sometimes drove an older, pink Cadillac, just as his college professor had anticipated.

The brothers' distinct personalities sometimes led to conflict. George, wearing a suit with a torn pocket, once walked into a colleague's office, interrupting a meeting he was having with Johnny.

"George, you look like crap," Johnny told his brother. "Why don't you buy a new suit and some new clothes?"

"If you'd pay me some of the money you owe me, I might be able to afford some new clothes," George shot back.

One day in 1952, a bookie in Chicago shared a tip with another of George Mitchell's investors, Louis Pulaski, about a natural gas field in north-central Texas, near Fort Worth. The bookie didn't know much when it came to energy; he usually just took horse bets for Pulaski, the owner of a Houston junkyard. But the out-of-town bookie told Pulaski that he had spoken with contacts in the drilling business and had a hunch that a 3,000-acre spot under the Hughes Ranch in Wise County would be a winner. Figuring the bookie might have a hot tip, Pulaski called George Mitchell right away, urging him to look into the field.

Mitchell was skeptical, and not just because the tip had come from a horse bookie eight hundred miles away. The acreage the bookie and Pulaski were excited about had already been picked over by locals. It even had acquired a damning nickname among veterans in the business: "The Wildcatters' Graveyard."

"I don't know, General, that field has been passed around for years," Mitchell told Pulaski, whom he referred to as "the General" after famed general Casimir Pulaski, the soldier of fortune who saved George Washington's life in the American Revolutionary War.

"It doesn't have much of a chance," Mitchell told him.

Pulaski persisted, though, and Mitchell agreed to check the field out. He got some help from a drilling contractor he had met in college, Ellison Miles, as well as another geologist who had studied the region. After some testing, Mitchell turned more enthusiastic and decided to try drilling the area.

The first well was a winner. So was the next one—and the next seven, too. Mitchell had hit a "stratigraphic trap," a huge underground reservoir of natural gas. The bookie was on to something after all.

To take advantage of the find, Mitchell and his company acquired 300,000 acres in the area in just ninety days. They paid three dollars an acre and kept buying land until they and their investors ran out of money. Large oil companies, or "majors," as they are known in the energy business, dismissed the acreage as worthless because it contained natural gas, not oil. But Mitchell was convinced he could tap profits. Some markets for gas, such as the Chicago region, were seeing growing demand. Natural gas prices were low—less than ten cents per thousand cubic feet, or the equivalent of about seventy cents in 2013's money—but Mitchell thought they were high enough to make money.

The Mitchell team would drill various rock layers in Wise County, including conglomerate sediment that was considered very "tight," or compressed. In other words, the pores of these rocks were so mashed together that it was hard to get gas to flow from it. The rock had such low permeability that it provided another reason for major oil companies to avoid the area.

But George Mitchell was willing to try a relatively new drilling technique he had read about in petroleum-engineering literature that held the possibility of loosening up this compact rock and getting the gas to flow. The technique, a way of "completing" oil and gas wells, or preparing them to produce energy, was called hydraulic fracturing, or "fraccing." It entailed fracturing the rocks, or breaking them up, by pummeling them with various liquids to free up the gas trapped in those rocks.

(Years later, hydraulic fracturing became known in the popular media as "fracking," with a "k" replacing the "c." From the beginning, industry members detested the word because of its closeness to the common expletive, not to mention a similarity to "fragging," the act of attacking fellow soldiers. "Fracking" also rhymes with "hacking," yet another word with a negative connotation. Energy veterans claim that "fracking" was coined by those with a bias against the industry. In truth, the word was first used in the late 1970s in the science fiction series *Battlestar Galactica* as a substitute for the curse word.)

The idea of violently fracturing rock to extract oil or gas traces back to an invention in the 1860s by Edward A. L. Roberts, who fought in the Civil War battle of Fredericksburg as a lieutenant colonel for the Union. During the heat of battle, Roberts noticed that when mortar shells fell into a narrow canal, a column of water shot up, sky high. It gave him an idea.

Until then, drillers had relied on black-powder explosives to coax oil out of stubborn wells, a method that often proved frustrating. John Wilkes Booth and some of his business partners once destroyed their company's Pennsylvania oil well while using black powder to try to speed the well's production. Booth eventually gave up on the oil business and grew angry about the South's defeat in the Civil War. Days after General Robert E. Lee surrendered, Booth shot and killed President Abraham Lincoln.

After returning home from the war, Roberts developed a torpedo mechanism that lowered explosive nitroglycerine capsules into wells.

These capsules managed to direct an explosion sideways, instead of out of the hole. That fractured oil strata around the hole much more effectively than previous attempts. The result: More oil could be extracted.

Roberts's invention was embraced by a U.S. oil industry that had begun just a decade earlier, in 1859, with the discovery of petroleum in Titusville, Pennsylvania. Drillers paid Roberts as much as $200 each time they used his torpedo mechanism, plus one-fifteenth of the value of the increased flow of oil they experienced. The fee was so high that a black market quickly emerged involving workers willing to employ Roberts's torpedo technique using the cover of night. Their secretive, explosive oil drilling under the stars was called "moonlighting," a term that soon became popular in the American lexicon.

By the 1930s, men working on rigs experimented with guns and ammunition, as some men are wont to do. An underground machine gun six feet tall became a popular way to open holes in casing and unlock oil. Swiss engineer Henry Mohaupt's bazooka projectile, invented as part of a secret U.S. Army program, also became a popular way to get oil flowing when it made its way to oil fields in the 1950s. The United States and the Soviet Union actually detonated nuclear devices to try to get oil and gas flowing in tight rock, though it wasn't a method that caught on, for pretty obvious reasons.

The former Standard Oil of Indiana (which later became Amoco) had first used high-pressure liquid to break up underground rock formations in Grant County, Kansas, in 1947. Some say the technique may even have been employed a bit earlier.

In the early 1950s, as George Mitchell focused on the compressed rock in Wise County, Texas, many rivals resisted fracking. It wasn't because there was any controversy attached to it. At the time it hadn't really occurred to Mitchell or many others that this activity might have any kind of environmental risks.

Companies avoided fracking because it was expensive and added time to a drilling project. They preferred the traditional method of looking for hydrocarbons: Drill a well, like a straw into the ground, and try to hit pools of trapped oil or gas capable of flowing to the surface without the "artificial stimulation" of hydraulic fracturing. After decades of low gas prices, companies were struggling to keep costs down, not increase spending on hydraulic fracturing.

But Mitchell didn't have much to lose, so he gave fracking a try, hoping to make the Texas fields yield oil or natural gas. Mitchell saw that fracking seemed to do a good job stimulating reservoirs that needed a little help to get going, a bit like giving the back of an old television a little bang. His company's efforts worked, and by the late 1950s the Wise County field had become their most important source of natural gas.

George Mitchell wasn't the nation's first fracker. Mitchell was mighty early, though. "The majors didn't bother with fracking, they didn't want to fool with it," he says. "I saw it as the new technology."

Over time, George Mitchell assumed greater control over the company, as Johnny focused on other interests. Soon Mitchell went on an acquisition binge, confident that gas prices would rise. But gas prices remained low and the company was often strapped for cash, forcing it to hold off drilling for long stretches.

Those delays sometimes grated on homeowners who had leased Mitchell land and expected regular royalty checks. After a number of leaseholders voiced their displeasure in 1956, Mitchell decided to deal with the unhappiness publicly. He invited the entire county to a barbecue and told three thousand residents that he was prepared to spend millions to drill. They just had to be a bit more patient. Most came away from the barbecue satisfied with the promise.[5]

An energy price decline in 1957 brought new stress, though. Mitchell Energy also fought with a rival and had difficulties shipping gas out of the area. The company's spending grew nonetheless, unnerving B. F. "Budd" Clark, a senior executive who was a native of New York City and a graduate of Harvard Business School. Pressure built within the executive ranks. The Mitchell staff didn't come to collective decisions so much as fight their way to them. There was screaming in the offices, in the halls, and pretty much everywhere.

It got so bad that Joyce Gay, who began as a secretary and eventually worked in public relations for the company, had to affix soundproof taping around the doors outside Mitchell's office so outsiders couldn't hear all the yelling. "George and Budd would scream in the hall until Budd had arrived at the four-letter word and then George would yank them all back in the office and shut the door," Gay recalls. "They used to joke and say it was management by decibel. . . . It was just their style, they'd scream and yell."

In the middle of an important meeting one day, one of the company's lead attorneys at the big Houston firm Vinson & Elkins fled Mitchell's office, joining Gay in her nearby office, she recalls. "I'm not going back in until they quit yelling," he told her.

It took about ten minutes before Mitchell and the other executives realized their lawyer had left, Gay recalls.

A way out of the jam was found in a pipeline being built by the Natural Gas Pipeline Company of America to take gas from North Texas through Oklahoma to the Texas Panhandle, where it would be transported to Chicago. The company agreed to a twenty-year contract to buy natural gas from George Mitchell's company at thirteen cents per thousand cubic feet, a price slightly above the market price. (The deal would be renewed in 1977.)

The contract was a lifeline for Mitchell Energy, which over many years supplied 10 percent of Chicago's natural gas needs. Mitchell's company continued to refine its fracking techniques, succeeding in bigger fields and benefiting from a program financed by the Department of Energy that split the costs of some expensive frack jobs, with each chipping in $1 million.

Over the years, George Mitchell bought out almost all his minority partners. Many were eager to sell, worried about the company's heavy debt load and cash flow issues. The strain affected Howard Kiatta, a Mitchell geologist, when he tried to buy a map one day in the 1960s.

"Just send me the invoice and we'll pay it," Kiatta told the salesperson.

"No, you work for George Mitchell & Associates. You send us the check first; if the check clears, then we'll send you the map."

"Why?"

"Because he doesn't pay his bills."

Mitchell felt the need to become involved in even the smallest decisions of the company, exasperating some employees. Jack Yovanovich, a senior vice president in Mitchell Energy's land acquisition area, recalls it this way: "George would come around almost three or four times a week, sometimes twice a day, and always tried to control every bit of leasing. Now you think about this: You're leasing lots that are less than, well, some of them, around one-tenth of an acre, some of them two-tenths, and George is coming around wanting to know what you're doing here and how you're doing it, why you're doing this and why you're doing that, and he was driving me and my guys crazy."

One day, Yovanovich told his boss, "Look, George, you've got to leave us alone." Mitchell agreed, after Yovanovich said he'd provide him with a weekly written report detailing what they were doing.

The company went public in 1972, providing the much-needed infusion of cash for newly named Mitchell Energy & Development, and George Mitchell was able to keep 70 percent of the voting stock. The 1973 Arab oil embargo sent energy prices flying and Mitchell's shares also climbed.

By the late 1970s, Mitchell had turned more relaxed and there was a bit less yelling and screaming. The period proved a fruitful time for the natural gas industry, with prices spiking and revenue increases following.*

By now, George Mitchell had ten children, a brood that became a matter of fascination and humor at his company. When Shaker Khayatt, a Mitchell Energy director, sat in Mitchell's office one day, he noticed ten different children's photos on the wall.

"George, that's very interesting. I never thought of taking pictures of my children at different ages," Khayatt said.

"What are you talking about?"

"Well, you have them young here and then about five years later."

"Wake up, they're all different," he told Khayatt.[6]

The growing family sometimes presented challenges for Mitchell. "When the family traveled together, even on short trips, the children would count off so that one wouldn't be left behind in the confusion," he recalls.

Mitchell, an Episcopalian, later confided that although he loved his kids, his wife, a devout Catholic, was the one spurring the family's

---

* Around that time, another George Mitchell, no relation, was elected as a senator representing the state of Maine, the beginning of his own illustrious career.

growth. After having three daughters and one son, "my wife kept saying she wanted one more girl," he says. Instead, Cynthia Mitchell had six more boys. "Finally, we gave up," he says.

As Mitchell loosened up, his employees developed a deep respect for him, partly due to the enthusiasm he brought to the job. At the company's headquarters, he roamed the halls, popping into various colleagues' offices, asking what employees were working on and how he could help. With gas prices higher, he was in a better mood, and his employees came to enjoy his surprise appearances in their offices.

Mitchell favored checkered pants and jackets that reminded some of the attire of a used-car salesman. He managed his company with an informality and dedication that inspired employees.

"George avoided all publicity," says Clark, the senior executive. "He was just work, tennis, work, tennis, work, tennis."[7]

Mitchell evolved into an unusual energy baron. A few years earlier, he had met R. Buckminster Fuller, a futurist and an early environmental activist. Fuller believed human sustainability was in jeopardy and that societies needed to turn to renewable energy sources, including solar and wind power. Mitchell became convinced of the need to pursue alternative energy options in addition to fossil fuels.

"It took me three or four days to understand what he was talking about," recalls Mitchell, who hosted Fuller at various conferences. "He made me believe we were really going to have trouble keeping up with society."

By the 1970s, Mitchell was reading the work of Dennis Meadows, a scientist who recommended ways to slow global growth to reduce the impact a growing population was having on finite resources. He also became an advocate for sustainable energy technologies and food sources.

Focused on preventing urban decay from taking place in Houston and elsewhere, Mitchell visited troubled neighborhoods around the country, such as Brooklyn's Bedford-Stuyvesant and the Watts area in Los Angeles. "All of our cities are in trouble," he said in an interview at the time. "The concentration of the disadvantaged and the flight to the suburbs of the middle-class whites—that's destroying all our cities."

Mitchell Energy purchased fifteen thousand acres of land twenty-seven miles north of downtown Houston and began building a planned city that would embrace his evolving views on the environment and sustainability. He spent evenings at home with his kids, playing with an elaborate electrical train set and designing forests for his new city.[8]

The company's bankers were dead set against his project, however, viewing it as a distraction and likely money loser. They refused to lend $10 million to get it going. Mitchell persevered, siphoning enough cash from his company to begin building The Woodlands, a mixed-income development.

Mitchell planted tall trees, built flowing waterways, and banned billboards in his new city. He decided that pine trees would blend with waterways and residents would be discouraged from cutting their grass, to create a natural look.

The Woodlands opened in 1974 and became the company's headquarters, though some investors and bankers continued to grumble about it. Mitchell himself grew frustrated when he tried to introduce wild turkeys to the development and one of his employees couldn't resist shooting and killing one of the birds.

B y the late 1970s, George Mitchell had much bigger problems than real estate and wild turkeys. His key Wise County field was beginning to run dry and his experts warned that the company had about ten years left before a severe decline in gas production would begin.

Mitchell Energy had committed to delivering one hundred million cubic feet of gas daily to the pipeline feeding Chicago. Mitchell couldn't see how his company could continue to meet its obligations, or where its future source of gas would come from, and he didn't know what to do.

"You could see it fading," he says. "I knew our gas stream was getting weaker. In a few more years I knew we'd be in trouble."

As oil and gas prices weakened and Texas real estate stumbled in the mid-1980s, pressure grew. It didn't help that demand for natural gas was lagging behind coal and nuclear power.

Mitchell and his men had spent years hunting the country for new sources of gas, in anticipation of their key field running dry. The efforts had largely met with abject failure, however. In the 1970s, Mitchell's team had explored a sandstone formation called the Clinton that extends through Ohio, New York, and Pennsylvania. Huge amounts of gas could be found in the region and it had close access to the energy-hungry East Coast, exciting Mitchell executives.

But the natural gas, called deep-basin gas, was below the level of most reservoirs and was tightly packed, making it a challenge to tap. While the Mitchell crew was experienced with fracking, their efforts were a waste of time.

"We nursed and piddled along in that area over a period of time and nothing really ever developed of any great consequence," recalls Jack Yovanovich, Mitchell's land-buying pro.

A promising spot in the Rockies turned into just as much of a disappointment. Mitchell chased other areas, from Whittier, California, to New Mexico, only to retreat back to Texas each and every time.

Mitchell was still searching for a way out of his quandary in 1981 when he read a research paper one of his veteran geologists, Jim Henry, had submitted to a geological publication. Mitchell was such a control freak that he demanded to see every paper submitted by employees to any periodical, so he got an early peek at Henry's research. The paper described a thick rock called shale that was found deep below Mitchell Energy's acreage in Wise County.

Rock layers below the surface are stacked on top of one another like pancakes; shale is one of the deepest pancakes in the ground. The layer in Texas was called the Barnett Shale, a name originating in the early twentieth century. That's when geologists mapping the area noticed solid, black, organic-rich rock in an outcrop near a stream. They gave it a name inspired by the nearby Barnett Spring, itself named for John W. Barnett, a farmer in Texas's San Saba County. Most of the shale formation was a mile or more below the surface, and it ran under much of North Texas, even below the city of Fort Worth.

Mitchell's employees had drilled through the Barnett Shale every once in a while, on their way to even deeper sedimentary rock formations. When they passed a drill bit through this shale, they often noticed gas "shows," or tantalizing indications that this rock likely held substantial amounts of natural gas. The gas "kicks" from the rock were promising signs, but Mitchell and his men usually ignored them, as others had. It just seemed too hard to extract gas from this tight rock so far below the surface.

Drillers had long deemed it fruitless to target other shale layers around the country, even when they seemed to hold gas. Indeed, in 1976, the Department of Energy's Eastern Gas Shales Project had determined that a substantial amount of gas was held in shale in Appalachian, Illinois, and Michigan basins. The study focused some attention on this rock, but it was considered too expensive to try to produce gas from it.

Those sedimentary layers were in the eastern part of the country and nowhere near Texas. So the project was of limited help to the Mitchell team. By 1980, drillers saw the Barnett Shale formation as little more than a marker indicating that another, more attractive layer called the Mississippian might be close by. There didn't seem much more reason to pay attention to this shale rock.

Jim Henry's paper was one of the first to detail the Barnett rock formation and to demonstrate that it might be full of oil and gas that actually might be extracted. Henry was proud of his paper, which effectively took the Barnett area off the industry's back burner. Even he didn't jump up and down at his results or tell his bosses to actually start drilling in the area, however.

"I made reference to hydrocarbons in the Barnett," Henry says, "but I was not advocating that anyone run out and drill these expensive wells."

Mitchell, desperate to find a new source of gas, was excited despite Henry's caution. If Mitchell somehow could tap this shale layer that his company already was sitting on, he figured he might finally have an answer to all his problems.

Drilling deep in the Texas fields was no more than a hunch, the best option he could come up with. Mitchell Energy already had invested a lot of money to build infrastructure to gather and process natural gas in the area. And the company held about 400,000 acres in Wise County. It made sense to at least see if this rock deep in the ground might produce gas, Mitchell decided.

"I thought maybe it would work in the Barnett because there are a lot of large faults, and a lot of movement in the earth's crust, and I thought that could lead to gas," he says. "We had a better chance of getting gas there than anywhere else at the time."

Mitchell was newly enthused about shale. But this rock—and its tantalizing potential—wasn't any kind of big secret in the energy patch. Most experts scoffed at drilling in shale, seeing it as the fool's gold of the energy business, a waste of time compared with much more promising types of rock below the surface.

Although oil and gas is said to lie in "reservoirs," petroleum deposits don't accumulate in anything resembling a subterranean pool. They're trapped within beds of various rock, hiding in the pore spaces within the rock. That helps explain the derivation of the word "petroleum"—from the Latin word *petra*, for rock, and *oleum*, for oil. For decades, drillers focused on the oil and gas in various layers of rock closest to the surface. By the 1980s, though, it had become harder to find large deposits of energy in those nearby layers. That's why George Mitchell became intrigued about the idea of going farther down, into shale.

Like other forms of petroleum, oil and natural gas in shale formed from the accumulated remains and excrement of plankton, algae, and other organisms that lived in large bodies of water, hundreds of millions of years ago. That origin explains why the term "fossil fuel" became popularized.

Over time, the organic-rich matter accumulated in various layers of sediment and was compacted and solidified, with new layers on top of it, just like a layer cake.

As the organic material was subjected to the intense heat and pressure found below the surface, it sometimes turned into oil; other times, when there was more heat deep in the ground and therefore more cooking, it became tiny bubbles of odorless natural gas. With more heat and time, some organic material turned into coal. The heating process also sent a lot of that oil and gas migrating upward toward the surface, like fat rising to the top of broth, until it was trapped in a higher rock formation with a seal preventing it from rising any farther.

Industry members had long called shale a "source rock" because it was understood to be the layer of rock that was the source of most of those oil and gas deposits found closer to the surface. Just as food is cooked in a kitchen and brought out to a dining room, oil and gas spent thousands of years "cooking" in these shale layers before it slowly made its way to reservoirs closer to the surface.

A lot of the oil and gas remained trapped in these layers, though. Indeed, there were few geologists who hadn't yearned to go straight to these layers of dense, often black rock to tap it for all its remaining energy deposits. Just as a hungry teenager would rather storm the kitchen than wait patiently at the dinner table, operators would have preferred to go straight to the shale than wait for its oil and gas to rise closer to the surface.

Fervor for this rock usually led to bitter disappointment, however. The first commercial natural gas well in the United States was in a shale formation in Fredonia, New York, in the early nineteenth century. But energy companies quickly moved away from this rock. It just didn't seem possible to easily extract oil or gas from shale.

Sure, shale had a lot of pores, allowing it to store oil and gas. But the rock was too tight and compressed. In other words, there weren't enough connections between those pores. Trapped oil and gas didn't seem to have necessary pathways for it to flow to a "wellbore," or a hole drilled into the ground to create a well.

Much of the oil and gas in shale eventually makes its way to shallower rock formations near the surface through natural fractures in the rock. But because shale is so compressed, this process can take millions of years. Geologists knew oil and gas remained in shale formations around the country and around the world, but it seemed much too expensive to try to go get it. The rock had such low permeability that it just didn't seem worth the time, cost, and effort. Whatever oil or gas shale held, it sure didn't seem to want to give it up.

"In the field the stance was that it wasn't economical, the formation wasn't viable and couldn't make enough gas," says Jay Ewing, an engineer who worked for a rival of Mitchell Energy in the early 1980s.[9]

Complicating matters, shale layers were just too far down—as much as two miles below the surface—adding to the difficulty and cost. Besides, it also wasn't clear how *much* oil and gas remained in shale, since so much of it had already flowed up to the surface. Most experts were dubious that there was enough pore space in shale to contain very much oil and gas.

"It was foreign to everyone's thinking that there could be that much gas in this rock," says Dan Steward, a Mitchell geologist. "We were all taught that there weren't enough holes."

Despite the broad skepticism about this rock, Mitchell told his team to go ahead and hydraulically fracture the Barnett Shale layer. Their goal was to connect pore spaces and try to create pathways for the gas to flow. The fracking involved pumping thousands of gallons of water and other substances into the rock to create small cracks to help the gas flow to the surface. It was a bit like keeping a thumb on a hose to amplify water pressure and create an impact, in the hope of creating pathways for the gas to escape.

Some of Mitchell's troops resisted this method of stimulating the rock, which they didn't think could work. Mitchell listened and acknowledged the doubts, but he pushed forward.

"Let's fracture it anyway," he told them.

Mitchell was determined to be the first to extract serious amounts of natural gas from shale and to do it before his existing fields ran out of gas, scoring a potentially historic windfall and shocking the industry in the process. He knew that if he was successful, the country itself might have a new way of discovering much needed energy resources.

In 1981, Mitchell Energy drilled its first well in the Barnett Shale in Wise County, the C.W. Slay No. 1. The results were good, though not remarkable. Most major oil companies weren't focused on the area and didn't notice or even care what Mitchell was doing.

Through the 1980s, Mitchell and his company weathered the ups and downs of a difficult industry, still trying to coax meaningful amounts of gas from the Barnett Shale. Profits fell to a meager $8.4 million in 1986 and rival independent energy companies went belly up. Mitchell trimmed expenses and reduced its budget for exploration in the Barnett. Only one area of the region was showing respectable production, and many of the company's engineers and geologists grew pessimistic.

By 1992, it was clear that the Barnett would have to replace Mitchell's natural gas reserves or they'd simply run out of gas. Mitchell pushed his men to make progress. They had tried what they called "massive fracks," costly attempts at fracturing rock relying on a technique the government had helped develop. This approach to getting gas to flow from the Barnett layer—pumping large volumes of fluid and sand down a wellbore to crack the shale and give it more permeability—had worked elsewhere in the state. They also used a "gelled water" formula that included one million gallons of fluid and nearly three million pounds of sand, to prop the cracks open and allow the gas to escape. It was as if the Mitchell team was giving the earth a massive enema.

Mitchell's crew got some gas to flow in, but it was an expensive undertaking and the results weren't especially promising. "We couldn't figure out how to make it work," he says.[10] The cost added up to more than $35 million, a huge sum for the company. "We kept trying, trying, trying."

Mitchell and his staff weren't getting anywhere. But word of his enthusiasm for the Barnett, and of his promising early tests, began to trickle out. The news piqued the interest of Sanford Dvorin, among the most unlikely wildcatters to risk it all trying to strike it rich in Texas.

Growing up in the Weequahic section of Newark, New Jersey, Dvorin was neighbors with future famed novelist Philip Roth. Dvorin didn't know much about the energy business. Yet from an early age the Jewish child of a meatpacker became excited by the idea of making a fortune in the oil patch.

"All I knew about oil was that it was huge and everyone needed it," Dvorin later told Ann Zimmerman of the *Dallas Observer*. "It seemed easier than the packing-house business. Boy, was I mistaken."[11]

A wildcatter is an independent operator who searches for oil or natural gas in areas that can be miles from the closest producing well. These men—and they almost always are men—are equal parts gamblers, salesmen, and geologists. Supremely confident, wildcatters drum up financing from banks or investors by describing how they will tap a gusher in a spot others have dismissed, ignored, or misunderstood. They repeat the pitch, no matter how poor their chances of success, until they have the funds to acquire acreage, drill a well, and wait for oil and gas to flow.

Wildcatters are responsible for discovering the majority of the nation's oil and gas. By the 1980s, huge gushers were becoming much less frequent in the United States and most of the biggest energy players had moved on to foreign locales. But the allure of the big strike, and the mystique of the wildcatter able to tap a lucrative well that the Goliaths of the industry had missed, captured Dvorin just as it had George Mitchell and countless risk takers before them.

Dvorin didn't know anyone in the energy world, nor did he have the first idea how to get started. He received a degree in engineering and moved to Texas in the late 1950s with his brother to be closer to the cattle market, not oil or gas wells. The brothers settled in Dallas and built the largest beef-boning operation in the state, Big D Packing. When his financial backers withdrew their support, Dvorin moved to Tyler, Texas, in 1968 and found himself surrounded by oil and gas workers. He was inspired to follow his childhood dream.

Now's the time, he thought.

For a while, things went well. Dvorin, heavyset and confident, put together oil and gas deals in north-central Texas and made enough money to raise a son and daughter with his new wife in Plano, an upscale bedroom community outside Dallas. In 1976, while searching for new drilling prospects in a local oil industry library, Dvorin decided to examine the history of drilling in the Dallas County area.

Most energy companies ignore urban areas, for the obvious reason that it's harder to obtain municipal approvals and persuade homeowners to lease land in more congested areas. George Mitchell once had considered drilling in the Dallas County area, but quickly reconsidered. The hesitance by larger competitors made Dvorin think there just might be some overlooked oil right there in suburbia.

There wasn't much in the Dallas database, but Dvorin found a well that was widely known among some early Barnett workers; it had been drilled two decades earlier by a company that later became part of the giant Mobil. The file didn't give details of where the well was. And all the Barnett activity was in Wise and Denton counties, where Mitchell was drilling, about fifteen miles away.

But Dvorin figured out that the well was under land that by then was across the road from Dallas/Fort Worth International Airport. "I had a feeling where it was," he recalls. The surface terrain near the airport "looked oily," he says.

Information related to the well suggested evidence of natural gas, Dvorin thought. But the well had been plugged, probably because the original driller didn't think the Barnett rock would produce very much, if anything. If experts were skeptical about Mitchell's prospects in the Barnett, they were even more dubious about this other area of shale.

"The Barnett was taboo," Dvorin recalls. "This was a rank wildcat well."

For nearly a decade, Dvorin didn't do anything with his research. Then, in 1985, while he was drilling northwest of Fort Worth searching for a layer of rock called the Mississippian reef, he drilled through the Barnett Shale formation. The "mud logger," a technician Dvorin had hired to monitor and record information brought to the surface, noted tremendous indications of gas and oil, much like those George Mitchell's crew had detected from the Barnett a few years earlier.

The logger ignored the findings, aware of the frustrating history of this rock. Dvorin was intrigued when he heard the results, though. He remembered the Dallas County well he had tracked down. And he heard rumors that George Mitchell was drilling in other Barnett Shale sediment near Fort Worth.

Dvorin sent his gas tests to researchers at Mitchell Energy's headquarters, hoping they'd take a look and maybe confirm that they in fact were drilling in shale. He reached one of Mitchell's petroleum engineers on the phone and casually asked how their Barnett Shale test was doing, acting as if Mitchell's interest in the Barnett was common knowledge in the business.

"The jury's still out," the engineer told Dvorin, inadvertently confirming that Mitchell had interest in the Barnett Shale layer.

The conversation buoyed Dvorin. He spent the next eight years testing various spots in the Barnett area, learning about the region. He didn't have enough cash to drill by himself in a proper way, so he tried to persuade other operators to partner with him.

Dvorin couldn't entice any companies, however, and an energy crash left him fighting to save his business, distracting him from the Barnett. He sold off all his production and his savings dwindled.

"For years, just years, it was, 'Do I buy clothes for the kids, or a new shirt for me?' Those kinds of decisions," Dvorin told the *Dallas Observer*. "Every day I gave my wife, Patty, a reason to run off."

Dvorin survived the downturn, as did his marriage, and he was back exploring the Barnett by the early 1990s. He focused on drilling spots near that original Mobil gas well in Dallas County, figuring that if he could make a well a success, transporting the gas would be easier if the well already was near a transmission line connected to a city.

One day, he got a call from an industry executive suggesting he hire an engineer who had opened his own practice. "There's someone in my office who you should meet," the industry member said. "He's a kook like you about the Barnett."

It was Jim Henry, the researcher whose study on the Barnett originally helped get George Mitchell excited about the region. He had left Mitchell Energy and now was on his own. Henry and Dvorin began working together, searching for productive wells in the area.

Dvorin desperately needed to raise new cash to make it work, however. In the fall of 1994, he pulled out a Dallas County Yellow Pages, went down the list of every outfit connected to the oil and gas industry, and sent them a letter, one that sounded as outlandish and unlikely as any get-rich-quick scheme.

"History is about to be made and you can be part of it!" Dvorin wrote.

The letter invited the lucky recipients to participate in the first commercially viable natural gas wells in Dallas County. Anyone with a shred of understanding of the energy business immediately threw the letter in the closest receptacle. Who drills under Dallas? If there was a real opportunity there, it sure as heck wouldn't be found in the morning mail, they all knew.

A few people wrote him back, just to tell Dvorin he was out of his mind. Michael Hart, a Dallas-based petroleum engineer with more than forty years of experience, told Dvorin, "Dallas is not worth a flip," he recalls.

"That was one of the mild ones," Dvorin says. "I got a lot of hate mail."

Dvorin was an oil and gas promoter, an archetype of the energy field.

He truly believed he was offering a once-in-a-lifetime opportunity and couldn't understand why others didn't see it the same way. But his unbridled zeal turned off some would-be investors.

They had good reason to be wary. The energy industry has long attracted charismatic salesmen spinning compelling stories. It takes charm to persuade investors to open their wallets to finance activity below the earth's surface and far out of sight. Dvorin's North Texas location was a notorious place where oil and gas promoters had spent years fleecing gullible investors. The region already had gained an unattractive nickname: "promoter's paradise." They flocked to the region because it wasn't hard for a wildcatter to bus a group of doctors and dentists from nearby Dallas and elicit a round of oohs and ahhs by lighting a flare and demonstrating some early production. Wily wildcatters would make sure to cash investors' checks before production from the wells died out. Dvorin was a true believer and wasn't out to rip anyone off, but potential investors had reason for suspicion.

He kept his hopes high, though. Government incentives had been put in place to encourage big companies to search for natural gas in formations previously believed unproductive. The incentives had helped Mitchell Energy, and they gave Dvorin encouragement that he'd eventually find a partner for the Barnett. Natural gas demand was rising and prices rose for a short while in 1996, as an industry advertising campaign touted it as the "clean energy source." At the same time, a technical paper by the Gas Research Institute had suggested that Mitchell's team was on its way to extracting gas in the Barnett, though it wasn't clear how much there would be.

Dvorin continued drilling on his own, using up the last of his life savings. He even persuaded his son, Jason, a financial consultant, to join the effort.

"Nothing in the oil and gas business is a sure thing. If there was, this was it," Dvorin said around that time.

George Mitchell and his gang weren't getting much of anywhere in the Barnett. But soon a few large energy companies wondered if he and early movers like Sanford Dvorin might be on the right track, at least. It was as if Mitchell had awoken some sleeping giants. Ray Galvin, a senior executive at Chevron, became determined to catch up and pass Mitchell. The odds seemed good for him and his company.

Few executives at Chevron were as popular as Galvin, a six-foot-two graduate of Texas A&M, George Mitchell's alma mater. Friendly and approachable, Galvin was a petroleum engineer who started as a grunt in the field. He encouraged his staff to share their ideas to improve Chevron's production, earning the respect of the company's engineers and geologists, some of whom resented that the executive suite seemed chock-full of MBAs, some with little industry background. It helped that Galvin was an upbeat personality who always seemed to have a smile on his face. He had come to Chevron from Gulf Oil, making him one of the few Chevron executives who hadn't spent his entire career at the company and providing him with a different perspective from the others.

By 1992, Galvin was president of Chevron's U.S. operations, the company's largest division. He sat on industry panels and gained an inkling of some of the progress being made in shale. That year, for the first time, the National Petroleum Council's estimate of long-term gas supplies pointed to shale as containing some gas reserves, causing a bit of a buzz among some and giving the Mitchell crew more encouragement. Galvin sensed that a technological breakthrough might be at hand. Chevron needed to begin exploring shale and other kinds of unorthodox rock, he argued within the company.

Galvin started a group to study and then undertake this "unconventional" drilling. Within Chevron, he preached that energy companies, like other kinds of businesses, were better off betting on improved technology than on the price direction of the products they produced. This seemed like the perfect way to put the theory to the test.

At the time, most of Galvin's colleagues in Chevron's corporate suite, as well as executives of other oil giants, were convinced that the most exciting exploration opportunities were anywhere but in the country's lower forty-eight states. Chevron was shifting its own focus to places like Kazakhstan, Nigeria, Newfoundland, and Angola, all but sneering at the United States.

The Barnett layer was, quite literally, in the backyard of at least one U.S. energy giant. The corporate headquarters of Exxon was in Irving, Texas—*right above* the Barnett layer. In fact, Mitchell and Dvorin were drilling just a few miles west. Executives were so focused on possible drilling spots worlds away, however, that they didn't pay much attention to what was happening under their noses. Most other companies figured that if oil majors weren't interested in the Barnett it couldn't be especially promising.

Despite the distinct bias against domestic drilling, Galvin managed to shift eight of Chevron's most talented up-and-coming stars—six engineers, one geologist, and one landman—to a new group to examine drilling for gas in tight rock like shale. Galvin segregated the unit from the rest of the organization, to try to ensure independent thinking, and he got them an annual budget of about $30 million, a sizable sum. By 1993, the group had grown to eleven pros.

"Go see if there are places where we can find an edge," Galvin told the team in one of their early meetings, "something big enough" to matter for Chevron.

The group quickly became convinced that drilling in shale could lead to huge amounts of gas. Chevron made headway using some fracking techniques in a shale area in Michigan and began leasing prime acreage elsewhere in Texas's Barnett Shale, hoping to succeed where George Mitchell was finding frustration.

"We knew Mitchell was trying different fracking treatments," Galvin recalls. "I was trying to position us for places where improvements in technology would give us an advantage."

Ray Galvin drew on the financial might of one of the largest energy companies in the world. And he had some of the industry's best engineers and geologists at his disposal. It seemed obvious that he was the frontrunner to make history solving the mysteries of shale drilling, not George Mitchell, who was encountering growing problems in the Barnett.

The Mitchell team had tried fracking with various types of fluid, but they just couldn't get much gas to flow. Mitchell's geologists, surprised by how hard the shale was, found that nearly half the mineralogy of the rock was quartz, considered the fourth hardest mineral, another reason some turned more downbeat about their prospects of getting gas out of the rock.[12]

Some of their problems were confounding. There were a few areas of the Barnett that didn't need to be fracked because the rock there had "natural fractures" that should have allowed gas to flow. But this rock actually proved harder to coax gas from, not easier.

Some at Mitchell Energy urged their boss to stop wasting time and money. Shale just didn't seem porous enough to produce much gas, they told Mitchell. He appeared to be jeopardizing his company's future. Getting blood from a stone is pretty impossible; getting gas from shale seemed nearly as difficult.

"People said if the Barnett was the best we had, then Mitchell was in deep doo-doo, though they used a different word for it," says Steward, the Mitchell geologist.[13]

It wasn't clear how long Mitchell could keep at it. Sanford Dvorin was hot on his heels, as were Ray Galvin and Chevron.

But it was an executive with a much more limited background in energy who seemed well on his way to becoming the nation's next oil baron. His name was Robert Hauptfuhrer, and he and his company had already mastered a radical and groundbreaking drilling technique that George Mitchell, Sanford Dvorin, and Ray Galvin knew little about.

CHAPTER TWO

On November 3, 1993, Mayor Steve Bartlett rose to introduce Robert Hauptfuhrer to hundreds of members and guests of the Newcomen Society gathered in the Grand Ballroom of the Adolphus hotel in downtown Dallas.

The group, dedicated to "achievement in American business," had come to the famed Beaux-Arts hotel, built by the founder of beer giant Anheuser-Busch, Adolphus Busch, to pay tribute to Hauptfuhrer, the head of Oryx Energy, the largest independent energy company in the United States.

"I take great pleasure in joining you this evening to honor an outstanding and well-deserved American enterprise, Oryx Energy Company," Bartlett enthused. "Bob Hauptfuhrer is a champion and advocate of free enterprise. . . . He is a visionary."

Hauptfuhrer strode to the dais, confidence oozing. Oryx had survived a shakeout in the energy business and its own problems and seemed on firm ground. The company was one of Dallas's five largest employers and Hauptfuhrer was a civic leader. Its success was due to Oryx's early embrace of a new technique called "horizontal" drilling that was showing great promise.

"Although we did not invent horizontal drilling, we've taken it further than most companies," he told the audience.[1]

Hauptfuhrer was a new king of the oil patch. Oryx wasn't creating fractures in rock to loosen up oil or gas and didn't really care what the Mitchell Energy gang was up to in the Barnett. With the drilling expertise Oryx was developing, fracking didn't seem to matter. Horizontal drilling was about to transform the energy business—and even the world—and rivals would be forced to play catch-up to the company.

What Robert Hauptfuhrer didn't know was how short his reign atop the energy world would be.

Born to an upper-middle-class Protestant family outside Philadelphia, Hauptfuhrer attended the William Penn Charter School before graduating from Princeton University's Woodrow Wilson School of Public and International Affairs in 1953. Six foot four, Hauptfuhrer played forward for Princeton's basketball team, was a Phi Beta Kappa member, and proved a quick study in challenging economics courses. After a stint in the navy, he received an MBA at Harvard Business School, graduating as a Baker Scholar, a distinction earned by those in the top 5 percent of the class.

Some people marry well; Bob Hauptfuhrer married very well. In 1963, he wed Barbara Dunlop, a daughter of Robert Dunlop, one of Philadelphia's most distinguished citizens. At the time of the nuptials, Robert Dunlop also happened to have been president of the Sun Oil Company, one of the most prestigious energy producers in the world. Later, Dunlop served as the company's chairman. Hauptfuhrer had met his future father-in-law when they both played golf for a team representing the Skytop Lodge, a resort in the Pocono Mountains that the Hauptfuhrer and Dunlop families regularly visited. Dunlop had suggested that Hauptfuhrer get in touch with his daughter, six years Hauptfuhrer's junior, who at the time was studying at Wellesley College, the women's college near Boston.

After finishing Harvard, Hauptfuhrer joined Sun, a perfect spot for an ambitious young man. The roots of the company, which sold gasoline under the name Sunoco, stretched back to the beginning of the modern oil industry in Pennsylvania in the late nineteenth century. That's when Joseph Pew, a schoolteacher turned oil and gas entrepreneur, helped start the Penn Fuel Company to supply energy to the city of Pittsburgh. In 1901, after the discovery of oil in the famed Spindletop oil field near the city of Beaumont, Texas, Pew was among the first to race southwest to tap his own wells. Pew and his brother, J. Edgar, began shipping oil from Texas to New Jersey under the corporate name Sun Oil Company, the beginning of several decades near the top of the U.S. energy business.

Hauptfuhrer spent nearly two decades working his way up Sun's ranks, excelling in various roles, including senior auditor and head of purchasing. In 1984, he took the helm of Sun's Dallas-based oil and gas exploration and production unit after receiving what he called a "wink and a nod" that he soon would become the company's president.

Bright and energetic, Hauptfuhrer worked with his staff to better allocate the unit's cash. Earlier in his career, he had hired University of Pennsylvania graduates to apply mathematical techniques to improve the analysis of geological basins, an unorthodox move that quickly set him apart from industry veterans. Now he streamlined the business by selling oil properties in Texas, California, and elsewhere and by embarking on a series of layoffs.[2]

There was growing unhappiness about Bob Hauptfuhrer's rapid rise at Sun, however, as his golden connections made him an easy target. Some veterans griped that he didn't deserve such a big job helming the exploration unit and that he didn't seem to have his heart in the business. During exploration meetings, some geologists noticed a blank stare on Hauptfuhrer's face. They figured he didn't care about the details of the discussion or couldn't muster much excitement about what they were doing. Hauptfuhrer's apparent lack of passion grated on some Sun pros.

Others resented Hauptfuhrer because he came from Sun's finance side, not its exploration or production businesses. Those guys were the ones actually finding oil, while Hauptfuhrer sat in his plush office talking about their discoveries while pushing numbers around, many felt. Antipathy for "business guys" is rampant within energy companies, much as Hollywood stars express disdain for the "suits" who make decisions at movie studios but can't act, write, or direct. For Hauptfuhrer, the label of "finance guy" was a stigma he couldn't shake.

"The view was that Robert Hauptfuhrer knew nothing about geology," says Kenneth Bowdon, who worked as a geologist in Sun's exploration department. Bowdon agreed with the criticism. "Hauptfuhrer wouldn't know oil if it dripped on him," he says.

For his part, Hauptfuhrer felt his familial connections served as a form of reverse discrimination that forced him to work harder to prove his value and advance within the company.

★　　　★　　　★

In truth, there was less to envy about Hauptfuhrer's position running Sun's exploration efforts than many of his critics realized. The 1980s were an awful time in the oil business and it became especially challenging as famed oil fields in California and elsewhere hit historic peaks. A glut caused by reduced energy demand after the 1970s energy crisis sent prices tumbling for most of the decade, pressuring profits. Growing supply from the North Sea and other regions outside the United States levied further pressure on prices.

Oil prices fell from thirty-six dollars a barrel in 1981 to about fifteen dollars by 1986, based on benchmark West Texas International prices. In 2013 dollars, that's a decimating fall from over ninety dollars to just over thirty. Some members of OPEC, the Organization of the Petroleum Exporting Countries, ramped up production, trying to offset lower prices, angering others in the cartel. The price collapse forced banks to shut down throughout America's Southwest, resulting in over $15 billion of bailout costs for the U.S. government.

Low prices made it hard for companies like Sun, which had to take a $260 million write-down in 1988 due to the falling value of its energy assets. Production also was a challenge for the company. In the mid-1980s, Sun began to develop the 14,000-acre Pearsall oil field in southern Texas. The area was part of the Austin Chalk region, which stretched from Texas into Louisiana and included areas near Austin and San Antonio. A few years later, however, production from Sun's wells sharply declined, causing frustration in Sun's executive suite.

If that wasn't enough to fray Hauptfuhrer's nerves, Sun's exploration and production divisions were notoriously competitive and uncooperative, causing infighting that Hauptfuhrer also had to navigate. "It was constant pressure," he said.

The stress took its toll on Hauptfuhrer, who left his wife and high school–age son in Philadelphia to take the Dallas job. He returned home nearly every weekend, a grueling lifestyle that he says led to a condition called alopecia areata, which resulted in hair loss all over his body.

On a trip to China in 1985, after a long evening of mai tai toasts with his hosts, who were selling oil drilling concessions for the first time, Hauptfuhrer felt sudden chest pain and was rushed to a pitch-dark local hospital in the middle of the night. There, he watched workers search to find the lights and struggle to properly place electrodes from an electrocardiogram on his chest.

"If I didn't have anxiety before, I had it then," Hauptfuhrer recalled.

He returned to the United States and suffered a heart attack at a Philadelphia hospital, resulting in quadruple-bypass open-heart surgery. It took months for him to recover.

Even as Hauptfuhrer struggled to run the division, something amazing was happening in the oil fields of Texas. One day in 1984, an oil field contractor brought an idea to Sun's production pros. Technology had advanced to the point where some drilling could be done in a sideways fashion, he said, rather than vertically, the traditional way wells had always been tapped. After years of experimentation, many of the kinks had been worked out and horizontal drilling now could be done in a more cost-efficient way, the contractor said.

He called this kind of drilling a "short-radius lateral." It was just a technical term that meant a drill bit now could bore horizontally over short distances. It had nothing to do with the hydraulic fracturing of compressed rock that George Mitchell's crew was focused on. Horizontal drilling was just an improved way to target oil and gas reservoirs below the surface.

Some horizontal drilling already was being done by a pioneering company called Union Pacific Resources, though the company could only drill about 150 feet at the time. The consultant advised Sun to give it a shot, too, adding that advances in computer imaging and improvements in seismic analysis of underground reservoirs made it a perfect time to try this horizontal approach.

For years, geologists had daydreamed of drilling long, narrow slices of rock below the surface that were thick with oil and gas, much as astronauts long to visit out-of-reach planets. The Sun production team instantly realized that horizontal drilling held the possibility of finally reaching this tantalizing rock.

They agreed to test horizontal drilling on one old, tired well in the

Austin Chalk area. Almost immediately, the loser well turned into a huge winner, as production soared.

It seemed great news, at least for the first day or so. But the glee of the crew quickly turned to panic—they realized that if Sun's rivals found out how much oil they were producing and how easily it was coming up, these competitors would do their own sideways drilling and Sun wouldn't be able to buy nearby wells from rivals at reasonable prices. They had to do something to mask their success or they'd be known for blowing a major opportunity.

"Shut it down!" a production executive screamed to several well hands.

When the company reported its production to the state at the end of the month, the results of the well didn't look out of the ordinary. The well produced with horizontal drilling had given up the same amount of oil that nine of Sun's old, tired vertical wells had produced. The big difference was that the horizontal well had produced all that oil in just two days! Hardly anyone knew about their breakthrough, though, because they had shut down the well so rapidly.

There was another, more devious reason for the secrecy of the Sun gang. Sun was such a fractious place that the production crew was determined to keep their colleagues in the exploration department in the dark about this newfangled drilling, lest they emerge as the company's new heroes.

But Bowdon, a junior geologist in Sun's exploration group, got a tip from a friend in the production group that something big was happening. Bowdon's job was to do the geoscience and recommend drilling locations, and he had been itching for a chance to pursue some huge wells, to make his own mark on the company. When his friend told him about the new way they were getting oil out of the ground, he immediately became energized.

"A lightbulb went off in my head," Bowdon recalls. "I could tell this was going to be disruptive technology, just like the computer."

The cat was out of the bag. The warring factions at the company would have to talk to each other about this new way of drilling. Bowdon's boss, Craig Bourgeois, managed to get approval from Hauptfuhrer and others to expand Sun's use of horizontal wells to new oil fields, not just the mature ones that had given the company fits. Bourgeois, Bowdon, and the rest of the production crew were full of hope. Even they had no clue how this drilling advance would change the course of history.

Once, it was a fairly simple thing to find petroleum—it was no more difficult than looking down. Far back into antiquity, bitumen, an oily, semisolid ooze, seeped to the surface through cracks and fissures in the earth. Bitumen was used as building mortar and is mentioned in the Bible. It was used to bind the walls of Jericho and bricks used for the Tower of Babel. Bitumen probably helped caulk Noah's ark and Moses's basket, according to Daniel Yergin's *The Prize*. The substance was ubiquitous, the WD-40 of its day.

Later, petroleum was used in ancient warfare, as noted in Homer's *Iliad,* and as an all-purpose medical remedy through the Middle Ages, especially in Europe. There, too, seepages were noted, making it relatively easy to gather petroleum. Sometimes peasants dug shafts by hand to obtain crude oil, which was turned into kerosene. A small oil industry developed in Galicia in Eastern Europe, and a pharmacist and plumber from Lvov helped invent a cheap lamp to burn the kerosene. This new kind of illumination caught on in Eastern Europe and spread through the rest of the world, Yergin notes.

Even during modern times in the United States, it wasn't very hard to find oil. Locals in the hills of northwestern Pennsylvania skimmed oily water off the surface of springs and creeks or wrung out rags and blankets soaked in it. They called it "rock oil," to distinguish it from vegetable oil and animal fats. The easy-to-find oil was believed to have healing powers, and was used as a cure-all for everything from headaches to deafness.[3]

A group of entrepreneurs in the United States and Canada soon became intrigued by the idea of using this flammable oil to illuminate, as well as to lubricate moving parts in automobiles and ships. They quickly realized that they would need a larger supply of it, and that they'd have to go belowground, rather than just skim the supply that made an appearance close to the surface.

To find oil, some of these entrepreneurs began adapting drilling methods used for many years to extract salt from areas below the surface. Boring for salt had been developed more than fifteen hundred years earlier in China, where wells went down as deep as three thousand feet. Drilling derricks later were used in Europe and elsewhere, also in search of salt.

In 1859, Edwin L. Drake drilled a well at a depth of less than seventy feet near natural oil seepages in the tiny town of Titusville, Pennsylvania, becoming the first to drill for oil in the United States. At the time, the drillers described what they were doing as "boring as for salt water," or just "boring." They may have used a dull name to describe it, but these early pioneers drilled the world's first commercial oil wells, though holes of several hundred feet were considered very deep in those early days.[4]

In the North Texas area where George Mitchell and his team later focused, early drilling was aimed at finding water, not oil or natural gas. Texas had few natural lakes and limited water sources, forcing early settlers to drill into the ground to keep burgeoning communities alive. Drilling devices and other innovations, such as rotary drilling, were employed in the state in the late nineteenth century, also in the search for water. When oil was struck it often was more of an annoyance than a cause for celebration. That all changed early in the twentieth century, when a series of remarkable oil discoveries shook the country, including the famed Spindletop gusher in 1901, and a growing need for fuel for automobiles and ships made it a more precious commodity.

As fields were discovered in Pennsylvania, Texas, Oklahoma, California, and elsewhere in the United States and around the world, the drilling methods employed were relatively simple—drill bits were sent vertically into the ground aiming for pools, or accumulations, of oil or gas trapped in small spaces in various "reservoir" rock. Poke a hole in and suck it out, just like with a giant straw.

That kind of drilling—straight down into the ground—sometimes led to gushers, and huge profits, of course. More frequently, this vertical drilling led to "dry holes," or wells with little or no oil or gas in them, as well as costly losses. Science was used to improve the odds of success, so drilling deep down in the ground in the hopes of hitting pockets of energy wasn't quite random. But vertical drilling too often was something of a hit-or-miss proposition, with exasperating dry holes regularly resulting.

"With traditional drilling it's like a door-to-door salesman, most people say no and then you get a yes once in a while," explains Jerry Box, who was a Sun Oil exploration executive. "There are emotionally really sharp peaks and really low valleys—the ups are damn exciting, but you have to have an emotional state to deal with it. If you can't deal with it, you don't belong."

Part of the problem with traditional vertical drilling is that among the numerous sedimentary layers under the ground brimming with oil and gas are those, like shale, that are very long and very narrow. They go on for miles and miles, but their "payzones," or the section with enough oil or gas to make it worthwhile to drill, are relatively thin. Traditional vertical drilling means penetrating only small parts of these wide horizontal energy deposits.

To get at other parts of these narrow reservoirs, multiple wells had to be drilled nearby, at additional expense. Even those couldn't always capture all of the horizontal oil and gas formations. With these wells, it was like playing a game of Battleship and only hitting parts of a huge ship, or missing the ship entirely.

By the 1970s, producers were realizing that there was less oil and gas to be found in the United States. American oil production peaked in 1970 at 9.6 million barrels a day and began dropping steadily. The challenges put pressure on drilling pros to come up with a better method, one that might reach those long, narrow reservoirs.

To access more oil and gas, some companies attempted to drill diagonally, by turning a drill bit on an angle of some kind. This kind of "slanted" drilling was a necessity for companies operating from a central platform in offshore waters, though some tried it onshore as well. In 1990, Iraq accused Kuwait of using a type of diagonal drilling to steal its oil, a pretext used by Saddam Hussein to launch a war against Kuwait. Diagonal drilling wasn't a perfect solution, however, because it didn't capture enough of those narrow oil and gas slices.

In 1976, William K. Overby Jr. and Joseph Pasini III, engineers working for a research arm of the Department of Energy, were searching for a method of drilling for methane gas in underground layers of coal. They achieved enough success to patent a method in which they started a drill bit "at a slant rather than directly vertically," according to the patent application, gradually curving it until it was at "a horizontal position." (The Department of Energy later provided funding for a number of experimental horizontal gas wells.)

This method, which came to be called horizontal drilling, a variation on diagonal drilling, was perfected through years of trial and error, usually with little fanfare in the industry, let alone among the general public. Consumers generally don't like to think about where the gasoline for their cars comes from, or the source of the natural gas that heats their homes, especially if it's a bit messy.

Nonetheless, the progress being made by the drilling tinkerers held immense promise. By sending a drill bit straight down as much as a few thousand feet, snaking it until it lay flat, and then boring horizontally, a larger area of a reservoir, or multiple deposits of oil and gas, finally had the potential to be unlocked.

When the consultant visited Sun's office in 1984, Kenneth Bowdon and the rest of the Sun team had a hunch that the nation's narrow, thin reserves finally might be within reach. Unlike George Mitchell's crew, the Sun group didn't have a lot of experience fracturing rock. They had no clue how to break up the dense limestone in the Austin Chalk. But they figured they didn't need to because the rock already had natural fractures that trapped oil in long, narrow slices. Drilling sideways seemed a perfect way of draining all the oil from these rock layers.

In 1986, Bowdon and Sun's exploration unit began using horizontal drilling in five wells in the Austin Chalk area. The results were immediate and stunning: a month's worth of oil in just three days, or 150 barrels an hour. Soon, a plume of smoke rose above the well, as oil escaped and was burned with the well's gas. The smoke rose so high in the sky—a full four hundred feet in the air—that it sparked calls to fire departments in four cities. When fire trucks reached Sun's drilling site, some trucks coming from as far as seventy miles away, sheepish Sun executives had to tell the alarmed men they had a surge of production, not a dangerous fire, on their hands.

This is amazing, Bowdon thought.

Then it dawned on Bowdon that their good luck might actually do harm to the company. It was the same conclusion that the rival production guys had come to when they met their own instant oil.

Craig Bourgeois, Bowdon's boss, ran to find one of the senior drilling specialists. "Don't tell anyone about this," Bourgeois told him. "Don't leave the site and make sure the guys don't go into town."

They knew they had a small window of opportunity. The terms of Sun's drilling permits meant the company had to share its monthly production results with the state, figures that were disseminated publicly. But they didn't have to reveal their results for 120 days. That provided enough time to do some serious leasing of nearby acreage without driving prices higher.

They didn't want rival companies to find out how much oil they were producing, and how quickly it was gushing. So Bourgeois ordered the drilling to slow. Roughnecks operating the rigs were told to sleep in their nearby trailers for a month, lest they go home and tell friends about the surging production. It wasn't easy to keep the secret—someone had built a stand on the edge of the property to spy on what the Sun guys were up to—so they tried to act as casually as possible.

Operating quietly, Sun began acquiring hundreds of thousands of unexplored acres in the Pearsall field, paying just about forty dollars an acre. When the company revealed the results of its early horizontal drilling, Wall Street investors cheered. Sun was getting over two thousand barrels a day from these new wells, some of the company's most productive wells on record. It was clear that horizontal drilling would be their ticket to glory and a way to boost the nation's propects.

Sun decided to spin off its exploration and production unit in 1988 as a separate company, while holding on to Sunoco's refining and marketing business. Hauptfuhrer was given the job of running the new entity. It was named Oryx Energy Co., for a type of large antelope with long, straight horns found in Africa and Asia. Rivals teased executives at the new company that the name sounded like a vacuum cleaner; Hauptfuhrer countered that it suggested "aggressiveness and pride."

"An oryx is one of the few animals that lions won't eat," he said. "And we knew there were a lot of lions out there."

Oryx immediately assumed the mantle of the nation's largest independent oil producer, or one that searched for oil and sold it to others but
didn't transport it or sell it to retail customers. Bob Hauptfuhrer and several experienced executives helmed the new company. They held the clear
lead in harnessing the exciting method of horizontal drilling that some in
the business were beginning to get enthused about. Hauptfuhrer's company seemed likely to lead a rebirth of American oil and gas production.

"There really wasn't anyone else doing it at the time," recalls Jerry
Box, the exploration executive.

Box began hearing accolades when he met with investors in New
York, Boston, and elsewhere, giving him more encouragement. "Hell, we
had come up with something that other people were imitating," he says.
"It was a big deal."

Horizontal drilling was hugely expensive. A single well cost more
than $2 million, compared with about $350,000 for a vertical well. But the
Oryx team managed to bring its costs down to $650,000 a well, making
horizontal drilling worth the added expense since so much more oil was
gushing out of the ground. They also improved how far their drills could
bore horizontally through the rock, going as far as four thousand feet
laterally, allowing Oryx to tap still more oil.

By 1989, the average well in the Pearsall field was producing thirteen
hundred barrels of oil a day of initial production, more than double the
results just a few years earlier. The oil would slow as the wells matured,
and then level off, but so much poured out in the first year that the company was on a roll. Some wells did even better, with initial production of
more than three thousand barrels a day and over two million cubic feet
of natural gas.

Oryx's success helped touch off a burst of leasing activity in the Austin Chalk area as rivals raced to emulate the company, prompting an
analyst at investment banking firm Petrie Parkman & Co. to proclaim
that the Austin Chalk area "may be the hottest land play in the lower 48
states."[5] At one point, activity got so heated in "the Chalk" that Dilley, a
town of about three thousand in the region, was overrun with prospectors "touting drilling prospects sketched out on their hands," according
to an article at the time in *Explorer,* a trade publication.[6]

The really crazy thing was that it was Oryx leading the pack, not bigger companies like Exxon, Mobil, and Amoco. Those companies had been focused on finding huge pools of oil abroad, not on improving drilling methods and squeezing out more from fields in the United States. Those "major" oil companies had been convinced that "elephant" fields in the United States were extinct. Now some were trying to figure out how to get involved in the Austin Chalk region.

Back in Houston, George Mitchell took notice of what was happening in the area. Mitchell Energy poured about $40 million into the Chalk region, one more spot that might save the company. But Mitchell got to the region a bit too late and ended up withdrawing in failure, another costly blunder.

By 1990, the energy industry was more convinced of the benefits of horizontal drilling. Few were as optimistic as Hauptfuhrer and his engineers, though. "We believe there are ten to fifteen, maybe as many as twenty other geological formations in this country that might lend themselves to horizontal drilling," Hauptfuhrer said at the beginning of the decade. "The experience on the cost side has been so good."[7]

Oryx pushed into natural gas exploration, viewing it as the energy of the future because it was more environmentally friendly than oil, coal, and nuclear energy.

On the surface, Oryx seemed to be firing on all cylinders. But within the company, there remained resentment about Hauptfuhrer. Some of his personal quirks also raised eyebrows. When Hauptfuhrer went to industry functions, employees noticed that he often looked disheveled, wearing badly wrinkled dress shirts that sometimes fell out of his pants. He favored dark suits with brown shoes, his socks drooping to reveal too much leg. Some employees rolled their eyes at their boss's "high-water trousers," which also showed an uncomfortable amount of leg. He tended to give data-driven, deliberate speeches that had some rolling their eyes.

"He was so unpolished you couldn't imagine why he was the one selected" to lead the company, a Sun veteran says.

Despite his upbringing, Hauptfuhrer often demonstrated a surprising lack of etiquette. On a buffet line with a colleague at the prestigious Dallas Petroleum Club, among the city's most exclusive private clubs, Hauptfuhrer once stuffed dinner rolls into his pockets, as if he was worried about the source of his next meal, an employee recalls.

"You didn't want to take him to an event with food," a Sun executive says. "You got to see what he was eating because half of it came out of his mouth. He was just oblivious."

Investors didn't seem to care about Hauptfuhrer's idiosyncrasies. They warmed to Oryx shares, sending the stock climbing as the company's production grew.

Hauptfuhrer felt just as confident about the company's prospects. In 1989, Oryx agreed to pay $1.1 billion to British Petroleum to buy a portfolio of international oil and gas properties in the North Sea, just off the British coastline, and in other nations.

Most major oil companies remained focused on foreign drilling areas, and Hauptfuhrer didn't want to buck the consensus, even though Oryx was making progress in the United States. He pledged to sell some U.S. properties to pay for the British Petroleum deal, sparking some grumbling by production pros focused on this country. To these skeptics, it was as if Hauptfuhrer didn't fully believe in the success his own men and women were seeing with their advances in horizontal drilling.

"Sure, we had to put on more debt," he said, "but I thought it was a pretty good deal, there were more unexplored areas offshore than in the U.S."

When Iraqi dictator Saddam Hussein ordered an invasion of onetime ally Kuwait in the summer of 1990, oil prices climbed and Oryx became a true Wall Street darling. Shares hit fifty-five dollars, doubling in just two years. *Fortune* magazine called Oryx the nation's "most admired independent producer of oil and gas" and "the leading practitioner of the industry's hottest technology, horizontal drilling."

That's when Hauptfuhrer made his critical mistake. In 1990, he directed Oryx to buy about 20 percent of the company's shares from the Pew Charitable Trusts, a group of seven charitable trusts representing Philadelphia's prominent Pew family, descendants of Sun's cofounder. The trusts had been itching to sell their shares because the BP purchase left Oryx with too little cash to continue growing its dividends, a payout that the trusts relied on. Charitable laws require trusts to distribute a minimum of 5 percent of their assets each year, creating a need for dividend-paying investments.

Oryx borrowed nearly a billion dollars to pay for the Pew's stock at a price of nearly $50 a share. It was a lot of money, but Oryx was on a roll, so Hauptfuhrer figured the company would pay the debt off eventually. He told colleagues that Oryx shares likely would tumble if the Pews sold their shares on the open market rather than to Oryx itself. Hauptfuhrer also worried that representatives of the Pew trusts, which had control of three of the company's board seats, likely would veto future spending on attractive projects if they remained stockholders. It was best if they took their cash and left, Hauptfuhrer figured.

Some employees backed the deal, worried that if Oryx didn't buy the shares they might fall into the hands of a rival, resulting in a merger and severe job cuts. Many had gone through layoffs at other oil companies during the industry's difficult period and didn't want to do so again.

But others worried about Oryx's rising debt levels. The company already had piled on debt to buy the properties from British Petroleum. Now it owed more than $3 billion, equal to nearly Oryx's entire market value. Credit rating company Moody's Investors Service cut its rating for Oryx's debt, citing the company's decision to borrow more money when its leverage was "already high."

Some employees grumbled that Hauptfuhrer was trying to help his father-in-law, who some believed was part of the Pew family. In reality, Robert Dunlop, the former Sun chairman, wasn't one of the Pews. But he was on the board of the Pew trusts and a director of the firm managing the family's money, raising other potential conflicts of interest in the buyback agreement.

Unbeknownst to most Oryx employees, the buyback resulted from a behind-the-scenes chess match that Hauptfuhrer appeared to have misplayed. The president of the Pew trusts, Dr. Thomas Langfitt, had indeed convinced Hauptfuhrer and Oryx executives that the trusts would sell their shares to a rival, threatening Oryx's independence, unless the company bought the shares from the trusts, according to a lawsuit later filed against the trusts related to the deal. The genteel Pew Charitable Trusts, known for their support of civic journalism and other like-minded causes, had embraced a tactic that Jack Willoughby in *Institutional Investor* magazine later likened to "greenmail," the hard-nosed tactic usually employed by corporate raiders.

"They put a gun to our heads," Hauptfuhrer recalled. "They said, 'If you don't buy us, we'll shop the shares to another company.' And I thought we were in the beginning of an important period for the company," so it wasn't the right time for a merger.

It turned out to be a bluff by the Pew trusts, however.

"Despite the fact that our own advisers told us that there was no third-party buyer out there on the horizon, we were able to induce them to believe that in fact there was," Peter Brown, chief operating officer of the firm managing the trusts' money, later asserted in court testimony, as reported by Willoughby. "And so they wanted to get control of the block."

Brown testified that Oryx developed a "genuine paranoia . . . about word getting into the market" that the trusts were considering selling shares to an outside company.[8]

"The president of the Pew trusts told our board that they intended to sell their block," Hauptfuhrer said later. "That may have been a bluff—in which case he lied."

Had Hauptfuhrer and Oryx rejected the buyout idea and told the Pew trusts to unload their stock on the open market, the selling likely would have weighed on Oryx shares. But the company would have been left in a healthier position to pursue horizontal drilling across the country. Instead, Oryx bought the shares back and found itself in a weakened state.

"Now we had a lot of debt," Hauptfuhrer acknowledged.

Hauptfuhrer assured investors that the company's growth, as well as firm oil prices, would enable Oryx to pay down the borrowing. He didn't think prices would keep soaring, but he also didn't plan on any kind of plunge.

But oil prices did plunge soon after the United States overran the Iraqi forces, ending fears of a global oil shortage. Prices remained low the rest of the decade, Oryx's profits dried up, and its shares crumbled. The company soon tottered under all its debt. Hauptfuhrer restructured operations and froze salaries, as production from the Austin Chalk area tapered off.

Searching for a way out, Oryx applied its horizontal drilling technology to other fields in Texas, to a promising layer of shale and other rock in the Bakken in North Dakota, and to one in Colorado called the Niobrara Shale. Oryx also focused on the Barnett, even as George Mitchell began to ramp up his drilling in the area. The Mitchell team watched what Oryx was up to with interest. Their data suggested that Oryx would find oil in its wells in the Barnett, a big reason Hauptfuhrer was so upbeat as he gave his 1993 speech in Dallas.

But Oryx's exploration efforts proved failures; the company ended up plugging and abandoning its wells in the Barnett. A big reason was that the company never mastered the fracking techniques necessary to extract oil and gas from fields like those in the Barnett and Bakken formations, where the rock wasn't as permeable as in the Austin Chalk region.

"We drilled almost every shale play, but we were just too early, we didn't have fracking capabilities," says Bowdon. "That was the magic ingredient missing for a technological breakthrough."

And while Oryx was expert at drilling horizontally, the technology wasn't at the point where it could be done for more than a few thousand feet, limiting its effectiveness.

It didn't help that Oryx had some bad luck picking drilling spots. "We were very close in the Barnett, we were just too far north," recalls Jeff Roberts, an Oryx geologist. "It felt like we hit a lot of long flyouts that were caught at the wall [rather than become home runs]. . . . It was very frustrating."

The horizontal drilling group soon found it harder to obtain funding for new wells. Oryx fired employees and sold assets, especially those related to domestic drilling.

"The price drop changed the economics" of the company's innovative drilling technique, Hauptfuhrer said.

All his efforts weren't enough to stem the fall of the stock, which dropped below fifteen dollars in 1994. Other Oryx executives, including one with a name straight out of a James Bond movie, Chief Financial Officer Edward Moneypenny, did little to strengthen the shares. On Wall Street, Hauptfuhrer went from beloved to detested.

"They're totally pathetic," Alan Gaines of New York–based brokerage firm Gaines, Berland Inc., told *BusinessWeek* in late 1994. For three years straight, Gaines, who profited by wagering against Oryx shares, gave Hauptfuhrer his "Jack Kevorkian award" for the executive most likely to aid a corporate suicide. "Lots of people thought oil prices would rise, but how many bet the ranch?"[9]

Internal rifts grew as employees fought for dwindling resources. Production executives demanded a bigger budget from Hauptfuhrer because they were generating crucial cash for Oryx. Exploration veterans countered that they were the ones finding new reservoirs, the key to the company's survival.

Hauptfuhrer also became a despised figure within Oryx, due to the layoffs and the company's declining fortunes. Between 1988 and late 1994, Oryx eliminated two thousand jobs, bringing its workforce down to twelve hundred. Awkward interactions with employees cemented the negative view of Hauptfuhrer, raising doubts about his future.

Hauptfuhrer ended up retiring from the company in the fall of 1994 at the age of sixty-three, a year before the end of his contract, a move Wall Street analysts cheered. After he announced that he would be leaving, some employees boycotted a party thrown in his honor.

Some of Hauptfuhrer's rivals had more sympathy for the executive. "Their timing was bad," Raymond Plank, chief executive officer of rival Apache Corp., said at the time, "there but for the grace of God go many oilmen."[10]

Hauptfuhrer's replacement, Bob Keiser, had earned his reputation working on international drilling projects. A year earlier, Oryx had risked $6 million of scarce cash on risky drilling in the Gulf of Mexico, a hot area at the time for energy companies, though it didn't stem the company's slide.[11]

In the fall of 1998, shares of Oryx slumped to just over ten dollars, as continued low oil and natural gas prices weighed on the company, a last, fatal blow. In October of that year, Oryx agreed to be purchased by a rival, Kerr-McGee Corp., of Oklahoma City. Before Christmas that year, Keiser convened the company's six hundred remaining nervous employees to discuss who would be kept on at the new company and who would be let go.

"Well, I know one person who still has a job," Keiser joked, according to three employees in the audience. "That's me." He said employees would have to deal with adjustments to their lives after the sale of Oryx, just like he would.

"I'm moving to Oklahoma City and my wife's going to have to find a new doctor," Keiser said, according to the employees. "I want y'all to go back to your offices, your bosses will tell you if you have a new job. By the way, Merry Christmas." (Keiser says he doesn't remember making those statements and that he knew he wasn't going to have a job at the new company. "If I said that, it was to lighten up the group," he says.)

Oryx employees turned to each other, mouths agape. Some tried to joke about Keiser's comments; others were too disturbed to say much. Soon, hundreds of employees would be let go. And Keiser *did* lose his job.

The fall of Oryx was an ignominious defeat both for a company that seemed on its way to greatness, and for an executive who seemed on the brink of remarkable achievement. Hauptfuhrer never sold the tens of thousands of Oryx shares he received during his tenure, meaning he suffered along with other shareholders. He did keep his portfolio well diversified, though, sparing him much of the pain.

Hauptfuhrer retreated to his ten-bedroom Pennsylvania-stone estate, built in 1929, outside Philadelphia, where he lived with the former Barbara Dunlop until his passing in August 2013. "It was disappointing," he said earlier that year, regarding Oryx's experience. "The price of oil went lower than we thought. We wouldn't have made the same bet had we known where they were going. That's hindsight."

Some of those at Oryx who believed in horizontal drilling, such as Kenneth Bowdon and Jeff Roberts, either were fired from Oryx or quit. Many continued to refine horizontal drilling methods at new companies or on their own, however, and dozens of these geologists and engineers would go on to help lease and drill millions of acres of land around the United States, playing a major role in the transformation of the country.

"Oryx management missed the most disruptive technology in the energy field in a century; they thought greatness would come if they followed the majors, when in reality the majors most often follow the innovators," Bowdon says. "While it is true Oryx was destroyed, the legacy lives on in dozens of guys that are still innovating and pushing boundaries."

George Mitchell was sure he could avoid the mistakes of Robert Hauptfuhrer and Oryx, and he was confident he and his men would figure out how to extract gas from the Barnett Shale.

He'd soon have to deal with an imposing and unexpected foe, however.

CHAPTER THREE

A s George Mitchell drove a brand-new mint-green Cadillac to meet a pair of reporters on a summer day in 1993, he was in good spirits.

The trim, balding seventy-four-year-old was set to give journalists from the *Wall Street Journal* a tour of The Woodlands, the planned city Mitchell had meticulously built outside Houston. By that time, the development had grown to thirty-six thousand residents. Just wait, Mitchell confidently told the reporters. The Woodlands soon would be home to a million people, he predicted, justifying the $600 million Mitchell Energy had invested in the project.

Energy prices remained low and the Texas economy was frail, but Mitchell couldn't have been more upbeat. Mitchell Energy still had its contract to deliver gas to Chicago at above-market prices, sparing the company much of the pain that rivals were facing. That summer, the company's shares hit twenty-seven dollars, giving George Mitchell a net worth of more than $700 million. His holdings included 60 percent of the shares of Mitchell Energy, along with seventeen hotels, restaurants, and shops in his hometown of Galveston, where he had spent $65 million of his personal money. It was the kind of wealth Mitchell never dreamed of as a youngster hunting and fishing to help his parents pay the bills.

George Mitchell had emerged as an energy tycoon for the new era. A local boy made good, he threw regular parties in his hometown of Galveston, including a Mardi Gras celebration in which he had his face playfully painted in traditional blue-cat whiskers. Mitchell spent another $17 million of his personal money to build a pavilion in The Woodlands in his wife's honor. He even lured the Houston Symphony as the centerpiece of the new structure. After musicians complained of the heat, Mitchell paid to install air-conditioning for the outdoor stage.

It didn't seem as if anything could stop Mitchell. When he returned to Galveston one day, however, a local dignitary named Jim Yarbrough approached him with concern.

Mitchell was spending gobs of money on The Woodlands and on Galveston, but neither of his pet projects was especially profitable. More important, most of his wealth was tied up in shares of his energy company. But Mitchell and his colleagues still hadn't figured out how to get gas out of the Barnett Shale, and there was no reason to think they ever would, raising questions about how long they could keep the company going.

"George, every night I light a candle for you and all your millions," Yarbrough told Mitchell.[1]

By the fall of the next year, Yarbrough's concerns appeared well founded. Mitchell Energy shares began to slip, falling by more than one-third to seventeen dollars. Some investors viewed his growing focus on real estate as a sign his energy business was running out of steam. Why else would Mitchell be spending all that time planning some idyllic community? Mitchell's gas production outside of the Barnett Shale formation had fallen by almost half over the previous decade. No one on Wall Street thought the company would get much gas out of the shale.

Doubts about Mitchell's drilling in the difficult rock were well justified. The United States is littered with regions that at various times seemed on the verge of becoming great gushers, only to result in dispiriting dry holes. The Barnett appeared just one more spot that ultimately would prove a huge disappointment.

For a long time, parts of Minnesota were viewed as chock-full of oil and gas, for example. The first reported discovery of natural gas in the state likely was in the late nineteenth century when a vein was struck in the tiny city of Freeborn, quite unexpectedly, emitting a loud "rushing and roaring" sound, according to reports at the time. The lucky landowner of the property milled about, likely trying to figure out what to do with the newfound gas. As he mulled his options, the landowner lit a pipe and suddenly "had his whiskers trimmed without the aid of a barber," according to an account in 1885 in a local newspaper.

News of the close shave sparked a rush of prospectors to the area.

Reports of oil and gas elsewhere in Minnesota fed the frenzy. Some wells were financed by reputable local businessmen and drilled by pros. Others resulted from prospectors relying on divining rods—forked-shaped branches or sticks that they claimed could indicate if oil or gas was nearby, sort of a faux metal detector of its day.

A few years later, a well-known "diviner" who had been paid $250 to travel to Minnesota from Ohio "passed over the ground and was taken by jerks and shakings so violently in certain places he could not endure the current," according to *The Geological and Natural History Survey of Minnesota*. Alas, when drilling commenced there was nothing to be found in the state. "Take the money to Las Vegas—your chances of coming home a winner are a lot better," said the former director of the Minnesota Geological Survey when interviewed in 1979 by a local paper about new petroleum exploration.

George Mitchell wasn't the first to roll the dice on unorthodox rock formations, either. Since the early 1900s, for example, interest has waxed and waned in areas of Colorado, Utah, and Wyoming jam-packed with so-called oil shale, an organic-rich rock that's a distant cousin of shale oil, despite the similar name. Oil shale holds high concentrations of kerogen, a precursor to oil that some liken to teenage crude because it hasn't experienced the millions of years of pressure needed to become oil.[2]

Rock in those U.S. states holds the energy equivalent of the world's entire proven oil reserves, government reports have said. Plus, it's close to the surface, making it easy to find. During World War I, *National Geographic* predicted that "no man who owns a motor-car will fail to rejoice" because oil shale surely would provide the "supplies of gasoline which can meet any demand."[3]

But fervor for this rock has petered out time and time again in the United States due to the imposing cost of heating it to get the kerogen out, as well as the environmental damage that usually results from this extraction.

All of those dispiriting experiences around the country helped explain why experts and industry members viewed efforts by George Mitchell and his men in the Barnett as quixotic and a tad amusing. Between 1988 and 1992, the world's largest oil companies spent over $150 billion on exploration and development outside the United States, about 50 percent more than they were spending domestically. It was a sign of how little of value was left in the country, and how much more attractive foreign locales were.

Good luck with that shale drilling, George, the Goliaths all but said. Let us know if you ever find anything. We'll be off in Africa minting money.

Mitchell knew all about the gloomy history of various regions of the country, and about how disappointing shale and other unconventional rock had been. But his early tests in the Barnett had been encouraging. And earlier in his career, he and his men had figured out how to fracture rock to coax natural gas from other challenging rock formations. He was determined to do it again, this time with shale in Texas.

Anyway, he didn't really have another good choice. "We had to make the gas separate from cells and flow to the pipeline, we were desperately trying to find the gas," he recalls. "Chicago was a good market for me, I wasn't about to lose it."

Mitchell's quest was about more than keeping his company afloat. Just as well-paid ballplayers with guaranteed contracts still throw tantrums after a benching or a bad loss, Mitchell's self-worth was tied up in his ability to find oil and gas fields. It was that way with most dedicated oil and gas wildcatters. That larger competitors were so dismissive of his efforts only stoked his fires.

Investors and colleagues became concerned that Mitchell was too distracted to make the shale effort succeed, however. He was guiding the Barnett efforts, but he was also involved in the minutiae of The Woodlands. He approved each of the over 150 wells his company drilled in 1993, even as he directed the planting of wildflowers along fifty-five miles of biking and hiking paths in the Woodlands project. He also took the time to select ice cream for the soda fountain of one of his Galveston hotels. (He chose Texas-made Blue Bell Creameries ice cream over Häagen-Dazs.)

"He's like a shade-tree mechanic who spends too much time on his hot rod and ignores the family station wagon," Thomas Lewis, a securities analyst, said in July 1993.[4]

Investors grew weary of the company's real estate projects, which accounted for less than 20 percent of earnings. Pressure grew on Mitchell to pick one business, stick with it, and finally see some results for all his dreaming.

Later that year, Mitchell's right-hand man, Don Covey, the head of the company's exploration and production division, died of a heart ailment on a flight to Europe. Covey had been Mitchell Energy's other ardent proponent of shale drilling, and his loss was another blow to Mitchell's efforts in the Barnett.

In 1994, Mitchell began to cede control of his company, handing the job of president and chief operating officer to industry veteran W. D. "Bill" Stevens, who emerged as Mitchell's heir apparent. George Mitchell remained Mitchell Energy's chairman and chief executive officer, but a passing of the torch had begun.

With the move, Mitchell, now seventy-five years old, suddenly had a new and unexpected foe in his lonely effort to unlock the mysteries of drilling in shale. His quest was on borrowed time.

When Bill Stevens retired from Exxon Corp. in 1992, it was an abrupt departure. He had spent more than three decades at the oil and gas giant and had reached the position of president of Exxon's U.S. operations, one of the top jobs at the company. He was popular with the rank and file and in the middle of a long, successful career.

But when the company's grounded supertanker, the *Exxon Valdez*, began leaking oil in March 1989, Bill Stevens's career also became ensnared. The oil tanker sent more than 250,000 barrels of crude pouring into icy Alaskan waters, one of the worst environmental disasters in U.S. history.

Within forty-eight hours of the spill, Stevens, who ran the division overseeing Exxon's crude shipments from Alaska, rushed to the site, trying his best to deal with the aftereffects. Stevens defended Exxon's response in testimony before Congress, though he acknowledged that the company's contingency plans for a big spill had been inadequate. Later, he was portrayed in a television movie adaptation of the disaster called *Dead Ahead*.

After the fiasco, Stevens clashed with Exxon management over cost and staff reductions, according to reports at the time, before retiring in February 1992, after thirty-five years at the company.[5]

Shortly after leaving Exxon, Stevens joined Mitchell Energy's board of directors, where he quickly gained the trust of George Mitchell and some family members. Stevens asked tough questions about how the company was spending its cash, upgraded Mitchell's decision-making process, and hammered board members with warnings about tough times ahead. "Do you understand what happens" if the company can't find new gas deposits and has to sell its remaining production at low market prices? Stevens demanded of the board. George Mitchell and other board members began to realize their situation was direr than they had imagined.

When Stevens was tapped as Mitchell Energy's president, chief operating officer, and head of exploration and production in 1994, investors assumed Mitchell Energy would refocus on oil and gas drilling and finally move away from real estate. They also expected Stevens, fifty-nine years old, to shift investments abroad, just like most large U.S. energy companies.

There was good reason to anticipate that Stevens would slow the company's shale-drilling efforts. As a member of Mitchell's board of directors, he had quietly campaigned for the company to slow its push in the Barnett, telling board members and other executives that trying to get gas out of much of the formation was too costly and unlikely ever to work.

Stevens told board members that much of the Barnett was "moose-pasture land" unfit for exploration. Sometimes he grew visibly frustrated when George Mitchell discussed plans to expand the company's shale drilling. At board meetings, Stevens didn't directly challenge the company's founder. Instead, he addressed his warnings to Todd Mitchell; Stevens seemed to hope the company's founder was listening when he alerted his son to the serious risks ahead.

Todd already held serious doubts of his own about the potential of the Barnett. Drilling in shale there was just too expensive to be worthwhile unless natural gas prices somehow took off, he had told his father. Stevens's alarm served to make Todd even more nervous about the company's fate.

Over time, the younger Mitchell came to see his job at the company as "describing to my father the dimensions of the box we were in," he says.

One day, after hearing more words of warning from Stevens, Todd Mitchell approached his father to discuss the latest rebuke from the company's president. "Dad, tell me what part of that is inaccurate?" he asked his father. George Mitchell waved his son off with a smile, unwilling to let his enthusiasm be dampened by a couple of worrywarts.

Most of the time, Stevens and Todd Mitchell kept their unhappiness to themselves. "I was pushing like hell to find gas and they didn't want to upset me," George Mitchell explains.

But other employees became aware of Stevens's growing displeasure with the work on the Barnett Shale. Within Mitchell Energy, the shale effort became viewed as another of George's pet projects, one unlikely to ever work. Some of those leading the work on the Barnett Shale, including Nicholas Steinsberger, soon would find themselves in difficult spots.

When Nick Steinsberger received a promotion to run the fracking effort in the Barnett in 1995, he was thrilled. A native of Indiana, Steinsberger was fascinated by engineering and geology at an early age. He dedicated a seventh-grade paper to the "THUMS Project," explaining how five oil giants drilled under Long Beach, California, in the 1960s, and the ways they replaced oil and gas under the islands with water to keep the city from shifting.

"I thought that was the coolest thing ever," he recalls. "I was really interested in how mountains and the earth are formed and things like that."

Steinsberger's parents didn't quite understand their son's growing curiosity. His father was a political science professor with a liberal bent, and his mother was a nurse. But they encouraged his interest nonetheless.

After studying petroleum engineering at the University of Texas at Austin, Steinsberger was hired by Mitchell Energy as an engineer, overseeing thousands of the company's wells by the early 1990s. In the fall of 1995, when Jim Addison, who had been in charge of fracking wells in the

Barnett Shale, was shifted to the company's headquarters in The Woodlands, Steinsberger was tapped to replace him.

By then, the quest to find gas in shale had become so unpopular
within the company that Steinsberger's new job was more of a booby
prize than any kind of reward. "At the time, being sent *away* from the
Barnett was a promotion," he recalls ruefully of Addison's departure.

The company already had drilled over two hundred wells in the area,
spending precious cash with little return. Gas prices were low, the industry was struggling, and prospects for extracting substantial amounts of
gas from the Barnett Shale looked dim. In one of his first high-level meetings that year, Steinsberger got a clear sense of the gloom that enveloped
management at the time.

"We might drill for a couple more years and then give up," an executive told him.

"Management was getting ready to walk on the Barnett project, they
were all pretty down on the whole thing," recalls Steinsberger. "The
entire industry was in the dumps."

Steinsberger, a soft-spoken man who looked even younger than his
thirty-one years, retained high hopes for what he and the team could
accomplish. George Mitchell was among the few executives who still
believed they could find a way to get large amounts of gas from shale. By
that time, however, Mitchell no longer was running day-to-day operations at the company.

It was Bill Stevens who had assumed greater importance at Mitchell
Energy. By then, everyone knew Stevens had a vastly different view of
what was going on in the Barnett than the company's aging founder.

As they discussed the company's drilling program one afternoon,
Stevens told Steinsberger that he probably was wasting his time, arguing
that shale is a "terrible" reservoir. "Stevens came from Exxon and he had
a big-company mentality that shales are a horrible thing," Steinsberger
recalls. "Majors didn't have the patience or time to make shale work—
they were all about out-of-the-park home-run wells in places like Angola."

For his part, Stevens says that "the issue wasn't can you hydraulically
fracture and drill wells in the Barnett, it was can you do it economically and
make money. We were in the business of making money, not drilling wells."

Steinsberger and his colleagues kept working on the Barnett wells, experimenting with different liquids and gels as they blasted the rock, trying to create fractures and pathways for the trapped gas to get out. The liquid they used to shatter the rock was mostly water. But they also included sand to prop open the rock's cracks and let gas flow, as well as a variety of heavy gels and chemicals, to transport the sand and loosen up the rock. The gel was made up of ground-up guar, a tiny and pricey bean harvested by dirt-poor villagers in northwestern India that acted as a thickening agent, carrying the sand to the wells.

The Mitchell group wasn't doing anything special. Others fracking tough rock around the country were using the same combination of water, sand, and chemicals in their own liquid blasts, though Steinsberger and his colleagues used new combinations of chemicals and additives to try to make the gas flow.

"We tried every kind of fracturing material," George Mitchell recalls. "Heavy water, oil, propane, ethane, and all sorts of things."

Steinsberger and the others succeeded in fracturing the shale, just as they had hoped. But too little gas came out. The solution didn't seem within reach. It seemed likely that Stevens or someone else would pull the plug on them.

"I knew we couldn't stay in the play forever," Steinsberger says. "I felt the pressure."

In 1995, Mitchell's key longtime customer, the Natural Gas Pipeline Company of America, bought out its twenty-year contract to buy natural gas from Mitchell Energy, two years before it was scheduled to expire. Now Mitchell no longer had a surefire customer at a guaranteed high price. The company still would sell natural gas to the Chicago area from its dwindling gas wells. Now, though, the price would be linked to lower market rates. The timing—with natural gas prices around two dollars per thousand cubic feet—couldn't have been much worse.

The news got still grimmer. In March 1996, a Wise County jury found in favor of eight plaintiff groups claiming that Mitchell's natural gas operations had polluted their water wells. The plaintiffs were awarded $4 million in damages and $200 million in exemplary damages, a stunning blow. Some thought the company was done, since Mitchell Energy was worth less than $900 million at the time. Employee morale sank, as did Mitchell shares, though the decision was later overturned.

Behind the scenes, George Mitchell had become convinced his company didn't have enough money to finance both his energy and real estate operations. He also realized they were no closer to figuring out how to produce natural gas from shale. He began discussing the idea of selling the energy business to focus on real estate. But the shocking lawsuit made the energy unit impossible to sell.

Mitchell decided he'd refocus on finding an answer in the Barnett. It was really the company's only shot left at avoiding a complete collapse, he decided, no matter what Stevens and some others believed. Mitchell stayed in touch with his engineers and geologists, trying to encourage them. He couldn't let go of the idea of getting gas out of shale.

At home with his family at night, he would ponder the effort and sometimes become downbeat. "I thought about it quite a lot—if it's rich with gas, why can't we make it work?" Mitchell recalls. "People told me I was wasting my money. It was discouraging. I thought there must be some way to figure the damn thing out."

Mitchell didn't have much time left, but he wasn't giving up. The price of natural gas rose in 1996, making their expensive drilling in the Barnett a bit more palatable—if they could ever find some of that gas.

Once upbeat and optimistic, Mitchell turned more frustrated. After the group gave up on one well in the region and explained their rationale to him in a Monday-morning meeting, he became irritated, the endless failures finally taking a toll on him. "Have you really succeeded in what you set out to do with this well?" he scolded the men.[6] After another failure, he cursed and ranted. He told us "in no uncertain terms with a number of expletives that we weren't that smart," remembers Dan Steward, a senior member of the Barnett Shale unit.

Mitchell never shared his growing private doubts with his men. Instead, he kept telling them that he was sure they'd find a way to access the gas. "We've done the tests, there's a hell of a lot of gas there," he reminded one engineer, trying to keep his spirits up.

★   ★   ★

One day, while supervising the fracking of a well in the Barnett region, Steinsberger noticed that the gel and chemicals that were part of the fracking fluid weren't mixing properly. As a result, the substance being pumped by the crew of the contractor hired for the job, BJ Services, was more of a liquid than the thicker Jell-O-like substance normally used to fracture the rock and open its pores so gas could flow. The faulty fluid was one more piece of bad fortune for George Mitchell and his men.

"It was just a mistake," Steinsberger says.

Steinsberger saw that the well's results were surprisingly good, though. To most of the Mitchell crew, it seemed a fluke. The well was producing a respectable amount of gas *despite* relying on faulty and watery fracking liquid.

Steinsberger wasn't so sure, though. Maybe watery liquid actually worked just as well at fracturing rock as the expensive gel everyone used, he thought.

He kept trying to find a way to reduce the company's costs, figuring that might discourage senior executives from pulling the plug on the expensive Barnett project. Steinsberger's predecessor already had cut some costs related to how they fractured the rock. Steinsberger wondered if there might be a way to trim expenses still further and maybe buy the group some time to find an answer to drilling in shale.

A few weeks later, Steinsberger went to an industry outing in Arlington, Texas, at a Texas Rangers baseball game. While enjoying barbecue and beer, he saw Mike Mayerhofer, a friend who worked for a rival company called Union Pacific Resources. They began to chat about work and Mayerhofer told Steinsberger that his company was drilling for gas in sandstone in a region called the Cotton Valley in East Texas using a fracking mix composed mainly of water, with most chemicals and gels left out.

Steinsberger was intrigued. It reminded him of the earlier well that relied on that watery fracking fluid. He scheduled a visit to UPR's offices to get a better understanding of how Mayerhofer and his boss, Ray Walker, were doing it. Mitchell wasn't drilling in the same area of the state as UPR, so Mayerhofer and his colleagues agreed to share their experience with Steinsberger.

Fracking had been going on in places like Kansas for over two decades with just water. It was frequently referred to as a "river frac" because the water and sand often came straight from a local river. Mayerhofer and UPR also relied on a fracking liquid composed of water with just a little sand. But they had added a small amount of polymers to the mix to reduce the water's surface friction and act as a lubricant so it moved faster. It was a high-pressure water stream aimed at breaking apart the dense rock.

Steinsberger returned to Mitchell's headquarters determined to use more water on Mitchell's wells and to cut back on costly chemicals and gels. It was a long shot, but maybe other wells could also produce gas with this cheaper fluid. If UPR was seeing success with this watered-down concoction, perhaps Mitchell also could, Steinsberger told his colleagues. Because the fluid was mostly water that was slickened with some of those polymers, Steinsberger and his colleagues named the new method "slick water" fracturing.

"We really need to try some slick-water fracks," he told production specialists at Mitchell's headquarters in The Woodlands.

Most everyone thought Steinsberger was out of his mind. It was one thing to use water on sandstone, like UPR was doing. But you can't pour so much water on shale, a different kind of rock, they reminded him. Shale has clay in it. It's almost like hardened mud. Everyone knew clay swells and absorbs water. When that happened, all those small fractures would be blocked, preventing gas from flowing, they told Steinsberger. Just look at what children's clay does when water is poured on it; the clay soaks it up.

Behind Steinsberger's back, some at the company expressed skepticism. Others did it to his face. Steven McKetta, a Mitchell petroleum engineer and the fracking guru for the entire company, was among the most dismissive. In a key biannual meeting with about a dozen senior members of the production unit, he let Steinsberger have it.

"I'll eat my diploma" if the approach works, McKetta told Steinsberger, according to two people who witnessed the put-down.*

It was a stinging rebuke that intimidated Steinsberger, the youngest

---

* McKetta denies being skeptical about Steinsberger's idea or telling Steinsberger he'd eat his diploma if it worked.

employee in the room. Steinsberger responded that there was little down-side to his idea—just time and some money—since they could always go back to using the usual gel.

"It's a stupid idea," McKetta retorted. "It's not going to work."

Soon, Steinsberger noticed that McKetta had stopped talking to him, a sure sign his career was headed in the wrong direction.

I'm really putting my neck on the line, he thought.

Mitchell Energy employed a variety of companies, including BJ Services and Halliburton, to frack, or "complete," the wells. Experts at these companies also were dubious. Without the gel, how are you going to get all that sand to the rock to prop open the cracks? they asked Steinsberger. You never want water to contact rock, experts reminded him. Yes, their fracking gel also had a lot of water in it, but it turned to a gelatin-like substance when it was pumped below the ground, where temperatures rose, preventing the water from contacting the rock.

The outside specialists had good reason to promote their gels—they could charge about $200,000 more for a frack job with gel than one with-out it. Anyway, everyone in the industry agreed that gels and chemicals were crucial to fracture rock.

Steinsberger didn't necessarily disagree with the conventional wisdom. But the old approach didn't seem to work, and he had to cut costs somehow. He figured it was worth a try to include more water, cut out the gel, and reduce the chemicals. He was just trying to save Mitchell Energy enough money so that his group could live to fight another day.

"We were still pumping chemicals, just not so much of them," he recalls. "They weren't working anyway."

Senior Mitchell executives were concerned they'd get less gas from the Barnett with this new approach. But they were prepared to live with reduced production from Steinsberger's new liquid mix if it could save them precious cash.

Steinsberger and his crew experimented with this new liquid, trying to ignore the critics. Instead of using gels to carry sand thousands of feet to prop open fractures in rock and make the fractures as large and long-lasting as possible, Steinsberger's radical idea was to use smaller amounts of sand than usual to create micro-fractures in the shale. Maybe gas could travel through little avenues and side streets to the surface, rather than on a superhighway.

Like traditional fracking fluid, Steinsberger's liquid mix also included various chemicals, such as polymers used for women's makeup, and hydrochloric acid, commonly used to clean swimming pools, to help make the gas flow. But the chemicals and sand would make up about 0.5 percent of the new mixture, rather than about 10 percent in traditional fracking fluid. The rest of the new mix would be plain old water. Their slick-water frack also was called a "light sand" frack, because their blend only had a light sprinkling of sand in it.

"The idea was crazy at the time. He had guts, no one else would have even thought of doing it," says a Mitchell executive. "If the oil business had a gonads-on-the-anvil award, he'd win."

Mitchell executives let Steinsberger try his new approach on three wells in May 1997. The water-heavy brew reduced costs dramatically, from $300,000 per well to $85,000, just as Steinsberger expected. Production of gas from the wells wasn't anything to get excited about, though. "Some people thought I was an idiot," Steinsberger recalls. "They thought it was a one-off thing."

Some Mitchell executives wanted Steinsberger fired for wasting time and money on his absurd idea. He began to think his days at the company were numbered.

Steinsberger's wife, Kathleen, became anxious about their future. A part-time nurse, she was caring for their two young children and they were trying to pay a mortgage on their home. Steinsberger told his wife that the energy industry was doing so poorly that if he got fired he'd probably look for a job as an engineer in another industry. Kathleen prepared for a move to another city.

Steinsberger grew dejected. Coming home at night, he headed upstairs to his bedroom, avoiding his family. He began having trouble sleeping, worried about his future. "It was a really low point for me, people weren't supportive," he recalls. "We already had gone through four rounds of layoffs in my fifteen years at Mitchell and another one was coming."

That year, Bill Stevens helped push through the sale of the division that had developed George Mitchell's beloved community, The Woodlands. The company used proceeds of $460 million to reduce debt, buy back stock, and fund exploration. Investors were thrilled that the real estate division was gone. But now the company was solely an energy provider and all of its eggs were in the Barnett basket, at least until Stevens figured out how to get rid of that, too.

Pressure built to make the drilling a success. At regular Wednesday-morning meetings in 1997 at Mitchell's headquarters in The Woodlands, George Mitchell told his men he wasn't satisfied with their results. "Push harder," he told Steinsberger and his colleagues at one meeting.

They already had spent $250 million on the area over sixteen long years but had little to show for it. Mitchell wasn't very happy at all. "If you don't think you're up to this challenge, I'll find people who are," he told Steward, the senior executive, one day. In meetings, he often cursed and ranted.

"He was hard on us all," Steinsberger recalls. "People were afraid of him."

Unbeknownst to the Mitchell executives, Ray Galvin and Chevron were making progress, even as Mitchell struggled. Galvin's group began drilling in "tight" rock areas in Michigan. Chevron was also among the first to drill in an area in South Texas called the Eagle Ford that seemed promising.

By 1995, Galvin had assumed the role of president of Chevron's North American production unit. He now was one of the five most important people at the company and was on its board of directors. His group was spending about $50 million a year, and he was committed to figuring out how to tap shale and other challenging rock around the country. Ray Galvin was in a perfect position to make history by being the first to crack the shale gas code.

Galvin began hearing rumors that George Mitchell was thinking about selling his company, perhaps frustrated at his inability to find the formula to unlock the shale. Galvin considered reaching out to Mitchell, but decided against it, confident that Chevron would succeed with its own approach. Galvin and his team were like sprinters closing in on the finish line, paying little attention to competitors such as Mitchell falling by the wayside.

Not everyone at Chevron believed in what Galvin's group was doing, however. When they began referring to themselves as the "nonconventional" gas group, others teased that they should be called the "noncommercial" group. Galvin's gang might have some creative ideas, the jokesters were saying, but they likely wouldn't generate any actual revenue for Chevron.

In 1996, an executive of the company's foreign division came for a visit. He met with geologists working for Galvin, including Kent A. Bowker, to try to recruit them for a job in one of Chevron's hot foreign locales. The executive didn't mince words; he told Bowker that their efforts to pursue drilling in the United States were pitiful.

"I feel sorry for you," he told Bowker. "There's nothing big left to be found" in the United States.

The foreign executive didn't mean to be insulting. He was trying to lure Bowker to the foreign unit. Bowker burned, though. He was a true believer in the possibilities of shale and other rock in the United States and he was sick of hearing that Galvin's group was a bunch of losers working on a dead-end project.

No thanks, I'm happy where I am, Bowker replied.

After stewing a bit, Bowker began examining earlier estimates of how much gas was in the Barnett formation. Mitchell Energy had shared some details of its work with both the Department of Energy and the Gas Research Institute, a nonprofit organization that helped develop new energy technologies. At first, George Mitchell had resisted revealing any of the details, worried that competitors might catch on to what they were up to. But others at his company managed to persuade him that the groups might be able to lend them some help. Besides, they had to file for various permits and submit reports to the Texas Railroad Commission, so much of the data was already available.

Work from the government and the GRI, building on earlier analysis of Appalachian shale, had confirmed to Mitchell's men that the Barnett held substantial amounts of natural gas. The data wasn't enough to drop any jaws, though.

As Bowker pored over the previously published numbers, he made a startling discovery: The work seemed flawed. "This doesn't look quite right," he told a colleague at Chevron.

Bowker noted that shale in the Barnett was deeper and under higher pressure than Appalachian rock that was used as a comparison by the government and Mitchell Energy. The techniques and equipment used by Mitchell also seemed faulty or outmoded. Bowker was sure Chevron's cutting-edge resources could give him a better sense of the true amount of gas in the Barnett Shale.

Bowker and the rest of the team didn't have much time to make it work, though. Their boss, Ray Galvin, was approaching sixty-five years of age, the mandatory retirement age for Chevron's managers. Galvin became aware that his slated replacement, Peter Robertson, wasn't nearly as sure that shale would produce much oil or gas. Galvin could see Chevron was losing patience with the group. Time was ticking away on their efforts.

Galvin loved what he was doing, and he believed that Bowker and his colleagues were close to figuring out a way to drill for gas in shale. But he began to realize it was probably too late for him to do anything about it.

On the last day of February 1997, Ray Galvin retired from Chevron. He worked until his very last day, but then had to vacate his office, removing the biggest supporter of the group's efforts. Galvin had a prime shot at becoming the first person to discover the perfect way to extract gas from shale. He was born just a few years too early, though.

After Galvin left the company, Kent Bowker and his colleagues at Chevron persevered, conducting their own tests of the Barnett. They discovered the amount of gas stored in rock in the region likely was in fact many times larger than the amount that even Mitchell had estimated. George Mitchell didn't even realize the size of the gold mine he was sitting on, Bowker told others at Chevron.

"At that point, I knew the Barnett was going to be much more successful than anyone else realized," Bowker says. "Exploration geologists live for moments like this, knowing where there's a huge hydrocarbon deposit before anyone else does."

The Chevron gang was like a group of fortune hunters, getting closer to a buried treasure they knew was much more valuable than even their fiercest rivals imagined. But they had to move fast to get the jewels before the Mitchell crew also figured out how much was in the Barnett—and before their new boss at Chevron closed down their nuisance of a group.

While George Mitchell's company had locked up virtually all the prime acreage in the Barnett area north of Fort Worth, Chevron acted quickly to gain control of about fifty thousand acres elsewhere in the Barnett area, in Johnson County, south of Fort Worth.

Peter Robertson, Galvin's replacement, paid little interest to the progress. The group was working on potentially groundbreaking techniques in shale exploration, but it was a long-term investment, one that made Robertson and others at Chevron uncomfortable.

Members of Chevron's drilling unit sensed that their effort was coming to an end. Some quit the company, discouraged at getting so close to such a major discovery. Eventually, Robertson disbanded the group.

Years earlier, Galvin's predecessor had handed him the reins and resisted second-guessing his decisions. He chose to do the same. "I didn't think I should poke my head in too much," he recalls. "My feelings about nonconventional gas were well known to Peter, and to all his direct reports. Shame on me if I wasn't able to do a better sales job."

Galvin also understood that Chevron needed to devote large amounts of money to deepwater spots offshore in the Gulf of Mexico and other areas it was much more excited about. "In Peter's defense, most of the managers of the company's business units weren't in love with a group getting up to $50 million a year to spend on projects that didn't have a guaranteed payout," he says.

Bowker, the geologist, didn't want to leave Chevron. But he worried that if he stayed around much longer he would be sent overseas to work on one of the company's core projects with one of the jerks who had been so dismissive of his group's work. Bowker's wife had just given birth to a baby boy and he wanted to remain in Texas, where he had dragged his wife from Princeton, New Jersey. He began to look for a new job.

"The handwriting was on the wall, people were starting to move out of the group," he recalls. "If I stayed at Chevron, I probably would have gone to West Africa."

For his part, Robertson says he doesn't regret the decision to disband the group. "Oil was at eleven dollars a barrel, it was a tough time," he says. "We were more focused on places like oil in Kazakhstan that turned out well, and we were cutting back on the U.S. We thought we could live without" unconventional oil and gas drilling.

Kent Bowker was bitter nonetheless. Chevron had seemed on the precipice of something great and now he was out of a job.

"Chevron didn't have patience for this," he says.

Geology was all Kent Bowker ever wanted to do, even though he grew up outside Detroit, far from the nation's oil fields. When he went to college, he expected to play football at Adrian College, a small Michigan liberal arts school. During the first day of practice, however, the brawny young man saw that he probably was better off focusing on his passion for academics. "I realized those guys hit too hard and I should probably stick with the studying stuff," he recalls.

In early 1998, Bowker got an interview at Mitchell Energy. Mitchell executives had discussed trying to hire him over the years, impressed by the knowledge he had displayed at industry meetings. As he chatted with a human resources executive at Mitchell during his interview, Bowker was asked if he had any questions about the company's policies.

He didn't really have any queries and he was uncomfortable making small talk. He knew a lot of the Mitchell geologists, had a good understanding of what they were trying to do in the Barnett, and was eager to get going.

The silence was awkward. Bowker knew he had to come up with something. The only question he could think to ask the human resources manager was one he thought was quite innocent. "Does Mitchell have a fixed workday," with set start and stop times, "or are there flexible start-ing times" like at Chevron? he asked.

The manager raised his eyebrows. Bowker knew he was in trouble.

"We don't have any of that flexible stuff," the manager said. "This is the kind of job where you're expected to come in when work is to be done," no matter the hour.

Bowker assured the manager that he was prepared to come in early or stay late, if needed. He was just being curious. A red flag had been raised, though, and the job was in doubt.

Later, the human resources officer approached Dan Steward, the se-nior executive in the group pursuing shale drilling in the Barnett, ex-pressing concerns about Bowker. The manager told Steward that Bowker wouldn't be a good fit for the company because he was asking questions about when the workday started, a sure sign he was lazy. "He won't amount to anything," he told Steward.

Steward seemed surprised. He said his experience with Bowker at Chevron suggested he didn't mind working long hours. Steward recom-mended that Bowker be hired, despite the flubbed interview. Kent Bowker had his dream job, but just barely.

At first, the fit seemed perfect. Mitchell was searching for gas in the Barnett, just as Bowker had been doing at Chevron. The Mitchell team believed shale could give up huge amounts of gas, also like Bowker. And Bowker lived a few miles from the headquarters of Mitchell Energy in The Woodlands; he was tired of traveling downtown to Chevron and now he had a ten-minute commute.

Best of all, he now worked for George Mitchell, an executive Bowker had long admired, rather than the suits at Chevron, who didn't get the significance of what they were doing in the Barnett. It seemed a match made in heaven.

Bowker's mother was disappointed by his move to Mitchell Energy. She and her friends had heard of Chevron, but Mitchell Energy was for-eign to them, she kindly noted to her son. And sure, the geologist before him at Mitchell who had focused on the Barnett formation had dragged his feet about working on the region and couldn't wait to get out of the assignment, creating an opening for Bowker. But he joined the company raring to go. Finally, he would be able to show that substantial amounts of gas could be extracted from the Barnett.

On his third day at the new company, Bowker walked to a nearby break room to buy a can of Coca-Cola. There, he saw Bill Stevens, the company's president. Bowker immediately got excited.

"Mr. Stevens, I just joined from Chevron," Bowker said by way of introduction.

Stevens, the former Exxon executive, responded with some good-natured ribbing about how Exxon was a superior company to Chevron.

Bowker didn't let the teasing douse his enthusiasm, telling his new boss that he was the new head geologist for the Barnett wells. "Mr. Stevens, I'm really looking forward to talking" to senior management about expanding in the Barnett, he continued. "I'm really excited—I have some ideas on how to improve" Mitchell's drilling there.

Stevens stuck his hand in Bowker's face. "Stop right there," he said sharply. "We've got enough of that stuff already."

Stevens told Bowker he was wasting his time focusing on the Barnett, Bowker recalls. There was no point drilling more wells there. You better find some other area to spend time on, he told the new hire.

Bowker walked back to his desk in a daze. He had thought Mitchell Energy was dedicated to figuring out how to make gas flow from shale. But here was the president of the company telling him that he had as little faith in the Barnett's possibilities as senior executives at Chevron had. "I was hired to study the geology of the Barnett, and the president of the company tells me on my third day he doesn't want me studying the geology of the Barnett," he recalls. "And he wasn't subtle."

Bowker told his direct bosses about his run-in with Stevens. They told him Stevens had growing clout at the company, and confirmed that he didn't believe in the Barnett's possibilities or what they were doing there. But George Mitchell remained a fan of their efforts and he still owned 55 percent of the company. Just try to ignore Stevens, they told him.

"But he's the president of the company," Bowker responded, unconvinced.

Bowker had joined the five-man Barnett group that included Steinsberger. The team was thrilled to add Bowker. He had more technical training in unconventional gas drilling—and was more enthusiastic about the potential of shale—than anyone they had encountered. By hiring Bowker, they also could draw on all the work Chevron already had done on the Barnett.

Some weren't quite prepared for how outspoken Kent Bowker was, however. Bowker was a serious geologist, but he usually came off as more cocky than confident. Some thought he had a chip on his shoulder or perhaps retained some anger at how Chevron had ignored his work. It was a disappointment he often shared with his new colleagues.

It took just a few days for Bowker to tell his coworkers that they were making a big mistake comparing the Barnett with Appalachian shale and using that as the basis for estimates they had made on how much gas was in the Barnett. Their approach was all wrong, Bowker said.

Chevron's data suggested the Barnett held much more gas than Mitchell believed, Bowker said. Something must have gone wrong with Mitchell's work, he told them. "You don't have access to the science we had at Chevron."

"So what, Bowker?" one of his new colleagues said to him.

"Don't you get it?" Bowker asked. "There's more gas than you think in the Barnett. . . . The prize is *bigger*."

One of Mitchell's scientists turned angry. "That's all we need, some smart-ass from a major telling us how dumb we are," he said.

Bowker said he wasn't blaming the company's analysis. Mitchell was relying on faulty data from GRI, the industry group, he explained. "We're all scientists and engineers, let's just look at the data," he said.

The scientist wasn't moved by the argument. All in all, it wasn't an auspicious beginning to Bowker's career at Mitchell Energy. "I just got there, I get yelled at by Stevens, and now I'm ruffling feathers and telling them they're doing it all wrong," he remembers.

It's not unusual for geologists like Kent Bowker to be dismissed or humored within large energy companies. Geologists are optimists, the dreamers of the business. They *know* there's oil or gas in a rock formation. It's their job to figure out how much of it is there and tell everyone to go get it. Failure can be a good thing for a geologist; it means they're aiming high and searching for a huge discovery.

Engineers are the realists of the oil patch, the ones less comfortable with ambiguity and failure. They're charged with getting energy out of the ground. They don't have much patience for geologists like Kent Bowker when they pound the table, convinced they've uncovered a huge gas reservoir. They're even more unlikely to listen when they've been working on an area for a while and a newcomer comes in to tell them how badly they've been botching it.

Bowker tried to make his argument a bit less forcefully when he met in the office of his boss, Dan Steward, and Steward's boss, John Hibbeler. Bowker said Chevron's data, based on cutting-edge science, showed there was a heck of a lot more gas in the area Chevron controlled south of Fort Worth than Mitchell was saying was in their acreage north of Fort Worth. His point was that if Chevron was right that there was a ton of gas sitting in shale south of Fort Worth, then Mitchell was underestimating how much gas it was sitting on from very similar rock north of Fort Worth.

"You guys are either drilling in the wrong spot or your experiments are wrong," Bowker said. "And I don't think you're drilling in the wrong spot."

It was as close to subtlety as Kent Bowker gets. And it seemed to work. Steward and Hibbeler proved more open to Bowker's argument. They realized that if there was in fact much more gas in the Barnett than they had previously assumed, the company was getting much less out of it than it realized. They'd have to find a better way to drill in the area.

"It was 'holy moly,' there's a huge amount of gas trapped in these rocks," Bowker recalls. "Not just some gas, but a world-class amount of gas."

If he was right, and if they ever discovered how to extract gas from shale, the Mitchell executives knew the Barnett could be a game changer for the company even for the country.

The analysis "changed many of our previous concepts in the Barnett," according to Steward, the senior Mitchell executive.

Bowker's work suggested that shale and other similar rock formations all over the country also might hold huge amounts of energy. He was given the green light to spend millions of dollars to do scientific work to prove that there was a lot more gas in the Barnett.

Other parts of the company were scraping for new cash, however. With natural gas prices still low, Mitchell fired employees, the latest in a series of painful layoffs. Remaining employees were on edge. The company now employed just about eight hundred people, down from thirteen hundred in just three years. But it still insisted on spending money on the Barnett. Some at Mitchell Energy thought their colleagues were throwing good money after bad and that all the wasted money might doom the company.

"I never got the feeling that other folks in the company resented us," Bowker recalls. "The attitude was more one of derision."

Mitchell and his men still weren't sure how *much* gas was in the shale. And at that point, Nick Steinsberger had only proved that the new, slickwater fracking fluid could save the company money, not that it could unlock much natural gas.

Steinsberger knew he had to figure out a way to get large quantities of gas out of the ground at a reasonable cost or they all were in trouble.

CHAPTER FOUR

Nick Steinsberger sat in his Fort Worth office, anxious and impatient.
It was August 1997 and Steinsberger felt the heat, but it had nothing to do with the intense Texas summer. He was watching daily results from dozens of wells Mitchell Energy had drilled in the Barnett area. He was most focused on three of them—the ones they had fracked with his radical, water-based mixture. The wells had been middling performers from the start, though, disappointing Steinsberger and his bosses.

The pressure had ratcheted up on Steinsberger. He watched the wells closely, hoping for some good news.

In late 1997, he noticed something strange. The wells that had relied on the gel substance used by Mitchell Energy and everyone in the industry to fracture rock showed a sharp decline in how much natural gas they had produced. That was to be expected after an initial surge of gas. But the wells fracked with Steinsberger's new, slick-water fluid were producing gas at more or less the same steady levels, without much of a decline.

The gas coming out of Steinsberger's wells wasn't that much, really. But each day that he checked, over several weeks, he saw little drop-off in the wells' production, surprising him. Most of his colleagues dismissed the results, and rivals at other companies still scoffed. But Steinsberger thought there was a slight chance they were on to something.

"They fooled us," Steinsberger recalls, referring to the wells that relied on his new watery fracking fluid. "They started slowly, but they kept on going."

The results were enough to get his boss, Mark Whitley, to approve trying Steinsberger's fluid on a few more wells. By then, Steinsberger had tweaked his methods, using more horsepower to pump the liquid. That seemed to create more fractures in the rock. He and his colleagues also learned to start with small amounts of sand and add a bit more as they fracked a well.

Steinsberger kept a close eye on the new wells during the spring and early summer of 1998, unwilling to give up on them. Each morning, workers in the field sent him a reading of the wells' production. When he couldn't wait for the next morning's reading, Steinsberger sometimes called a pumper or foreman on-site at one of the wells for a fresh afternoon reading. Sometimes he drove thirty miles to a well to see the results himself. He was like a father pacing in a hospital waiting room, awaiting the birth of a first child.

One day, while Steinsberger was examining results from a well called S.H. Griffin No. 3, he was taken aback. From the start, this well was special; it managed to produce nearly one and a half million cubic feet a day, making it among the best performers of the three hundred wells Mitchell had drilled in the Barnett. Steinsberger and his colleagues had waited for the well's production to decline, as it had in other wells. After ninety days, though, the Griffin well kept churning out natural gas at the same remarkable pace. No Barnett well had *ever* produced even one million cubic feet of gas after ninety days, and this one was doing much more.

"It was unheard of," Steinsberger says.

Steinsberger kept his excitement in check, worried the results might tail off. He tried to distract himself with other work. When he checked back after another thirty days, though, his eyes grew wide: The Griffin well was still at it, producing huge amounts of natural gas, like an ever-flowing river.

This is unbelievable, Steinsberger thought.

He began telling colleagues about the wonder well. It was a flow of natural gas the likes of which they'd never before seen. Finally, the Mitchell team had justification and reward for the hard work they'd put into drilling in shale.

"This was the aha moment for us, it was our best well ever in the Barnett, and it was a slick-water frack," Steinsberger recalls. "And it was my baby!"

The Mitchell crew began to see extraordinary results from other wells relying on Steinsberger's fracking fluid. He and his bosses came to a startling conclusion: Their slick-water concoction wasn't just cheaper than the chemical-and-gel fluid they had been using. It somehow was more effective as well.

All the water pouring into this rock didn't make the clay swell after all, despite the warnings of the experts. That was because there wasn't as much clay in the Barnett Shale as there was in other types of shale. Instead of swelling when the high-pressure liquid hit it, this brittle shale shattered like glass, allowing gas to flow.

The gel used in their old formula had gummed up fractures in the rock, preventing natural gas from flowing. But the water-based liquid seemed to go out in every direction in the rock, creating complex mini-networks of cracks, each enabling gas to flow to the surface. Steinsberger and his colleagues even got gas from natural fractures in the shale, and from the Upper Barnett, a layer of rock above where they were drilling.

They had discovered the right fluid to fracture rock, the secret sauce for drilling in shale. And it had come about by happenstance, when too much water got into their earlier mix and they were just trying to save some money.

By September 1998, Mitchell's engineers had switched to relying on Steinsberger's slick-water method to fracture rock for all their wells, dropping gel-fracks entirely. They decided to increase their activity, drilling a well every two weeks, to take advantage of their discovery.

George Mitchell received his own daily reports about the wells, and he too saw the startling amounts of gas suddenly being produced. "You get excited, you think it's working," he says.

Sitting in his office in The Woodlands, Mitchell tried to contain his enthusiasm, though, worried the results wouldn't pan out. Other executives also didn't get too worked up. That year, a book of interviews with company executives was published, *How Mitchell Energy & Development Corp. Got Its Start and How It Grew*, written by Joseph Kutchin. It had one mention of what they were working on, a throwaway line by George Mitchell simply describing the Barnett's location.

There was good reason George Mitchell and others were so subdued. In late 1998, shares of Mitchell Energy fell below ten dollars, down by over half in just a year and by over 60 percent from the summer George

Mitchell eagerly showed off The Woodlands to two reporters. The company had piled on so much debt to afford all its spending that lenders wouldn't offer more. Skepticism abounded about the company's efforts to tap shale. Gas was being produced from the Barnett, but it wasn't a lot, and it wasn't making them any money. When Stevens and Todd Mitchell argued that the company needed to rein in its spending on the Barnett, few at the company's 1998 board meeting disagreed.

Elsewhere, financial markets were rocked by a debt default by Russia, an economic slowdown seemed possible, and natural gas prices appeared likely to stay low for the foreseeable future.

George Mitchell was dealing with mounting personal problems, though his employees didn't realize it. Over the years, he and his family had resisted selling his company's shares. To finance big-ticket items, such as his investments in Galveston and various pledges to a group of charities, Mitchell had used shares of his company as collateral to borrow millions of dollars from a group of ten banks.

Now, as Mitchell Energy shares fell and the collateral dropped in value, banks levied pressure on George Mitchell to pay the loans back or add new collateral. He didn't have the means, though. The family met and decided to reduce its spending, to try to deal with the sudden cash crunch. Mitchell was on the verge of foreclosure from the banks.

"It was a serious time for the family, we had to get our act together," Todd Mitchell says. "We were hustling."

George and Todd set up meetings with various charities, including the Houston Symphony, to say that the family needed more time to make good on its pledges. "We don't have a choice," Todd told them apologetically.

By not meeting their pledges, the Mitchells had violated the contractual terms of their agreements with the charities. Now they risked being sued by the charities, a move that would surely draw ugly publicity.

"It was embarrassing, I had to go and talk to people and say our circumstances had changed," Todd Mitchell recalls. "We didn't have an option."

Cynthia Mitchell, who had grown up poor, saw the growing strain on her husband. Each night, when George came home from work, he seemed tense. Cynthia began to go to extremes, cutting back on her spending. One day she stopped writing checks to help pay for her grandchild's college education. Todd, who ran the family's finances, drove right away to his parents' home to assure his mother that she could still spend some money despite their new restrictions.

"Mom, it doesn't mean you can't help your granddaughter go to college," he told his mother.

Family members noticed out-of-character behavior by Cynthia. The behavior confused them and saddened them. Later, Cynthia would be diagnosed with Alzheimer's disease. At that point, though, all George knew was that something disturbing was happening to his wife, and it added to his strain.

Mitchell retreated to his office in Houston, infrequently visiting the company's headquarters, which remained in The Woodlands. He was seventy-nine and dealing with a personal ailment of his own. He had been diagnosed with prostate cancer and was undergoing a brutal regimen of treatment, even as his wife struggled in the enveloping fog of her Alzheimer's disease.

In the Barnett, the team had discovered the perfect way to fracture and get natural gas from shale, succeeding where countless engineers and others had failed. Extracting gas from this über-difficult rock was the most important advance in decades in the energy industry. Few investors, rivals, and industry experts cared very much, though. Almost no one had noticed the breakthrough. Most didn't think there was that much gas in the Barnett anyway. It seemed like a lot of wasted time.

Kent Bowker knew he had a chance to change the way most people viewed Mitchell Energy's efforts in the Barnett. He just had to convince his own company's senior executives first.

It was the late spring of 1999, and Bowker, after a year of hard work, finally thought he had proved that the Barnett held huge amounts of gas. He had shared his results with his senior managers, who said they had relayed them to George Mitchell. Bowker doubted his bosses fully understood his research, though, or could properly explain its significance.

"I felt like the only person in the country who knew how much gas was in the shale," he says.

When Bowker was invited to present his research to George Mitchell and other senior executives at Mitchell's headquarters in The Woodlands, he knew he had his chance to show the world how much natural gas was hidden in the Barnett.

As he walked through the entrance of Mitchell Energy's office, he was energized. He had his presentation down cold. At Chevron he had been taught how to make high-level presentations to top energy executives, giving him added confidence.

If Bowker could prove to the executives that the Barnett held huge amounts of gas, Mitchell Energy could devote more resources to find it all, changing the company's history, Bowker thought. Perhaps he could even get the industry, Wall Street, and the world to take notice. If his presentation went awry, though, Bowker knew they'd miss a historic opportunity, just like Chevron had.

It wasn't going to be easy. Bowker knew that some of the executives he was about to face remained skeptical about the work in the Barnett. A colleague who knew that Mitchell had a short attention span and wasn't a very good listener had given Bowker one last piece of advice: "Keep it short or you'll lose George."

Bowker joined a group of men as they walked down a long hallway to Mitchell Energy's fifth-floor conference room. Everyone was dressed a little better than usual, some wearing ties and jackets. Soon, fifteen executives had crowded around a long oak conference table. Technical staff members also joined the meeting. Nick Steinsberger and Dan Steward were there, as was an outside expert who had been hired to double-check Bowker's work.

During his career, Bowker had been in the conference rooms of a few energy companies. He had told friends that those with art on their walls were companies that cared more about balance-sheet maneuvering than finding new oil or gas deposits. They were the fakers of the business, he had said.

As he looked around the room at Mitchell Energy, Bowker saw dozens of maps on pushpin walls.

That's a good sign, he thought.

George Mitchell walked in from an adjoining room. He was with his son Todd, who wore a backpack on his shoulder, like a true geologist. Bowker's sprits climbed still further; he figured Todd Mitchell must have been invited because his father expected to hear some big news.

Todd's a scientist, he'll understand this! Bowker thought.

The odds were more stacked against Bowker than he realized. Todd Mitchell had spent years as a behind-the-scenes naysayer about the Barnett. And he hadn't been invited to the meeting by George Mitchell. Rather, the invite had come from Bill Stevens, the company's president, of all people. Stevens wanted another skeptic at the table to counteract any potential enthusiasm by George Mitchell about whatever Bowker was about to say.

"My father has a tendency to ignore obstacles," Todd Mitchell says. "Bill wanted me to understand what was going on. . . . Stevens saw me as someone who could communicate with my father."

Bowker began his presentation by showing why earlier tests of the Barnett had relied on improper pressures, among other mistakes. He focused his attention on George Mitchell, who was sitting at the middle of the table wearing a sports jacket that looked at least a decade old. Bowker tried to ignore Stevens, who was wearing a stylish suit and sitting to Mitchell's right.

Bowker cited his research to explain why there likely was much more gas in each ton of the Barnett Shale than previously estimated. All the company had to do was apply Nick Steinsberger's new fracking liquid to their Barnett acreage and they could tap a true gusher of gas. Mitchell Energy had a unique, even historic opportunity to get much, much more out of the ground, Bowker argued.

George Mitchell didn't show any kind of reaction. Neither did his son. It wasn't clear Bowker was getting through.

Bowker rattled off some more numbers, such as an estimate of the amount of natural gas in each ton of rock in the Barnett.

George Mitchell looked impatient. Some of Bowker's terminology was best suited for technical journals. Mitchell was an old pro, but it was still too wonky for him, at least at that moment. He was at the edge of his seat, hoping to make a decision on whether to go full steam ahead in the Barnett. He didn't need a dissertation on the rock's characteristics.

"What does that mean?!" he interrupted in frustration. "Give it to me in square miles."

Mitchell wanted to know how much gas was in each of the five thousand square miles that Mitchell Energy controlled in the Barnett. He needed a simple number representing how much gas the company was actually sitting on per square mile. Mitchell didn't care how much gas was in each ton of rock.

Bowker hadn't really thought it through in those terms. He borrowed a calculator and began punching up figures as the executives in the room watched and waited. After a long minute or so, he looked up. He said he and his team believed there were 185 billion cubic feet of gas for each square mile in the area. That was *more than four times* Mitchell's previous estimates. Mitchell was extracting only about 7 percent of the gas stored in the Barnett Shale, not the 30 percent the company thought it was getting, Bowker argued.

Bowker was finished. The consultant in the room was asked for his view on the presentation and he concurred with Bowker. There was a hushed silence as everyone looked at George Mitchell, waiting for his reaction. A smile formed on his face that grew bigger and bigger. It was dawning on the septuagenarian what kind of treasure his company was sitting on. Mitchell stretched his arms out wide and looked excited.

"This is huge!" he exclaimed.

Todd Mitchell flashed Bowker a smile. He too was convinced by the presentation and agreed with his father's enthusiasm, a sign that management would go all in on the Barnett.

George Mitchell now understood that the Barnett Shale held more gas than almost anyone believed possible. And thanks to Steinsberger's work, they now had a method to get it out. The company, and perhaps even the nation itself, had an unmatched opportunity. And it was happening just as more experts were wringing their hands about where new sources of oil and gas would come from in the United States.

Mitchell realized he'd have to lease a lot more land in the area if he was going to seize this prime opportunity. But the only way they could do it at reasonable prices was if no one else found out how eager Mitchell Energy now was to expand its drilling. He spoke up again, this time with more urgency and seriousness. "This is the biggest secret in the history of the company," he told the group. "No one can hear about this!"

The group left the room in high spirits. Bill Stevens exited through another door by himself, heading in a different direction.

Sanford Dvorin, the stubborn wildcatter from Newark, was making his own progress. Dvorin was working to unlock gas from shale, just like Mitchell. He was just as certain the Barnett area had a world-class supply of gas, if only someone would let him drill in their backyard.

By 1996, Dvorin had focused on drilling in the city of Coppell, a Dallas suburb five minutes from Dallas/Fort Worth International Airport. The city was close to a natural gas transmission line, making it easier for Dvorin to sell gas if he ever found some of it. Coppell was just fifteen miles away from Mitchell's wells, another reason he thought he had a sure thing. If George Mitchell was excited about the area, it must have real promise, Dvorin figured.

Coppell was an up-and-coming city with little interest in energy drilling. Its young and affluent homeowners already enjoyed brick homes worth a quarter of a million dollars, paving-stone intersections, and turn-of-the-century lampposts.[1] The city had little need for gas wells, and little need for Sanford Dvorin. Besides, striking oil or gas in suburbia seemed preposterous. Most were certain that if there were reservoirs to be found, someone already would have done so.

After a year, though, Dvorin met Bill and Adelfa Callejo, Coppell residents who agreed to let him drill on their land. The Callejos were well-known attorneys in Dallas who had devoted a lifetime to sticking up for outsiders; Adelfa had spent three decades as an activist defending the rights of Mexican Americans and others after becoming the third woman to graduate from Southern Methodist University's law school. The Callejos also happened to own 130 acres of open land zoned for light industrial development. At the time, the acreage was being leased for cattle grazing, making it perfect for drilling.

The Callejos agreed to lease their mineral rights to Dvorin and even formed a close relationship with him, rooting hard for his success. As a royalty partner, the family figured to do well if Dvorin discovered gas on their land, of course. It was about more than money, though.

"We were *mishpucha*," Bill Callejo, a New Yorker of Puerto Rican Catholic descent, told the *Dallas Observer*, referring to the Yiddish word for family. "We established a good family relationship, and the business came later."

They fully appreciated their odd pairing with Dvorin. "You have a Puerto Rican Yankee, a Mexican Texan [Adelfa Callejo], and a Jewish oil operator from New Jersey," Bill Callejo told the *Observer*. "What a trinity. . . . We're giving it a legitimacy it claims in TV shows."

It took Dvorin another year to persuade Coppell officials to let him drill on the Callejos' land. Early on, he couldn't find any financial backers. He was a lone, quirky wildcatter with a limited track record, and he was focused on questionable acreage. Dvorin had to put all his net worth into drilling his one well, which he dubbed Callejo No. 1.

To ease the financial pressures on Dvorin, the Callejos agreed to forgo their lease payments in exchange for a bigger cut of profits if Dvorin discovered gas. Soon he found a small company called Foundation Drilling & Exploration willing to finance his efforts, allowing him to proceed.

In early 1997, Sanford Dvorin and his son, Jason, began drilling. Their work caused an initial roar that disturbed some neighbors, but others began to root for their hometown Don Quixote. "I never prayed to God for money, only for the health of my family," Adelfa Callejo said at the time. "I know if the gas well comes in, I will benefit. But I asked God to please let it happen—for the Dvorins. They have so much at stake."

Even the Mitchell team was pulling for Dvorin. Mitchell executives never wanted the bother of drilling in the highly populated Dallas–Fort Worth region, but they figured if Dvorin could extract gas from his slice of the Barnett, investors might become excited about their own drilling, boosting Mitchell Energy's slumping shares. They figured they might even learn something from Dvorin.

Dvorin's early attempts at fracturing the rock met serious problems, just like George Mitchell's had. After some trial and error, though, he reduced his gel and increased the water in his mixture, just like the Mitchell team. He began to see some progress.

Often, Dvorin gathered his drilling team to go out for lunch at a local steak restaurant, where he picked up the bill, according to Jim Henry, who lent a hand to the effort. Other times he set out silver trays of barbecue for members of his crew who had worked through the night.

"Sanford liked to ask how your wife was, how your kids were, there would be plenty to eat, and he'd pick up the tab," Henry says. "Some provincials might have been a little suspicious of a New *Joisy* Jew who spoke with a thick accent, but he was a generous and goodhearted character who became popular."

Dvorin and his family lived in a trailer on location, sometimes munching on bagels and whitefish salad from a Dallas delicatessen as they watched the fracking crew do their thing.

"They were a poor excuse for bagels," Dvorin says.

When politicians or members of the Mitchell team visited, the Dvorins handed out T-shirts reading "The Well in Coppell." The Mitchell engineers shook their heads, befuddled by their neighbor's unbridled optimism.

Dvorin and Foundation received permission to drill additional wells nearby. As they made progress, some of their field workers began to notice unfamiliar men roaming around, trying to get on their property. The men seemed to be attempting to ascertain how much progress Dvorin was making.

The same thing was happening near George Mitchell's wells. In past drilling eras, such skullduggery often resulted in gunshots aimed at pairs of fast-retreating feet. The ever-confident Dvorin had a different approach to his uninvited guests.

"I waved to 'em and yelled over, 'Hey, what do you want to know?'" Dvorin remembers.*

---

* Undercover activity was commonplace in the Barnett, part of the legal cat-and-mouse game played by wildcatters and drilling companies. An unrelated lawsuit later filed against an active investor in the Barnett, Trevor Rees-Jones, suggested that Rees-Jones's company, Chief Holdings, LLC, had hired "spies," "moles," and other "contacts brazen enough to scale barbed-wire fences and sneak onto private property for precious information." Other times they would conduct "daring trespasses and reconnaissance at competitor well sites," according to the suit, filed by a disgruntled backer of Rees-Jones.

At least once, Rees-Jones called the intelligence he received from the field a "Special Ops Re-

In April 1997, a year or so before Mitchell began making real progress in the Barnett, Dvorin received an early indication that he'd struck gas. Not only that, but it seemed to be in substantial amounts.

"This has got us so excited we can't see straight," he told a reporter, while pointing to a copy of early test results on the well. "That's a big well. A very big well."

Dvorin was sure he was on the brink of uncovering life-altering amounts of natural gas. He leased new properties in the western part of Dallas County and set plans to drill a hundred new wells. He figured the financing would come as word spread about his amazing discovery. Within months, he got a few wells going, and they also produced gas, just as he had predicted. He had somehow succeeded in extracting natural gas in Dallas County for the first time, despite all the skepticism.

As he watched the wells pump out gas, though, Dvorin's enthusiasm slowly began to ebb. His wells produced gas, just as he had predicted, but the amounts weren't huge. His costs were soaring and it was proving more expensive to drill in the city than he expected. Natural gas prices weren't keeping up, adding to his strain. He kept at it, but he wasn't making nearly enough from his wells to offset their high costs. He had yet to master the horizontal drilling that would be crucial to that part of the Barnett, further handicapping his efforts.

"I thought he'd produce a lot of natural gas," says Jim Henry, the former Mitchell Energy executive. "His ideas turned out to be geologically sound and people who came in after him did very well. He was just a little bit ahead of his time."

Soon, Dvorin found himself in a messy legal dispute with his backer, Foundation Drilling. Dvorin didn't have enough capital to hold on to his acreage and he had to give up his leases, a disastrous turn of fortune. He remained as the operator of the Coppell wells, but eventually was fired from that job, too.

---

port," according to the lawsuit. Sometimes the intelligence gathered had to do with internal estimates of the value of certain acreage. After appearing to uncover details of a high valuation another company, Devon Energy, had placed on its holdings in the Barnett, one of Rees-Jones's contacts sent him an excited e-mail: "Either someone at Devon is smoking crack or we're gonna be rich." The case was eventually settled.

"When the smoke cleared the only ones who made money were the lawyers," Dvorin says. "It got so nasty I almost sat down and cried."

At one time, Dvorin had leased five thousand acres in the Barnett at an average of fifty dollars per acre. Less than a decade later, the same acreage would sell for $22,000 an acre, or $110 million. Around that time, the area under the Dallas/Fort Worth International Airport that Dvorin was set on drilling was leased for nearly $200 million, as big companies like Chesapeake Energy became convinced the Dallas area was full of natural gas, just as Dvorin had tried to argue to investors.

By then, Dvorin and his son, Jason, had moved on from the drilling business.

"I'm seventy-three years old, I don't want to spend my time looking into the past and cursing," he said in an interview in early 2013.

Sanford Dvorin lost his big lead in the Dallas–Fort Worth area and could only watch as Mitchell and others made progress. Rather than become a rich Texas oil and gas man, Sanford Dvorin became an altogether different archetype of the energy pitch—a wildcatter who came close to a huge find but couldn't quite pull it off.

"The saddest part is we were right, and took heat, but we didn't profit from it," he says. "I didn't make millions, but I sure got screwed out of millions."

Today, Dvorin lives in a 2,000-square-foot home in a Texas suburb and drives a 2003 Cadillac. He and his son roam the region searching for attractive acreage to try to flip to exploration companies, a challenging endeavor.

"If it was easy work the Girl Scouts would be doing it," he jokes.

Dvorin holds out hope that he and his son will strike it rich yet, perhaps by discovering the next hot drilling location or by operating wells for someone else. He's also developing new and improved production tools, as he waits for his lucky break.

"Sooner or later, the pendulum might swing back," he says.

Mitchell Energy's shares slowly began to rise in 1999, as the outlook for natural gas prices improved. Relying on Bowker's data, Mitchell ordered the company to step up its leasing in the Barnett. The company also drilled more wells and did another round of fracking on old wells, realizing there was a lot more gas to be found than they had first thought.

Stevens grumbled when he heard about Mitchell's desire to lease even more land, several people at the company recall. One day after a board meeting, Stevens approached Todd Mitchell and another executive with an urgent request. "Go talk to your dad and tell him to stop buying" acreage outside a core area of the Barnett. He continued to argue that the core held promise, but the rest likely wasn't worth the time and drilling expense.

For his part, Stevens downplays his disagreements with others at the company. "We were all working for the same end result, and depending on your role in the company, you may have had different perspectives," he says. "I was balancing a lot more than the Barnett," including other parts of the business.

The leasing continued, but it was clear the acreage would have to produce larger amounts of gas soon or someone would pull the plug on the efforts in the Barnett. It didn't help that by October 1999, George Mitchell was a hands-off figure at his company. He had turned eighty years old, and while his cancer was under control, Cynthia was showing early signs of Alzheimer's disease and her illness weighed heavily on him.

Mitchell's company had survived a brutal period for the business, but he knew they'd have to come up with huge amounts of cash if they really wanted to take full advantage of its advances in the Barnett. It was cash they didn't have.

Reluctantly, Mitchell toyed with the idea of selling his company to a rival and slowing his pace. Bowker's data on the size of the Barnett seemed to make the company's acreage look attractive, and Steinsberger's improved fracking methods made it seem likely that a burst of gas could be extracted from the rock. If Mitchell could get a hefty price for his company, he decided he'd agree to a sale.

Mitchell hired Goldman Sachs, the preeminent investment banking firm, as well as Chase Manhattan Bank, to make a sale happen. Mitchell executives agreed to the difficult step of allowing competitors to come to their offices to spend a month reviewing their assets and proprietary information. George Mitchell told his bankers he'd only agree to sell his company if someone paid a price that was a premium above the Mitchell Energy's weak share price. "We'd done too much work not to do it for a premium," he says.

No one else seemed interested in his company, though. Rivals repeated the same thing: Mitchell's team was doing interesting work in the Barnett and there was a lot of potential. But the Barnett was a new play and they couldn't project how big it might get or if Mitchell's production would last. After all, short-term production rises were common in the business, but they often died down. Companies such as ExxonMobil, Conoco, and others had spent years ignoring Mitchell's work in Texas's Barnett Shale area. They sure as heck weren't going to start betting on Texas fields and this newfangled drilling now.

They still don't get it?! Mitchell thought.

Mitchell Energy had earned less than $100 million in the fiscal year that ended January 31, 2000, and the company had lost nearly $50 million just the year before. Mitchell executives spoke of how much gas was in other areas of the Barnett. But they hadn't actually drilled in those newer areas yet, potential suitors noted. The latest survey by the U.S. Geological Survey from 1998 had judged the Barnett Shale to have just ten trillion cubic feet of gas reserves. It was a puny and stale estimate that didn't reflect Mitchell's growing production, but it was all the industry had to go on at the time.

"We turned our noses up because we didn't think it would work," says Larry Nichols, then the chairman and chief executive of Devon Energy.[2]

It didn't help that natural gas prices remained under pressure in 1999, averaging just about $2.30 per thousand cubic feet, a popular measure of gas prices. The low prices doused the spirits of even the most optimistic rival.

Kent Bowker had another reason why the company was being ignored. "We were a bunch of goofy mothers in a little office in The Woodlands who weren't intellectual geniuses," he recalls. "We were just guys from Oklahoma State and Texas Tech with a task—to keep Mitchell in business. . . . We weren't PhDs in white lab coats or Harvard first-in-their-class yahoos."

With each rejection, George Mitchell became more enraged. The Internet boom was near its peak; any kind of company with a dot-com in its name was going public at an eye-popping valuation, but Mitchell couldn't find a single taker for his own company that was producing the energy for all of it.

"I can't believe it!" he screamed one day.

Mitchell took each rejection personally. Soon he was throwing temper tantrums around the office, railing about how companies with no assets were getting all the attention. "But we have the Barnett!" he said. He couldn't let it go, turning angrier with each rebuff by a potential suitor, and with each billion-dollar IPO by a new dot-com company. "This is ridiculous!" he screamed to his son Todd, who tried to calm his father.

In April 2000, Mitchell Energy announced it wasn't going to sell itself after all, an embarrassing public retreat. George Mitchell was devastated. "I just couldn't convince them," he recalls.

Mitchell realized there might be a silver lining to the brush-off from would-be buyers. If the industry was still skeptical about the Barnett, it meant one thing: They would have less competition leasing more acreage in the area.

The Mitchell team leased even more land. To avoid tipping off competitors to their new buying, brokers weren't told who was doing all the leasing. Before long, Mitchell had leased more than 180,000 acres in the Barnett, including areas Chevron had given up on a few years earlier, bringing Mitchell's total holdings to nearly 600,000 acres.

Surely the added acreage, new production, and other moves would make Mitchell Energy a more attractive acquisition target, Mitchell and other senior executives figured. Plus, natural gas prices were finally showing signs of recovering.

Their plan seemed like it could work—if the company didn't split apart at the seams while it waited for a suitor, a scenario that seemed increasingly likely. One day that year, management sent an e-mail to employees alerting them to an appearance on business cable television network CNBC by Mitchell's president, Bill Stevens. As Mitchell employees tuned in to watch the interview, their jaws dropped as Stevens described the shale in the Barnett as "black tombstone."

"That's how he talks about our biggest asset?" a Mitchell employee said to a colleague.

Stevens may have meant it as a joke, and Mitchell's men were being a bit sensitive. Shale actually does look like a tombstone, due to how compressed it is, and the comparison was common in the business. But to some, Stevens's comment was one more insult after years of hard work on the Barnett.

Stevens did other things to get under their skin. In a meeting in 1999, a geologist proposed that the company, which had begun to drill a few wells using the new horizontal method, ramp up its horizontal drilling in the Barnett. Stevens became annoyed. "I will be dead before Mitchell drills another horizontal in the Barnett," he told the geologist, ending the meeting.

After Stevens left the room, Kent Bowker turned to another geologist in disbelief. "Now we have to kill Stevens," he said.

The company never did drill another horizontal well in the Barnett.

"It was in jest, of course," Bowker remembers.

A year later, in an interview with the trade publication *Oil and Gas Journal,* Stevens added to the resentment when he talked about how easy it had become for the company to find gas. "I laugh and tell my exploration people, 'I don't need geologists. We don't find [gas], it's there.'"

Stevens likely meant his comment to suggest the company was an attractive acquisition candidate because its shale acreage didn't need sophisticated geology to produce gas. But his detractors took it as yet another insult. When he read the interview with Stevens, Dan Steward, the senior member of the Barnett team, went home and told his wife he wanted to quit.

"We have four kids," she told him. "Suck it up."

George Mitchell began hearing complaints about Stevens. It pained him that his geologists and engineers were chafing. But he and his son Todd remained big fans of Stevens and appreciated that he had made tough decisions that George Mitchell couldn't bring himself to make, such as the sale of The Woodlands. To the Mitchells, Bill Stevens was a perfect bad cop to George Mitchell's good cop.

"He's not the most warm and cuddly person, and my father's obsession with the Barnett was clearly a point of tension," says Todd Mitchell. "He took over a company in crisis . . . we felt he put the company on firmer ground."

The hard feelings within the company soon didn't matter very much, because Mitchell Energy's gas production was surging. In the summer of 1999, when the company had tried to sell itself, it had been producing nearly 100 million cubic feet of gas each day. By the summer of 2001, however, production was approaching a remarkable 300 million cubic feet a day. It soon passed 365 million cubic feet, a stunning increase of 250 percent in two remarkable years, one of the quickest and most unlikely discoveries in energy history. And it all was thanks to the groundbreaking work of Mitchell's small team in the Barnett, which somehow figured out how to extract gas from shale.

Around the Mitchell headquarters, even those responsible for the gusher were amazed at how their hard work was finally paying off. "Production grew like a hockey stick," says a member of the exploration team. "It was way past theoretical."

Word spread throughout the region and the oil and gas industry, and rivals demanded to know what was happening at Mitchell. "People began to wonder what the hell was going on" at the company, Mitchell recalls.

In late September 2000, the Oil Information Library of Fort Worth held a symposium to discuss the Barnett Shale. The library was hurting for cash after a decade of troubles in the business and was eager to raise some money. A recent tornado had severely damaged its building, another reason they hoped the meeting would be a success.

Organizers set up seats for 125 attendees, optimistic that industry members would come to pay the $195 entrance fee. On the day of the meeting, though, over 200 people stormed the library, including representatives of major oil companies. Officials had to set up speakers and monitors in the halls so the overflow crowd could hear the presentations; others were turned away. They all wanted to know how Mitchell was producing so much gas.[3]

They weren't the only ones. Larry Nichols, the chairman of Devon Energy, another independent producer with little pedigree, had turned down an opportunity to buy Mitchell a year earlier. But as Nichols read data showing how much natural gas Mitchell suddenly was producing, he was astonished. He couldn't figure out what was happening or how Mitchell was doing it.

"I challenged our engineers as to why this was happening," Nichols says. "Why was Mitchell's output going up?"[4]

Devon officials reached out to the Mitchell Energy team to better understand what they were doing in the Barnett. Over a matter of months, the Devon executives became converts to the possibilities of drilling in shale layers. Nichols and Mitchell formed a personal bond, two independent energy producers focused on the United States rather than foreign spots.

By then, Cynthia Mitchell was in the full grip of Alzheimer's disease and George Mitchell was even more eager to sell his company, as were his children. Trying to spark a bidding war, Mitchell reached out to companies such as Anadarko Petroleum.

"If you fracture these wells you can make two to three times the gas," Mitchell told one potential suitor. "We know what we're doing."

Most remained skeptical, though, figuring that drilling in shale layers was a short-term wonder. The companies still wouldn't pay a premium above the price of Mitchell Energy shares, one more insult to George Mitchell.

"Anadarko didn't believe we could separate the gas" from the rock, Mitchell says. "People didn't believe what we were saying."

Mitchell had produced a miracle, but it was as if no one believed their eyes.

"At that time," says Nichols, "absolutely no one believed that shale drilling worked."[5]

In August 2001, Devon Energy agreed to pay $3.1 billion for Mitchell's company and to assume $400 million of its debt. The price was 20 percent higher than Mitchell's stock price at the time, just as George Mitchell had hoped.

"We convinced Devon," Mitchell recalls. "They got it."

Many doubted Nichols and Devon could ever justify the expensive purchase of such speculative acreage. Some close to Mitchell also seemed unconvinced the Barnett ever would truly work out. After the deal was announced, Bill Stevens, the longtime skeptic, turned to Dan Steward, saying, "Boy, we hit a home run there."

"He said it with relief," Steward recalls.

George Mitchell and his team had cracked the code to the Barnett and had proved that difficult rock like shale could be a gold mine. Texas's Barnett region would become the nation's largest onshore natural gas field, representing about 6 percent of the entire nation's energy supply in 2013. George Mitchell, who believed natural gas in shale could quench the nation's desperate thirst for energy, would inspire a group of wildcatters to wonder if other areas with shale could produce similar gushers.

After the deal was completed, George Mitchell was worth nearly $2 billion. He and his son Todd would go on to lease new acreage in other shale plays around the country, expanding their fortune. George remained a maverick, however, donating millions of dollars for clean energy research, among other causes, while urging a crackdown on companies that weren't protecting the environment.

To Mitchell, an abundance of gas from fracking and drilling in shale seemed a great way to give the nation time to find dependable renewable energy sources. "We've got to sequester" carbon dioxide, "because it's a serious problem," he told the *Fort Worth Star-Telegram* in 2008. What "we need to figure out now is how we get through the gap from fossil fuels to renewables."[6]

Bill Stevens enjoyed his own successes after Mitchell. Two years after the sale to Devon, Stevens joined the board of EOG Resources, an up-and-coming oil and gas explorer set to do revolutionary things of its own in the nation's energy fields.

Those responsible for the energy industry's most important breakthrough in nearly a century didn't fare nearly as well as Mitchell and Stevens, however. Just days before Christmas in 2001, Dan Steward, Kent Bowker's boss, met with Devon's executives to go over some budget items related to the newly merged company. During the conversation, Steward's new bosses surprised him by saying he wouldn't be needed after the merger was completed. Steward had lent crucial support to the work in the Barnett, but he was allowed only to stay a few more months at the new company before he had to vacate his office.

Nicholas Steinsberger, who had discovered the perfect mix of liquids to extract gas from shale, received a salary of just over $100,000 the year Mitchell was sold. He never received any bonus for his breakthrough, though over time he cashed in another $100,000 of stock options. Steinsberger lasted a year or so at Devon before he also left, beginning a successful career working with other energy companies. Steinsberger saw his methods copied by countless companies around the country, though he also had to field questions from his liberal-minded parents back in Indiana, who began hearing from their friends about how fracking was affecting the environment.

Kent Bowker made about $120,000 the year Mitchell Energy was sold, his usual salary. He cashed in about $20,000 of stock options after the sale but never received any kind of reward for calculating the true size of the Barnett's gas deposits, an advance that paved the way to a lucrative sale of the company.

A few months after the merger was announced, Bowker was interviewed by a Devon executive charged with helping to decide which Mitchell Energy employees would be retained. As Bowker chatted with the executive, explaining his role at the company, the Devon executive seemed to grow tired. A few minutes later the executive fell asleep right in front of Bowker.

Bowker realized he wouldn't be given a meaningful role in the new company, so he quit to join a smaller company active in the Barnett.

"It was time to go," he says.

In nearby Oklahoma, two young men watched with special interest what George Mitchell had accomplished. Aubrey McClendon and Tom Ward weren't convinced a new era of drilling in difficult rock had arrived. They were among those who doubted that Mitchell's success in the Barnett could be replicated elsewhere.

McClendon and Ward had their own reason to bet on a new era of energy for the nation, however. They were so confident they could strike it rich that they decided to wager it all on that belief, with shocking results.

THE
RACE

CHAPTER FIVE

Aubrey McClendon and Tom Ward weren't convinced George Mitchell had discovered anything special. Sure, Mitchell and his crew had figured out how to access large amounts of natural gas from shale deposits over in Texas. But McClendon and Ward ran an energy company in Oklahoma City, one they had founded a few years out of college. It wasn't clear to them that Mitchell's water-heavy fracking techniques could be applied to rock elsewhere in the country, at least not at a reasonable cost.

Instead, McClendon and Ward were excited about horizontal drilling and other emerging methods of finding oil and gas. Mitchell's team hadn't spent much time experimenting with horizontal drilling, but McClendon and Ward were quick converts. They figured the Oryx team and other early practitioners were on to something really special.

By early 1999, McClendon and Ward were sure that natural gas prices were heading higher. Fewer companies were drilling for gas, reducing supply, even as signs of new demand were emerging. McClendon had been tipped off that a huge utility was ramping up its use of natural gas and that others soon would follow, news that cemented their bullish view.

"It's a classic supply-demand situation," Ward told McClendon one day.

McClendon and Ward seized on a plan: If they could lock up wells throughout the country that were already producing natural gas, and then employ horizontal drilling and other newer techniques to extract even more gas from the acreage, a historic fortune could be theirs. They decided to spend some serious cash to buy as much land as they could, all over the United States, as quickly as possible. They were determined to buy like no one before them to take advantage of what seemed like the opportunity of a lifetime.

There was only one problem: Neither McClendon nor Ward had quite as much money as they needed. It was a lifelong problem for the pair.

Tom Ward grew up in Seiling, a town of one thousand residents in northwestern Oklahoma. The Ward family was among those who remained in the state during the Great Depression and the Dust Bowl periods, devastating eras when nearly a half million "Okies" fled for California and elsewhere looking for work. Others sent children to live with families or friends in other states, hoping they'd find a better life.

The Ward family stayed around to farm the state's often challenging land. Searing memories of those difficult years left a deep impression on the entire clan. Years later, members of the extended family cultivated vegetable gardens near their homes to ensure that they always would have something to eat. Most barely made enough to get by.

Tom Ward's grandfather William Ward had an especially tough time providing for his family with odd jobs around town. By the time Bill Ward was thirty, he was turning to alcohol for comfort, to the dismay of family and friends. A friendly man when sober, Ward turned belligerent as soon as he began drinking.

On Sundays, Bill Ward often stumbled into church with the service well under way. He'd slump into a seat in a first-row pew as tears fell from his eyes. The young preacher, Orville White, often stopped midsermon to ask if there was anything the congregation could do for him.

"Sing 'Amazing Grace,' " Bill usually responded, referring to the Christian hymn of forgiveness and redemption. The song had special meaning—it had been sung several years earlier at his mother's funeral.

The congregation, with abundant sympathy for their struggling neighbor, would launch into a rousing rendition of the hymn, as Bill Ward sat in his seat, listening and crying. Then he'd receive a blessing from White and take his leave, well before the service had concluded.

Bill Ward's wife, Reva, was a kindhearted and popular waitress at a local restaurant who made easy conversation with customers. Each evening, she came home braced to deal with her inebriated husband. Later, Reva would sit with her grandson Tom and add up her day's tips, using the coins to teach him how to count. Reva Ward refused to take a day off, even when she felt ill, instilling a strong work ethic in Tom.

"You can be sick at home or sick at work," she often told her grandson, "so you might as well be at work."

At an early age, Tom Ward recognized that his father, Jody, also was an alcoholic. For a while, Jody managed to run the family's horse-training business, even while drinking prodigious amounts of ten-year-old Old Charter bourbon. He got to the point where he was consuming a half gallon in a single day. Jody Ward brought his bottle to work, replacing the cork stopper with each debilitating swig, in full sight of his son.

When he was sober, Jody Ward was kind and good-natured, much like his own father. He just couldn't manage to quit drinking. Jody Ward's reputation, and the stories told of other Ward men, were such that at least one local woman was warned not to marry into the troubled family.

"He drank every hour he was alive, straight out of a bottle," Ward recalls of his father. "It was embarrassing."

Tom's mother's family had a bit more wealth, and they helped the family build a brick home, one of the nicest in town. Nonetheless, the family expected Tom, the youngest of four children, to help his father and his older brothers in their work. At the age of eight, Tom began cleaning stables and doing other chores while his father trained horses for a nearby track.

Over time, Jody Ward's heavy drinking took a greater toll and he was unable to handle a full workload. At a young age, Tom began helping his older brother Ronnie run the business. On weekends and each day before and after school, Tom cooled down the horses by walking them around a track, a challenging task for the small boy because the racehorses often turned nasty.

"They were some mean horses," Ward says. "They often tried to hurt me, it was like life and death."

In his forties, Jody Ward was diagnosed with cirrhosis of the liver, and he died of a heart attack at the age of forty-eight, when Tom was sixteen. The sudden death forced Ronnie to return from college to take full control of the business, even as Tom's responsibilities grew.

To escape the family dysfunction, Tom began devouring books, finding solace in the tales of Ernest Hemingway, Edgar Allan Poe, and O. Henry. Undersized compared with his schoolmates but quick and ultracompetitive, he excelled at sports, another diversion from his family life. Tom was an outfielder on his high school baseball team and a star point guard on the basketball team, ran track, and starred as a halfback and defensive back on the football team. He sometimes carried several less determined opponents across the goal line, neighbors recall. Decades later, some in Seiling still recount a game late in Tom's senior year when he somehow scored six touchdowns to lead his team to victory.

As he emerged as a star athlete, Tom became more popular and confident. The Wards weren't the only local family plagued by alcoholism, and neighbors viewed it as an addiction and a disease, lending ample sympathy and understanding to Tom and his siblings.

Without a father to provide guidance, Tom turned to various coaches or spent afternoons at the home of Orville White, the local preacher. Tom frequently sat with older members of the town at a local diner, drinking coffee and listening to their stories. He relished the attention and respect they accorded him despite his youth and troubled background.

Gasoline was cheap, so most Saturday nights Tom and his friends piled into their cars and drove around town, joining a slow procession of automobiles packed with teenagers who waved and honked at neighbors. They'd head to the local Dairy King for ice cream, much like the classic drag scene in the movie *American Graffiti*.

Tom and his friends dabbled with alcohol, but he steered clear of drugs and serious drinking, worried they might affect his athletic performance. "Sports were my one thing, the one thing I loved, and in my mind drinking would hurt me," he says.

A visit to Seiling by a charismatic church revival helped transform the young man. Ward watched in awe as a traveling pastor discussed the concept of forgiveness and the support Jesus could provide. As Ward watched the pastor speak in tongues and saw neighbors baptized with the Holy Spirit, he experienced a religious awakening.

He later told friends he had been "humbled" by the experience. Within weeks, he was rethinking his behavior and devoting himself to religion. The pastor "was an instigator, I felt a calling," recalls Ward. "I made a conscious decision to follow Jesus."

Seiling was too small to field its own American Legion baseball team, so it formed one with Waynoka, a town thirty miles away. After one game, Ward met a pretty local girl, Sch'ree Ferguson, who was selling tickets for a cake raffle. Ward and Sch'ree began dating and he discovered a sense of security and comfort within the Ferguson family, which showed a warmth and love for one another he had never witnessed.

"My family was more stoic, they were Norwegians," Ward says. "When you have hard times it's hard to show love, you just live."

Ward began studying the Bible and attending church services with his girlfriend and her parents. He became a part of Sch'ree's family and joined their traditional Protestant church. "They became a stabilizing factor," he says.

Ward didn't have much interest in his schoolwork. With graduation a year away, he couldn't decide whether to stay in town and pursue what he viewed as an attractive opportunity to become a truck driver and make four dollars an hour, or to try to get into a local college. He chose to apply to college, gaining acceptance at the University of Oklahoma in Norman, one of only a handful of students in a high school graduating class of about forty to attend college.

Each weekend, Ward drove a used 1974 Buick Regal back to Waynoka to see Sch'ree, who was a year younger and still in high school. They married after she graduated, when Ward was nineteen. At the time, it wasn't unusual for couples in the state to marry before their twentieth birthdays, but Ward seemed more eager than most to start adult life.

"I think I may have matured at a younger age than others because of my life circumstances," he recalls. "I was ready."

While Ward attended college classes, Sch'ree worked at a local flower shop, helping to pay for her new husband's education. "Our goal was to have a fireplace," Sch'ree Ward says. "That was a big deal for us."[1]

At the University of Oklahoma, Ward majored in petroleum land management, inspired by his mother's brother, a land broker who was the wealthiest person in the family. School wasn't much of a thrill for the young man, though. "It was drudgery," he recalls of his college years. "I was working all the time. . . . I worked at a local horse farm. I had no fun in college, got through in three-and-a-half years and took the last three hours by correspondence."[2]

After graduating in 1981, Ward joined the land brokerage started by his uncle and recently purchased by Ward's brother as a "landman," or someone who leases mineral rights from homeowners on behalf of exploration companies hoping to drill for oil or gas under the homeowners' property. The career move made sense. Ward had studied to be a landman, and the energy business was booming. He and his young wife moved to a town near Seiling to start a new life.

But the business hit a tailspin in the early 1980s, as oil and gas prices plummeted. There wasn't enough work and Ward found himself out of a job. He tried doing land brokerage work for himself but couldn't make enough money doing that, either.

In the summer of 1982, at the age of twenty-two, Ward was cutting wheat in a desolate Oklahoma field, the only job he could find. He and his wife had a toddler at home and lived in a 900-square-foot house that Sch'ree also used as a day-care center to make some money. It didn't seem likely that they could turn things around, and Ward grew dejected.

After several months, he decided that he had had more than enough of cutting wheat and that he wanted to try to make a living in the energy business once again. He just had to figure out a way to do it.

At home at night, Ward recalled that his mother's father regularly rejected offers from energy companies hoping to lease land he owned in western Oklahoma. His grandfather turned the proposals down in hopes of eliciting better terms. The companies generally refused to sweeten their offers, though. After Ward's grandfather declined to drill the land himself, he was forced to lease it at the best rates paid to others in the same field.

This forced leasing was legal—the law in Oklahoma was such that a landowner could either lease his land to those wishing to produce oil or gas on it, or drill the acreage himself. To this day, a landowner can't hold up drilling simply by refusing to lease his land. It's a common measure shared by many states and aimed at preventing recalcitrant landowners from standing in the way of energy production, something akin to the government's ability to acquire land by means of eminent domain.

As he recalled his grandfather's experience, Ward had a brainstorm: He'd find areas in the state where major companies were having luck extracting oil or gas. Then he'd contact local landowners who had rejected lease deals from these companies, like his grandfather had. Ward would offer a bit more for the drilling rights than the landowners had been offered. He and his investors could make a profit, even by paying more for the drilling rights of the holdouts, because this was acreage that was close to wells already pumping oil and gas, making it close to a sure bet.

In the fall of 1982, Ward obtained a list of recalcitrant residents and began making calls. He developed a persuasive pitch and met almost immediate success. Before long, he was coloring in maps of the best drilling areas around the state and wooing nearby holdouts, making $83,000 his first year, more than double what he had made working with his brother in land brokerage.

"It was the one good idea I've had in my life," Ward says.

In time, Ward began doing his own drilling on the acreage he leased instead of selling the drilling rights to others. He continued to focus on areas where big oil companies already had found success. By 1984, he and Sch'ree had enough money to move to Oklahoma City, where Ward hoped to be closer to the real action of the oil and gas business.

Soon, Ward noticed someone else in town employing a similar strategy. His rival was just as aggressive with landowners and oil companies as he was, and just as successful, but the young man stood out from others in the business. His name was Aubrey Kerr McClendon, and it was obvious that he was destined for great things.

Born in 1959 in Oklahoma City, just three days before Ward, McClendon came from true Oklahoma and energy nobility. His great-uncle was Robert Kerr, the former Oklahoma governor and powerful U.S. senator. Kerr also was the cofounder of oil and gas pioneer Kerr-McGee Corp., the company that would purchase Robert Hauptfuhrer's Oryx Energy. McClendon's mother, Carole Kerr, was Robert Kerr's niece.

Aubrey's father, Joe, arrived at the University of Oklahoma with limited means and worked his way through school, taking time off to serve in World War II and the Korean War. He met Carole in college, where she was the president of her sorority and an outgoing campus leader. After graduating, Joe McClendon joined Kerr-McGee and rose within its ranks, as if the oil giant was the family business.

Aubrey McClendon grew up in Belle Isle, an upper-middle-class suburb of Oklahoma City, and his family lived in one of the few two-story homes in the neighborhood, a 2,400-square-foot house. The McClendon family was comfortable, though they weren't wealthy. Friends were well aware of Aubrey's rich pedigree, however.

"Aubrey was big-time, even back then," recalls Chuck Darr, a friend from grade school. "He was a Kerr, we all knew it."

As a boy, McClendon was friendly and popular. He also demonstrated early signs of a sharp intellect. At Belle Isle Elementary School, he regularly read more books than other kids, wowing classmates when his teacher made a regular tally. Later, he received straight As at Heritage Hall Middle and Upper School, a private college-prep school regarded as among the best in the city. McClendon also had a fierce competitive edge. At nine years old, for example, while playing a game of backyard football, he broke his good friend's collarbone with a hard tackle into a house, a friend recalls.

Future energy tycoons often show an early love for science, geology, or engineering. McClendon's passion was geography, a hint at a focus later in life. On car trips with his parents, he brought along maps to trace the route as his father drove. "Some people are born with art or science influence; I've always liked places and geography," says McClendon.[3]

During one year in high school, McClendon grew eight inches to reach six foot two, enabling him to start as a wide receiver on the high school football team. Athletic, tall, and slender, with a passing resemblance to actor Richard Gere, McClendon dated an attractive cheerleader, Mary Anne Brown.

McClendon proved a natural leader. He was senior class president, co-valedictorian of his class, and led a number of clubs. The high school team was called the Chargers; McClendon earned the nickname of "head Charger."

McClendon was such a smooth talker that he could persuade others to follow him, even if it meant doing something dangerous. During high school, he founded a drinking club called "The Soused Seniors of '77." Classmates recall that he printed T-shirts to build camaraderie and persuaded fellow students to drink eight full ounces of hard liquor in a single shot to gain entrance to the club.

"He convinced us all to chug eight ounces," one friend recalls. "He was very articulate."

Aubrey McClendon may have come from Oklahoma royalty, but true wealth seemed just out of reach. Joe McClendon spent thirty-five years at Kerr-McGee, but he was on the dull and dreary side of the business. He worked for Kerr-McGee's refining and marketing division, which operated the company's decidedly unglamorous filling stations and sold various other products. Joe McClendon never searched for oil and gas and he never became rich.

"As a kid, with him, I didn't spend time on oil and gas rigs," said Aubrey McClendon. "I spent time looking at dirty bathrooms in gas stations."

Joe McClendon had ample time to spend with his kids and coach Aubrey's sports teams. He was a relaxed presence around the house and something of a neighborhood comic, say McClendon's friends, who remember Joe with a constant smile on his face.

Carole McClendon, a schoolteacher, could be a different kind of parent. Aubrey's friends recall her as a stern, strict disciplinarian who demanded respect and pushed Aubrey to excel. "You didn't want to mess up in front of her," Chuck Darr, the old friend, recalls. "I was afraid of her growing up."

Dan Jordan, an energy industry executive who got to know McClendon's family, says Aubrey's mother was cut from a different mold. "The Kerrs are tough . . . they eat their own," he says.

McClendon grew up with more money than many of his friends. But it wasn't enough to afford everything he wanted, so Aubrey worked for his spending cash. He sold holiday cards door-to-door, delivered newspapers, and started a lawn-mowing service, competing with a boy three years older named Shannon Self. The energy business was booming, the local economy was strong, and McClendon could charge as much as ten dollars for an hour's work on a lawn. Each week during the summer, he rode his bicycle to deposit his earnings in Penn Square Bank, saving nearly $3,000 to help buy his first car, a 1977 black Oldsmobile Cutlass. (His parents kicked in the other $3,000.)

"The eight years I was in high school and college, the price of oil went from three dollars a barrel to thirty-nine dollars a barrel," McClendon recalls.

McClendon traveled to North Carolina for college, attending Duke University. He was a history major and was serious about both his studies and his social life. Friendly and personable, he was an active member of the Sigma Alpha Epsilon fraternity. He served as social chair for his pledge class during his freshman year and rush chair during his junior year.

"Athletes and non-Athletes, party boys and geniuses. . . . It was a collection of good guys from across the nation," McClendon later told a Duke University publication. "We studied hard, we played hard."

McClendon was tenacious at forming social connections. At a frat event one time, he struck up a conversation with another undergraduate, Ralph Eads, peppering him with questions about his background. They would forge a lifelong friendship that would come in handy when Eads emerged as one of the nation's most prolific energy bankers.

"Aubrey had this energy and intelligence, he did his schoolwork and then read a book or magazine, he was always reading," Eads says. "Who even knew what the *Economist* was? We didn't have any idea."

McClendon's inquisitiveness left an impression on Eads. "He always wanted to know everything you knew, as rapidly as possible," according to Eads.

Even when McClendon failed, he somehow emerged a winner. He didn't always have much success organizing formal events and one time found himself without a date at a fraternity party. He saw an attractive brunette, Katie Upton, who also seemed to be alone. Upton's date, a fraternity brother of McClendon's, had become ill earlier in the day when McClendon brought him to his first barbecue dinner.

"That opened the opportunity for me and I took it," according to McClendon.

It wasn't a plan, of course, but McClendon took advantage—he approached Upton and struck up a conversation. It turned out that Upton was from a small town on the shore of Lake Michigan and was the granddaughter of Frederick Upton, cofounder of home appliance giant Whirlpool Corp. Soon they were dating.

McClendon seemed on such a sure path to success that classmates took to emulating him. He participated in a Big Brother program at Duke, arranging for fraternity brothers to mentor local underprivileged boys. He often was seen around campus with his assigned brother, Terry, on his shoulders. McClendon encouraged his fraternity brothers to join the program, and John Landa Jr., who was a freshman during McClendon's senior year, says that's why he also enlisted. "Aubrey was a role model for many of us, he was a guy who had it all together and someone we could learn from," Landa recalls. "He was kind, considerate, and fun to be around."

McClendon graduated in 1981 and prepared to move to Dallas to work at Arthur Andersen, the big accounting firm. The nation's economy was struggling, as crude oil prices saw a thirteenfold increase since the 1973 Arab oil embargo, but the Southwest was a pocket of strength, thanks to the local energy business. Just before McClendon left for the accounting job, though, his uncle Aubrey Kerr Jr. called to offer him a job as a staff accountant for his small oil and gas company in Oklahoma City, Jaytex Oil & Gas Company.

"He offered me $3,000 more than Arthur Andersen, so I thought it'd be fun to come back to Oklahoma City," according to McClendon. Having by then married Katie Upton, he convinced his new wife to move to his hometown.

McClendon was no natural-born wildcatter, like George Mitchell. He only took the job because it was the best one available. But McClendon had spent some time thinking about the oil business in his senior year at Duke. That's when he came across an article in the *Wall Street Journal* about two young men who had made a killing in energy.

"They sold their stake . . . and got a $100 million check. I thought, 'These are two dudes who just drilled a well and it happened to hit,'" McClendon later told *Rolling Stone* magazine. "That really piqued my interest."

After nine months in his uncle's accounting department, McClendon shifted to the land group, which acquired and leased land. Like Tom Ward, Aubrey McClendon had found his calling. Land combined his love for history, maps, and math. He even had to do some detective work when he tracked down who owned which below-surface minerals.

It didn't take long for McClendon to bump up against the hard realities of the oil business, however. Jaytex ran into trouble, much like Tom Ward's first employer did, and soon McClendon also needed a new job. Katie McClendon might have been a Whirlpool heiress, but her trust fund wasn't huge and no fortune flowed to the McClendon home.

The 1980s were among the worst periods in the history of the domestic energy industry, amid a glut of oil and slowing demand. An estimated 90 percent of oil and gas companies went out of business and the bulk of the industry's petroleum engineers left to try their luck in more promising businesses. Some called it the industry's "gulag" era. It was difficult to persuade investors to finance attempts at discovering new oil and gas deposits. The most they would do was invest in drilling of already proven energy deposits that carried less risk and had fewer capital requirements. The days of the American wildcatter seemed over and the job market seemed closed to McClendon.

"You couldn't get a job at a bank, oil and gas companies weren't hiring and there really weren't any jobs in town," McClendon later told Oklahoma City's *Journal Record*. "I didn't want to leave town so I thought, 'Well, I'll go hang out by myself for a couple of years and see what I can learn about the business.'"

In 1982, McClendon bought a typewriter and some maps, rented an office, and turned himself into an independent landman, trying to lease small parcels of land in areas where the biggest exploration companies were already drilling. Charismatic and outgoing, he proved a natural. It helped that competitors had vanished, many after suffering huge losses from the fall in prices. McClendon had the added advantage of being able to drop his well-known name on leaseholders.

"He could go out there and blend right in because he was an Oklahoman and because of his family name—the Kerrs," says Ted Jacobs, who sometimes competed with McClendon.

McClendon searched for leases in prime areas of the region, persuading landowners to lease acreage that he flipped to larger oil companies, just like Tom Ward was doing. McClendon dug up information about successful wells and became familiar with the area. Then he'd find available leases and cobble them to together to enable a more lucrative sale to a larger company.

"What Aubrey would do, in essence, he was out there picking up a lot of crumbs . . . getting in everybody's deal," Jacobs says. "He put enough crumbs together to where he was able to put together a whole loaf of bread."

In the spring of 1983, Ward and McClendon began bumping into each other while vying for the same oil and gas leases, many near Will Rogers World Airport. One day, McClendon called Ward with a proposal.

"Why don't we just join forces?" McClendon asked.

Ward agreed. They decided to keep their companies apart, with offices in separate locations, but to work as fifty-fifty partners with only a handshake binding the twenty-three-year-olds. For the next six years, Ward and McClendon put together deals, often holding on to the leases instead of flipping them, while mustering enough cash to participate in drilling done by others.

McClendon and Ward were "non-operators," or individuals who participated in the drilling and production of others rather than doing it themselves. It was a profitable business, but not a way to get rich. McClendon was making less than $30,000 a year in some of those years, friends say, and Katie often made more in her job as a real estate agent.

By May 1989, McClendon and Ward were ready to search for oil and gas themselves, hoping to score some real wealth. McClendon asked his old lawn-mowing competitor, Shannon Self, who had become a lawyer, to help them incorporate.

The organization didn't seem built to last. McClendon and Ward continued to maintain separate offices, with McClendon in a suburban brick building and Ward in a nearby building of his own. The young men made important decisions after a fax or phone call, rather than conversing in the same room. They began with $50,000 in cash, eight employees, and no oil or gas reserves, a fragile position in a horrible energy market. Most of their money was gone on their very first day of business as start-up expenses quickly piled up.

McClendon later said he and Ward didn't want to name the company after themselves because times were so tough there was a good chance of failure. McClendon, the geography buff, named the company Chesapeake Energy after the Chesapeake Bay, an area of the country he'd always enjoyed visiting.

They were equal partners, but McClendon—passionate, outgoing, and more comfortable with investors, lenders, and the finance side of the business—became the company's chairman and chief executive officer. Ward, more introspective and thoughtful, enjoyed focusing on operations and working with the men and women in the field; he became Chesapeake's president and chief operating officer.

Their families didn't have much spare cash to invest in their deals, nor did McClendon and Ward have great insights into the next great energy plays. They did have a unique way to entice investors, however. In 1980, the U.S. government had introduced a tax credit called the Nonconventional Fuels Tax Credit, or Section 29. The measure, enacted amid concerns about American dependence on imported oil, was aimed at spurring expensive production from shale and other challenging rock, as well as from coalbeds.

McClendon and Ward made sure to target areas qualifying for these tax credits and to emphasize the tax benefits to doctors, lawyers, and other potential investors in Dallas and elsewhere. They managed to raise $1 million, which they used for Chesapeake's first well, the Newby 1-1 well in Grady County, Oklahoma, drilled on June 27, 1989, as well as for another well nearby.

Shannon Self, McClendon's lawyer, had a contact with the Belfer family of New York City, prominent philanthropists and donors to the Democratic Party. Soon, McClendon sold the Belfers on becoming investors in Chesapeake's drilling deals. (The Belfers later sold a group of energy holdings to Enron and watched their investment drop in value by nearly $2 billion when the utility filed for bankruptcy protection in 2001 amid accusations of fraud.)

It helped McClendon and Ward that many veteran wildcatters still were licking their wounds from years of troubles, leaving them unable to spend much cash to compete with Chesapeake. The new company often acquired dominant positions in oil and gas fields and told those holding acreage in the area that they had to come up with their share of money to participate in Chesapeake's drilling or sell their acreage to Chesapeake. It sparked grumbling among landowners who resented having to fork over scarce cash in an economic downturn to drill with young operators with little track record.

McClendon and Ward were good at finding oil and gas and rarely hit dry holes or those with little production. They were even better at discovering ways to make their drilling profitable. Because fewer companies in the state were searching for energy, the pair received aggressive bids from drilling service companies. They also could play hardball with those hired to drill their wells and provide other services. When Chesapeake hired a company to provide drilling fluid or something else for a site, it asked as many as eight providers for bids, while rivals offered similar jobs to as few as three companies.

McClendon and Ward only paid dirt-cheap prices and made it clear they would pay their bills at a slower pace than others.

"I told you that you wouldn't get paid for five months," McClendon told one company that tried to get Chesapeake to pay its bills because it said it couldn't meet its own expenses. "Sorry, a deal's a deal."

McClendon was up-front with most everyone, making sure contract terms were clear, and some service companies appreciated the business. Others took exception to the tactics, though.

"It was a love-hate thing," says Dan Jordan, who ran a drilling company that Chesapeake used for many of its wells. "They were just brutal to service companies, they beat them down. But Chesapeake was the biggest game in town, they had balls [to drill] when no else did, so they could get away with it."

Some in the industry, accustomed to a friendlier way of doing business, resented the young men. "In New York, Aubrey would have blended in, but in Oklahoma he stuck out—he was young, brash, and brutally honest," Jordan says. "People talked trash about both of them, they couldn't stand them and were envious . . . but Aubrey and Tom were trying innovative, fresh" approaches to drilling wells.

McClendon and Ward kept leasing more acreage, regularly spending more than Chesapeake had. It forced the company to borrow big sums, leaving it in a precarious position. "It was near-death on a daily basis," Ward says.

All the debt might have made others uncomfortable, but McClendon had outsized aspirations for the company and for himself, friends say. Borrowing heavy amounts of cash to get his hands on land was the only way he was going to meet those goals.

As for Ward, he figured that if Chesapeake failed to make its debt payments, he and Sch'ree would just move back to their hometown region and live a quiet, contented life. This safety net gave Ward courage to borrow more money. He also felt he didn't have much choice but to rely on copious amounts of borrowed money. "I didn't have much to start with in life, so without leverage I wouldn't have anything," he explains.

The pair saw their run of good luck end in the early 1990s amid weak oil and gas prices. One day, Jordan drove to Chesapeake's offices to try to get paid. Jordan was operating eight drilling rigs for Chesapeake and was owed one and a half million dollars by the company. He knew McClendon and Ward were slow to pay their bills, but some of them were more than seven months late and Jordan's own expenses were piling up.

"Tom, I know it's a hard time for you guys, but I need something right now," Jordan told Ward.

"I feel your pain," Ward responded. "But we don't have enough to buy a cup of coffee. We have a plan, though."

Ward told Jordan that he and McClendon hoped to take their company public. By selling shares to investors, they might be able to raise enough cash to pay their bills and keep Chesapeake going, Ward said. They had to clean up the company's balance sheet first, though, or investors wouldn't be interested in Chesapeake stock. Ward and McClendon asked Jordan, along with others waiting to get paid, to turn their receivables into long-term notes. That way, Chesapeake could make its balance sheet look healthier while pushing off short-term expenses it couldn't otherwise pay.

In 1992, Chesapeake's tax accountants at Arthur Andersen wouldn't give the company an "unqualified" opinion saying its financial statements were sound, apparently due to the company's heavy debt, which prevented a public offering. Chesapeake had to switch to PricewaterhouseCoopers to get an opinion that enabled the company to pursue an IPO.

The plan to dig out of the mess worked, though, and Chesapeake managed to go public in February 1993. McClendon and Ward placed the six hundred or so wells they had interests in into Chesapeake while retaining the right to purchase a two-and-a-half-percent stake in each new well the company drilled.

"There was no grand ambition, we just needed the money," Marc Rowland, by then the company's chief financial officer, says of the sale of shares.

From the start, McClendon and Ward were a bundle of contradictions. McClendon was a friendly presence in the office. He exuded optimism and was well liked by employees, who called him by his first name. Friends from both grade school and college received a quick return phone call when they reached out to him, and McClendon quietly helped a number of them when they hit tough times. Ward was just as generous, writing big checks for various charities while encouraging employees to spend time on volunteer projects. At times during the day, employees found him studying the Bible in his office.

When it came to business, though, McClendon and Ward were less kindhearted. By the time their company went public, the pair faced lawsuits from aggrieved parties, some of whom claimed that they were mistreated or misled in various land deals. Just weeks after Chesapeake went public, one of these cases came back to haunt them. The dispute also marked their first battle with Harold Hamm, another determined and ambitious Oklahoma energy man.

Back in 1988, Ward had approached Hamm with a proposal. He told Hamm that Chesapeake had a deal with an operator named Ralph Plotner Jr. for some attractive acreage in western Oklahoma. At the time, Hamm ran a company that drilled wells for others, in addition to a midsized exploration and production company called Continental Resources. Ward asked if Hamm's company would be willing to drill the wells for Chesapeake's project with Plotner.

Hamm was reluctant. It was a difficult period for the industry and he needed to make sure he'd get paid for any drilling. Hamm was wary of working with Plotner, a 300-pound local energy explorer with an outgoing personality, a high school diploma, and a checkered and tragic past.

Nearly a decade earlier, Plotner had been accused of forcible oral sodomy and attempted rape of a female acquaintance. The accusation was based on the woman's allegations, as well as on paint chips from a door in her home that allegedly were found on Plotner's wristwatch. Plotner vehemently disputed the charges, arguing that the woman had tried to blackmail him.

During the resulting trial, Robert Webb, an examiner from the respected FBI Crime Laboratory in Washington, D.C., who had scrutinized evidence in the case, flew in to testify. Webb walked into the courthouse wearing black cowboy boots, affecting a western look, perhaps in a bid to win favor in the eyes of jurors.

"He *was* the epitome of what an FBI agent should look like," Fred Whitehurst, who worked with Webb in the FBI lab, later told *GQ* magazine. "He was good-looking and very fit from being a triathlete."

Webb seemed an expert, but some had doubts about his testimony. "I considered Bob a friend, but when it came to things like paint . . . he didn't know what he was talking about," Whitehurst told the magazine.[4]

Plotner was convicted on both charges. He maintained his innocence and the rape charge was later reversed on appeal. But the oral sodomy charge stuck, and Plotner served nearly five years in prison. It proved a crushing experience. He was in a jail with a leaky roof and an infestation of cockroaches and rats. Later, he watched an inmate take a baseball bat and beat another inmate to death.

During his incarceration, Plotner's wife divorced him and he lost most of his possessions, including a 7,500-square-foot mansion. Plotner couldn't bear to let his five-year-old son, Kyle, see him in a prison, so he barred the boy from visiting, a move that brought him intense pain. Several years after Plotner left prison, Kyle, then seventeen years old, committed suicide after joining the local goth scene. Plotner blamed the tragedy on his own troubles with the law and time in jail.

After leaving prison, Plotner managed to rebuild his career by working on small oil and gas deals. He happened to control acreage in western Oklahoma that Ward and McClendon took an interest in, resulting in the call from Ward asking Hamm's company to drill wells. Hamm was hesitant, but his drilling company was dealing with slowing business, so he accepted the job and work commenced.

After a well was drilled, but before pipe was placed in it, McClendon and Ward's enthusiasm for the project seemed to wane, according to three people involved in the deal. Soon, Plotner was told that for technical reasons McClendon and Ward no longer had interest in the acreage.

"When we set pipe they elected not to go in with us," Plotner recalls. "I thought that was strange."

When Hamm tried to get paid for his company's drilling from an escrow account, the money wasn't there, he says. Hamm, who was owed about a half million dollars, turned furious, blaming McClendon and Ward for transferring the money out of the account, though they said they were blameless. McClendon and Ward argue that they weren't the operators so they weren't responsible for paying Hamm. McClendon also denies that there was an escrow account.

Hamm fumed nonetheless.

"It was a tough time for the business and that was a lot of money for me," Hamm recalls. "When you don't get paid for drilling wells it's pretty upsetting."

In the view of Plotner's attorney, Charles Watts, the Chesapeake executives likely got cold feet about Plotner's acreage after agreeing to the deal. "They probably had interest in a lot" of land at the time, Watts says, "and they backed out when the acreage dropped in value. . . . They were trying to hedge their bets with this deal."

Hamm was so upset that he moved to sue McClendon and Ward. Before the trial began, the parties agreed to a settlement requiring Chesapeake to use Hamm's drilling company for work on a number of future wells, Hamm says. Hamm couldn't get over how McClendon and Ward had treated him, and he vowed not to do any more work with them.

"After that, we steered clear of them" for many years, Hamm says.

Plotner suffered more serious consequences from the failed deal. He was left with acreage that dropped in value. Eventually, he was forced to seek bankruptcy protection.

Plotner also sued McClendon and Ward, alleging that they had committed fraud by misrepresenting their interest in his land. As the proceedings began, McClendon and Ward's attorney raised Plotner's disturbing past, an apparent attempt at discrediting the Oklahoma oilman, Plotner's attorney says. The judge wouldn't allow Plotner's prior convictions to be shared with the jury, however, ruling that they had no relevance to the dispute, the attorney says.

When McClendon took the witness stand, dressed in a sharp suit, his attorney asked if he was upset about Plotner's charges of alleged fraud.

"Yes, I am," McClendon responded, according to two people who were in the court. "No one has ever accused me of" acting improperly in a deal.

The response gave Watts—Plotner's attorney—an immediate opening. He says he grabbed a tall stack of legal papers and began to slowly and deliberately cite details of previous lawsuits McClendon and Ward had faced since starting in the energy business.

"I liked Aubrey and Tom and I still do," Plotner now says, "but Aubrey was a real pompous smart-ass . . . he tried to out-slick" the jury.

The suit's outcome was still in doubt when Chesapeake went public. At the time, it didn't seem likely that Plotner would win any kind of sizable payout. Chesapeake's prospectus said the company was not party to any legal proceedings that management believed would have a "material adverse effect." Even if some payout was to be made, it was expected to be the responsibility of McClendon and Ward, not Chesapeake itself.

But on February 12, 1993, a state court jury awarded Plotner $2.2 million, including interest and attorney expenses. Equally surprising was that the judgment was against one of Chesapeake's units, in addition to McClendon and Ward.

"The jury liked Plotner and they didn't like Ward and McClendon," Watts recalls. "The jury thought they were lying."

McClendon and Ward say they simply decided not to back Plotner, as was their right.

"He thought I lied to him but I didn't," Ward says.

McClendon adds: "Tom and I were the innocent guys on the sidelines."

"We didn't do what they claimed but it wasn't seen that way," is how Ward explains the verdict.

The legal decision sent Chesapeake shares reeling; it wasn't clear how the young company would come up with the money. McClendon and Ward volunteered to pay the entire legal bill themselves, even as they appealed the decision, a move some said they didn't need to make. Chesapeake shares kept falling, though, as shareholders launched a lawsuit claiming the company's IPO prospectus hadn't provided adequate warning about the ongoing litigation, a suit that eventually was settled.

By the end of the year, Chesapeake was trading for under four dollars a share and the company had the dubious distinction of being the worst-performing IPO of the year.

A few years later, Plotner received more vindication. Whitehurst, the FBI agent, became a whistleblower and leveled serious complaints about the work of the FBI Crime Lab that had been involved in Plotner's rape case. A subsequent Justice Department investigation produced a scathing report about the lab's shoddy work and about contaminated evidence in a number of other cases.

A lawsuit brought by Plotner against the FBI was dismissed by an Oklahoma City federal judge, however. Soon his bad luck returned. One night in the summer of 1997, while Plotner was at home with his young wife, Margarita, he turned distant and despondent. She didn't realize it, but it was Plotner's late son's birthday and he was having a tough time dealing with the memories flooding back. Plotner hadn't shared his loss with his new wife, preferring to suffer in silence. Margarita kept asking him to talk to her about why he was so down in the dumps, but he resisted. Eventually, he began yelling at her. Soon their neighbor, a cop, called the police to report a disturbance.

When the policemen arrived, they didn't find evidence of violence, according to both Margarita Plotner and Plotner's attorney. The officers did discover a safe containing a collection of hunting guns handed down for generations in the Plotner family. Plotner was arrested for gun possession, a violation of his parole.

Days before he entered a minimum-security prison in Dallas, Plotner's mother died, adding to his sad plight. During his two years in prison, he gave power of attorney for his oil business to an aunt. She soon clashed with Margarita. Eventually, Margarita grew so unhappy that she divorced Plotner, though they remained close. In 2013, after Plotner suffered a stroke, Margarita nursed him back to health.

"He's a really nice guy who's been through so much," she says.

Plotner's attorney, Charles Watts, experienced his own unexpected postscript to the Chesapeake lawsuit. As the company appealed the award, Watts joined a new church in Oklahoma City. With reluctance, he agreed to join a church feet-washing service. It was a religious rite tied to the day before Easter and shared by some Christian denominations, but one that Watts was dreading.

During the rite, Watts, already uncomfortable having his feet washed by a stranger, looked up to see Tom Ward standing over him. Watts had just ripped into Ward and McClendon in court, calling them all kinds of insulting names as he defeated their effort to overturn the award to Plotner.

Ward didn't seem to hold a grudge, though. "May I wash your feet?" he asked. He put out a hand to Watts and said, "Welcome to the church."

Watts still believed Ward and McClendon had committed fraud against Plotner. But the warm welcome was a surprise, and Ward and Watts soon became friends. Later, Watts wrote a letter to McClendon thanking him for his work in helping Oklahoma City grow and apologizing for his role in the case.

C hesapeake was on the ropes once again, due to the Plotner award, and the company's shares were nearly worthless. McClendon and Ward had reason for some optimism, though.

For one thing, they had done a good job of acquiring acreage, such as in the Sholem Alechem oil field in southern Oklahoma, a productive field with an odd name and a colorful history. Some locals believed it was named after a Native American expression. It's more likely that it was a tribute to Bill Krohn, a popular and outgoing reporter for the *Daily Ardmoreite* in the city of Ardmore.

Krohn had been an indefatigable journalist who traveled to oil fields to visit workers, breaking news on new oil discoveries and chronicling the 1920s oil rush. When he saw a roughneck, he'd call out the traditional Jewish greeting of "shalom aleichem," or "peace be unto you." Krohn became a well-liked figure, and his greeting caught on. He even popularized the traditional Jewish response to this greeting: "aleichem shalom," or "unto you, peace."

When a newcomer heard the odd phrase and gave Krohn a look of confusion, the reporter would buy the fellow a raspberry soda at a nearby confectionery store and explain it all to him. Krohn even started a Sholem Alechem society in the lobby of an Ardmore hotel, which was nothing more than an excuse for oilmen to get together, drink liquor, and smoke cigars. Legend has it that Krohn was with a group of men at a new well one day when oil suddenly shot to the sky. They named the field Sholem Alechem in tribute to their unusual friend, who later in life turned in his pen and became a wildcatter.[5]

The Chesapeake team wasn't really sure how the Sholem Alechem field acquired its name. All they knew was that it was churning out oil, helping the company regain its footing.

Chesapeake became aware of the progress companies like Oryx Energy were making drilling sideways through bedrock to extract oil from tired old fields in Texas. McClendon and Ward were young and open to new technology, and they instantly recognized that horizontal drilling could revolutionize production. They also embraced an improved mapping technique called three-dimensional seismic imaging.

Soon, McClendon and Ward saw promising results in the underdeveloped limestone formation straddling the border between Texas and Louisiana, the Austin Chalk. At first, Ward's team drilled at sixty-five-degree angles. Eventually they figured out how to properly turn their drill bit horizontally, scoring success in areas including the First Shot field in the formation.

Major oil and gas companies ignored the area, leaving it to a group of upstarts. Union Pacific Resources, an offshoot of the Union Pacific Railroad, was the first to seize on using horizontal drilling in the Giddings field in the Austin Chalk area. Over the next few years, the company produced more natural gas in the United States than giants like Exxon, Conoco, or Shell.

The limestone in the Austin Chalk region was naturally fractured. As a result, oil and gas came up quickly when it was drilled in a horizontal fashion. It didn't need to be hydraulically fractured, unlike the Barnett Shale elsewhere in the state, where Mitchell Energy was drilling around the same time.

"Back then we had twenty-seven rigs drilling, and if an oil well didn't come in at one thousand barrels per day, we didn't think it was worth much," according to Darrell Chmelar, who was a production supervisor at the time for UPR.[6]

McClendon and Ward locked up some acreage ahead of Union Pacific, scoring instant winners with its early Austin Chalk wells. Chesapeake's Navasota River field would go down as the most lucrative in the company's history, producing six hundred billion cubic feet of gas from just one hundred wells.

Most other companies were reeling amid the industry's continuing struggles. Many were firing experienced drilling engineers. Ward hired some of the best of them to help push farther into the Austin Chalk region. New members of the team found ways to improve Chesapeake's horizontal drilling, allowing the company to capture more oil and gas. As they bought more acreage in Texas, moving closer to the Louisiana border, Ward was like a general sweeping across enemy territory, picking up defectors along the way.

By early 1994, Chesapeake's wells were showing huge production and Chesapeake shares were soaring. They reached nearly seventy dollars a share in November 1996, up from less than five dollars at the beginning of 1994. Chesapeake was the best-performing stock in that period and the company had a market value of more than $1 billion.

These achievements weren't enough for McClendon and Ward. Looking across the state's border into Louisiana, they spied hundreds of miles of fresh land that seemed fit for drilling. They were well aware that rock can change in quality every few miles, making the new push no sure thing. But Chesapeake had met early success drilling just over the border in South Louisiana, so it made sense to keep going under the assumption that nearby fields would be just as productive.

Their growing interest in the Austin Chalk began to leak out. Industry veterans, such as Morris Creighton, reached out to lend a hand. Creighton had worked as a landman for the Hunt brothers, who had famously tried to corner the silver market in the 1970s and also held oil interests. In the past, Creighton, a Texan, had purchased mineral rights around the country, from Michigan, North Dakota, and Montana to Wyoming, Texas, and Louisiana. But he was struggling to find work amid the industry's tough times when he heard about Chesapeake's ambitious drilling plans in the Austin Chalk region.

When they met, Creighton, well over six feet tall, towered over Ward, who was not quite five foot seven inches, wore round glasses, and bore a resemblance to Jason Alexander, the actor who played George Costanza in the television comedy *Seinfeld*. It was Creighton who felt intimidated, though. Ward spoke quickly, had more energy than anyone Creighton had ever met, and grilled him for information about the fields he had worked on and how horizontal drilling was evolving.

"I'm from a part of the country where you speak slowly, I was just wishing he would slow down," Creighton recalls. "Tom was a fireplug, he looked right at me and got right to the point and you could feel the excitement in his voice."

There were no newspapers or Web sites tracking new drilling techniques, and operators like Oryx were trying to keep their advances quiet. Ward knew the best information came from field workers, so he sent Creighton to do some reconnaissance work. Creighton chatted with rig hands, roughnecks, truck drivers, and the like at local bars and other spots. He asked which companies had made progress with new ways to drill horizontally and who was leasing new acreage, reporting it all back to Ward.

"He wanted to know everything and anything I knew and everything I could find out about the newfangled technology of horizontal drilling," Creighton recalls.

Late one afternoon, Creighton received a phone call from Ward. In an urgent tone, Ward said he had heard that Occidental Petroleum was leasing land in the Austin Chalk region in Louisiana, the same area Chesapeake was eyeing for new wells.

"We need you to go to Rapides Parish to see horizontal drilling there" and potentially lease some land, Ward told Creighton. Ward was worried that Occidental might be locking up choice acreage before Chesapeake had a chance to do much leasing.

"Sure, I'll be there in the next few days."

"You don't understand," Ward told Creighton. "I want you to go *now*."

Creighton got into his pickup truck and drove nearly eight hours, arriving in Louisiana at close to midnight. The area was nearly deserted and he didn't know how to find his rival's wells. He located a sheriff's substation and asked the sheriff if he had any idea where the new wells might be located. The sheriff took pity on the lost Texan and led Creighton to a well site deep in the pitch-dark woods.

Soon, Creighton saw a trailer. He knocked on the door but didn't get an answer. No one seemed to be around. Through a big window, though, Creighton saw a large computer monitor that had been left on. On the screen were full production details of the well. Creighton grabbed a pen and wrote it all down. Then he found a nearby pay phone and placed a collect call to Ward, who was waiting with some colleagues by the phone, in the middle of the night.

After Creighton read the well's results, Ward relayed a quick new order: "Get thirty men there!"

Creighton's mission was to buy up as many leases as possible, as quickly as he could, as Chesapeake pushed farther into Louisiana. McClendon and Ward ordered the purchase of over a million acres, as they prepared to establish hundreds of well-site locations.

"I thought we could march across Texas and Louisiana," Ward says.

Ward spent so much time at the office, quarterbacking the push, that he built a Murphy bed to pull out of the wall. That way he could work until 1 or 2 a.m., crawl into bed for a few hours of sleep, shower, and start work anew by 5 a.m.

After a well blowout forced a crew to evacuate, Ward flew to the site to hand out toothpaste, deodorant, and other toiletries to displaced field hands, a gesture that inspired them to work even harder.

By late 1996, Chesapeake's shares were soaring. Investors were so excited they were willing to overlook the fact that the majority of the company's reserves were "proved but undeveloped," an industry term for reserves, like those in Louisiana, that the company was confident would produce oil even though they weren't yet producing anything.[7]

Once production from the Louisiana wells began, though, a curious thing happened. The wells began with a burst but quickly saw sharp production declines. Soon, Chesapeake's Louisiana crew wasn't getting its paychecks. Tempers rose and some of the workers in the fields accused Creighton of gambling away their money.

"It was an ugly time," Creighton said. "I had sixty-five guys in the field all starving to death."

The men didn't want to direct too much anger at Chesapeake. Jobs were scarce and they didn't have a contract with the company to give them any security. Finally, a meeting was arranged at Chesapeake's offices.

"Y'all are doing a great job, you're picking up a lot of leases," Aubrey McClendon told Creighton and a few others. "Y'all have any suggestions on how to do things better out there?"

"Yeah, I got a suggestion," one of the captains on the project said. "Pay us."

McClendon looked confused. He didn't seem to realize the Louisiana crew hadn't been paid. He looked to his lieutenant and former classmate at Duke, Henry Hood, for an explanation.

"Well, I think we're a little bit behind right now," Hood said sheepishly.

By the end of the day, McClendon got the men paid. But McClendon and Ward watched with dismay as production kept declining. Not only that, but the Louisiana wells produced huge amounts of water that had to be disposed of.

It turned out that the Louisiana fields were very different from those in nearby Texas. The rock in Louisiana didn't have as many fractures and it was more expensive to separate the energy in its pores. By the end of 1997, Chesapeake had abandoned almost the entire area, forcing the company to record a painful write-off of more than $200 million.

"I thought the play extended, but it didn't," Ward says.

Chesapeake was hit by shareholder lawsuits once again, this time charging that it had overstated the value of its Austin Chalk acreage. The company fended off the suits, but was on the ropes once more.

McClendon and Ward retreated to Oklahoma, vowing to use their remaining cash on more conventional natural gas formations. They'd avoid drilling for oil and focus on simple, boring gas fields. "We had been the darlings, and we had been the goat, and now just wanted to be a regular company," according to McClendon.[8]

Chesapeake had taken on so much debt to finance its land acquisitions and drilling that the company owed a billion dollars. But McClendon had made sure most of it wasn't due for several years, giving Chesapeake breathing room. Within six months, Chesapeake had spent $800 million to buy a series of companies with about eight hundred billion cubic feet of gas equivalent reserves, as it reinvented itself as a traditional gas exploration company.

The purchases didn't help much, though. Anemic oil and gas prices kept pressure on Chesapeake shares, as the Asian financial crisis and OPEC's inability to keep a lid on its members' production flooded the market with surplus energy. By February 1999, Chesapeake shares traded at a measly seventy cents each, giving the company a market value of only about $75 million.

When McClendon and Ward met with veteran landman Larry Coshow to try to recruit him, Coshow was skeptical: "Man, I've been reading y'all's balance sheet and y'all are *broke*."

McClendon and Ward couldn't believe it. They were sure Chesapeake owned natural-gas-producing assets that had more value than the market was according them. The tumbling stock brought special pain for the Chesapeake cofounders; they each had borrowed millions of dollars from banks to purchase boatloads of their company's shares. Now they, too, were nearly broke and the banks were making noises about calling in the loans.

The U.S. economy was on a roll, so bankers gave McClendon and Ward some time to meet their payments. McClendon and Marc Rowland, the chief financial officer, flew to New York to ask some big-name investors, like Carl Icahn, to buy Chesapeake. After an hour's discussion at Icahn's office, he came to a decision.

"Your bonds aren't cheap enough," Icahn said.

His point: Chesapeake wasn't yet enough of a bargain, even at seventy-five cents a share, because its debt wasn't trading at the dirt-cheap prices that usually got Icahn excited.

Other investors were even more dismissive. It was 1999 and technology, Internet, and biotechnology shares were all the rage. Natural gas was about the last thing anyone wanted to invest in. Gas prices were below two dollars per thousand cubic feet and oil prices were around twelve dollars a barrel, close to the lowest levels of the decade. Experts anticipated that a glut of energy and continued limp demand would keep prices under wraps for years to come, just as it had in the 1980s.

At the time, fewer than five hundred oil and gas rigs operated in the country, down over 10 percent in a year, as companies went out of business or stopped searching for gas with prices so low. That could only mean less supply in the years ahead, McClendon and Ward figured, no matter what the experts said.

Meanwhile, demand for natural gas seemed to be growing, as a decade's worth of low prices spurred utilities and others to use more cheap gas and less coal and other energy sources. The Department of Energy's Energy Information Administration produced a study saying the country needed thirty trillion cubic feet of natural gas a year to meet growing demand from a new generation of power plants. If demand for gas was growing and supply was shrinking, prices were destined to climb, the Chesapeake executives concluded.

To test their hypothesis, McClendon and Rowland, got on a plane to San Jose, California. There, they met with Peter Cartwright, the chairman and chief executive officer of Calpine, one of the fastest-growing power companies in the country, and Ann Curtis, Calpine's executive vice president. McClendon and Rowland wanted to see how serious Calpine was about its announced plans to build a slew of new power plants fueled by natural gas aimed at helping Calpine become the world's biggest natural gas consumer and electricity generator.

During the meeting in Calpine's dining room, the Calpine executives seemed distracted. Every five minutes or so, one of them got up to check their company's soaring stock price, Rowland recalls. When they returned, they emphasized how Calpine was determined to grow its power production by expanding its use of natural gas.

"Our business model is that gas will be" in the range of two dollars or less for a long time, Cartwright told them. "There's so much cheap gas" that Calpine could generate huge profits building new plants relying on inexpensive natural gas, Cartwright said.

McClendon and Rowland were floored. The Calpine executives were going to build enough capacity to make use of fifteen billion cubic feet of gas a day—one-third of what the entire country was consuming at the time. If Calpine even consumed close to that amount and others followed their lead, prices would leap, McClendon and Rowland agreed.

"These guys are really betting on cheap gas forever," Rowland told McClendon as they flew back to Oklahoma City.

Back at the office, Ward grew excited when he was apprised of the meeting. Everyone else in the business seemed stuck in a mind-set of weak demand and falling prices, they thought. But the United States seemed to be transforming into an economy more dependent on natural gas than ever.

McClendon and Ward had an opportunity to make a fortune. They crafted a strategy to buy the best natural gas assets in the country as quickly as possible. They felt compelled to act quickly, before others caught on that prices were headed higher. They suspected they could use newfangled horizontal drilling techniques to help locate gas.

Around the same time, Harold Hamm, their former legal combatant, was making plans of his own to change the country's energy equation. Like McClendon and Ward, Hamm was convinced that energy prices were heading higher.

Hamm thought McClendon and Ward had it all wrong, though. Oil wells—not natural gas—held real promise for the country. Hamm even had a hunch where he might find enough crude to change America's direction, as well as his own.

CHAPTER SIX

Formula for success: Rise early, work hard, strike oil.

—J. Paul Getty

Harold Hamm had a lot in common with Aubrey McClendon and Tom Ward. Hamm was born in Oklahoma, just like the Chesapeake Energy cofounders, and he also ran an energy company in the state. By 2000, all three men suspected that America held vast energy resources that might be tapped profitably, despite the skepticism of major oil companies and industry experts.

McClendon and Ward set out to find huge quantities of natural gas, but Hamm was fixated on a different energy source: oil. Hamm wanted to discover enough crude to leave a mark on the nation and even the world. He had his exploration team scour various U.S. locations for the nation's next big oil field. He even told friends the United States had a chance to achieve energy independence.

"There's oil everywhere, man," Hamm told Roger Clement, his chief financial officer, referring to an overlooked rock formation in the North Dakota region. "There's oil everywhere."

If there happened to be loads of untapped oil in the country, few gave Harold Hamm much of a chance to find it. Hamm's company, Continental Resources, was based in Enid, a small, second-rate city about a hundred miles from Oklahoma City.

At one point in 1999, Continental was losing so much money that it had to fire exploration pros and slash the pay of all of the remaining employees. Things got so bad that year that Hamm tried to persuade the government to impose tariffs on oil imports from OPEC member nations, arguing that they were improperly keeping a lid on prices. The effort drew bemusement and some ridicule from industry giants.

Smaller oil and gas producers in Oklahoma looked up to Hamm, and he was an advocate for his fellow "independent" producers within the industry. But the top brass at energy powers in Houston, Dallas, and Oklahoma City hardly gave a thought to Hamm and his dreams of huge oil discoveries.

"Harold wasn't exactly lighting the world on fire," says Roy Oliver, an oil industry veteran in Oklahoma. "He was just another good, hard-working guy who was grinding it out and making money. No one accused him of genius. . . . The smart money would have bet against him, not on him."

Hamm's vision of better times for his company and his country was dismissed as unrealistic, if not half-baked. If someone was going to become a new oil titan, it surely wasn't going to be Harold Hamm. Meeting Hamm sometimes reinforced the doubts. He had a humble upbringing, didn't hold a college degree, and spoke in a slow country drawl. Earlier in his career, he had raised eyebrows by struggling with his speech.

"He talked like a hick," says Allan DeVore, Hamm's attorney and a longtime friend. "People thought he was a dumb country bumpkin."

Hamm knew he was more than just another poor small-town boy with an unrealistic vision, no matter what the skeptics believed. His genial personality masked an outsized ambition and fierce competitiveness that only friends and colleagues fully appreciated.

"He *was* a country bumpkin," Devore says. "But he was far from dumb."

By 2000, Hamm already had accomplished more than most expected. The achievements helped explain why he was confident that he could find more oil in the United States than his acclaimed rivals and perhaps even make an indelible mark on the country.

Harold Hamm's father, Leland Albert Hamm, was a descendant of British colonist Thomas Buckland Hamm, who fought for America in the Revolutionary War. Leland enjoyed few victories in his own life, however. He lost his father at the age of three. After two short marriages, he married his third wife, Jane Elizabeth Sparks. The couple rented a farm in a small Oklahoma town called Agra, where they raised cattle and horses. One year, several of the cows died when their feed was inadvertently filled with splinters, according to Harold Hamm's older sister Fannie. Leland couldn't make much of a living with his remaining cattle and eventually sold them.

The family moved to a small farmhouse five miles north of the tiny, one-stoplight town of Lexington, Oklahoma, where Harold was born in December 1945, the youngest of thirteen children. Leland and his wife worked as sharecroppers on farms in the region, eking out a meager living and relying on federal food aid.

The first recollection of many boys is a summer ice cream outing or perhaps an early ball game. One of Harold's first memories was helping his parents pick tomatoes in a nearby field in the middle of the summer and feeling intense heat emanating from the ground through his flimsy shoes. As a five-year-old, Harold stood and watched his family's home burn down in an electrical fire. There was a silver lining in the accident: Neighbors pitched in to help the family rebuild, donating Harold's first new pair of shoes.

"It was kind of exciting" to wear new shoes for the first time, Hamm later told a colleague.

The Hamms' new home was more of a glorified shack; it had two bedrooms, no electricity, and no running water. The parcel of land became even more cramped when Leland donated about half of it for the construction of a small church, the General Assembly and Church of the Firstborn.

Neighbors remember Leland and Jane Hamm as a quiet, religious couple who didn't associate very much with townsfolk other than members of their church, which had a strict doctrine and believed in faith healing rather than relying on doctors.

Leland Hamm didn't have free time for hobbies, but he did seem to have a fondness for children. He had a son from one of his previous two marriages, in addition to the thirteen children with Jane. Years later, after Jane died and Leland married for a fourth time, he would have one more child. (In all, Leland had fifteen children of his own and was stepfather to at least nine others. At his death, he had seventy grandchildren and forty-two great-grandchildren, according to a local obituary.)

In their home, Leland was a calm and easygoing presence who discussed values and religion with his kids and sometimes served as a volunteer preacher at the church next door. "My father was very active, honest, and smart, but he didn't have much get-up-and-go," says Fannie.

Jane Hamm was more demanding. She taught the Hamm kids to make their beds each day, sweep and wash the floors, and do other chores to keep the household running smoothly. Jane also tended to the family garden and made sure her children were clean and well fed, even when there was little money.

"My mother was easier to get upset, she expected more from us," Fannie recalls. "She's the one who raised the family. . . . Because of her we managed, she taught us hard work."

Hamm, a respectful, clean-cut boy, didn't spend much time dwelling on his family's plight or the fact that he had to go to a neighbor's home to watch television, partly because few others in the area were much wealthier. Leland regularly gave away what little extra the family had to even poorer neighbors. Often, one or two worse-off neighbors lived in the Hamm home. Indeed, Lexington was a rural, hardscrabble town of about two thousand residents, and many locals were in more or less the same straits as the Hamms. More prosperous Oklahoma City was just forty minutes or so by car, but it felt a world away.

"I didn't know I was poor, it was a happy childhood," Hamm says. "I was like a lot of others, I thought, supporting my parents and working."

Hamm had his quirks as a child. Early on, he acquired the nickname of "Buzz" because he seemed to enjoy making an odd buzzing sound while playing with neighbors, an old friend says.

But some who grew up in the area remember Hamm as a quiet loner who could be insecure and withdrawn. He did have some unusual pets to keep him company, though. He helped raise coyotes in the family's backyard, brought a black calf to 4-H competitions, and doted on a pet crow. He kept the crow in the house, his sister remembers, and even managed to teach it to speak a bit. One day, he took the bird outside to get some fresh air. When the pet seemed to become hot in the Oklahoma sun, Hamm brought it inside and put a cold washcloth over it, accidentally suffocating the bird.

There wasn't much time for playing, though. Hamm's parents worked for local landowners, raising crops and livestock for them, and the Hamm children understood that they needed to help the family make ends meet. In the fall, the Hamms frequently grabbed belongings and followed the cotton-picking trail, traveling as far as the cities of Blair and Altus, nearly 150 miles away, and even into Texas. From a young age, Harold pulled, chopped, and hoed cotton, gathered watermelons, and took on other difficult labor. "I'd pull the cotton, put it in the middle of the road, dad would scoop it up, then we'd repeat again," Hamm says.[1]

As they traveled and worked, Harold and his family lived in tiny homes owned by various landowners near the fields. "Money was tight for everyone, we all had to work," Charlie McCown, a neighborhood friend, says. "But Buzz had a remarkable work ethic, he'd jump in and be energetic and never complain."

Hamm and his siblings usually couldn't begin their school year until the work slowed, typically after the first freeze or around Christmastime. Hamm wasn't the only child of a sharecropper who got a late start each year, but that didn't make it easier for him. "It was hard to catch up," he says.

In his sophomore year of high school, Hamm was a top-flight pitcher on the town's baseball team, thanks in part to a baffling curveball, a pitch one of his older brothers taught him. In a tournament one year, he pitched back-to-back games with a throbbing shoulder without telling his coach of the pain. "I was very competitive," he acknowledges.

Hamm learned to persevere without complaining, a lesson he took from Jim Hunter, a shop teacher in junior high school. "Jim had a lot of scars, no hair, missing teeth and pains . . . all courtesy of the Nazi guards in a prisoner of war camp during World War II," Hamm says. "Jim refused to become a negative person and taught me some life lessons that I never forgot."[2]

Halfway through his junior year of high school, Hamm's family moved to the city of Enid, 130 miles away. Hamm's arm still ached from the baseball season, so playing ball after school wasn't especially appealing. He also knew his family could use any cash he could generate, so he decided to seek a proper job.

At the time, Enid was enjoying an early 1960s oil-fueled mini-boom. The city was the headquarters of Champlin Refining Company, started by

Herbert Champlin, a local banker. Champlin hadn't wanted to get into the risky business of searching for oil, but his wife pushed him to invest an initial $25,000 grubstake. The company eventually became the nation's largest privately held "integrated" energy company, or one that does everything from exploring and producing oil to refining and selling it.

Hamm began working at a big Champlin filling station that was open twenty-four hours a day and had a sizable business servicing trucks. He pumped gas, fixed tires, and washed trucks, making a dollar an hour while earning class credit.

Enid was a hub of activity, and Hamm got a glimpse of oil drilling when he delivered diesel fuel, lube, and other material to fracking crews in nearby Hennessey. At the time, some crews were using river fracks, the progenitor of Nicholas Steinsberger's slick-water fracks, to improve production results. At that young age, Hamm "learned that you could fracture the rocks" with this new technique, he says.

Hamm met oil industry veterans and was struck by their generosity and outsized personalities, so different from the humble men in his hometown. On one of his first days at work, Charles Potter, who ran the Champlin distributorship, opened his billfold and asked the young man if he needed any cash. A few months later, Hamm drove his service truck to fix a flat on the huge tire of a portable rig operated by a squat, gruff man named "Tough" Cunningham, who ran a company called Tough Drilling. Hamm struggled with the tire, and as Cunningham approached, Hamm braced for a round of abuse from his customer.

"Hey, boy," Cunningham began, as Hamm turned even more nervous. "I'm going to send some help—don't try to do that yourself."

Cunningham took a liking to Hamm and began looking out for him. Hamm spent time with other men in the oil fields, listening to their stories and learning how they discovered energy. The veterans enjoyed sharing their perspective with the young man and a few became early mentors.

All around Hamm, men were striking it rich as the local oil industry thrived. They spent their money as quickly as they made it, eager for even larger strikes, exhibiting a cockiness and a true zest for life he had never seen before. Hamm was smitten by the uninhibited, energetic men.

"The oil people there were different," he says. "They were charismatic and bigger than life, I had never been around people like that."

He was struck by an encounter with a group of bankers outside a Denny's restaurant. Standing in the parking lot, he watched the bankers debate who had eaten a piece of pie with lunch, and who was responsible for a larger percentage of the meal's bill.

"Oil and gas guys would have just grabbed the check," Hamm says. "They were generous to a fault and were willing to teach me."

Hamm's father had cautioned him about the rough men in the oil business, but Hamm warmed to them. "It was a shock to be in that environment, to find out there are good people in the oil field," he recalls. "I never told my father about their language, though."

Hamm's days were grueling. Most days, he'd head to the service station after school and work from 2 p.m. until 11 p.m. Then the young man would go home and do his schoolwork. He would get up early, usually arriving at school dead tired, to start the day again. On Saturdays he'd work twelve straight hours, logging an average of sixty hours of work a week.

One morning, Hamm walked into school half asleep, and was directed to the auditorium, where a school assembly was getting under way. Soon, John Frank of Frankoma Pottery, a pottery company in Sapulpa, Oklahoma, famous for its sculptures and dinnerware, walked in wearing a white smock. Frank sat in front of the stage and began pedaling a potter's wheel, slapping a lump of clay like it was a baby.

As he formed a beautiful vase, Frank addressed the one thousand students watching him. "Pottery's my thing," he said, explaining that he had become successful because he loved his job. "You'll do better if you find what you're passionate about."

Later that day, as Hamm fixed yet another flat tire, he began thinking about what he was excited about. He knew it wasn't pumping gas into trucks, but he couldn't come up with anything better.

"It bothered me for days," he says. "I didn't know what I had to be passionate about."

Then it occurred to him that finding crude was his thing. He remembered the fun the oilmen seemed to have spending their money, as well as their acts of kindness and generosity. "I wanted to be like them," he remembers. "I figured if you can find one [huge well], all this wealth was there for the taking."

Hamm decided he'd learn as much as he could about the exploration business. He wrote a senior-year high school thesis about J. Paul Getty, Bill Skelly, and other architects of Oklahoma's oil industry.

One day late in his senior year, as he was working at the truck stop, a man who ran a nearby hauling company drove up to him.

"What are you gonna do after high school?" the man asked.

Hamm didn't know. He couldn't afford college and realized he'd need to make some money. The man offered him a job and Hamm jumped at it. Soon he was working for a man named Johnny Geer who ran a company that serviced the oil industry. Hamm was given a tank truck to haul water to oil and gas drilling sites. He wasn't searching for oil, but at least he was working with wells.

Holy cow, this is what I want to do, he thought.

"I fell in love with working in the oil field," Hamm says. "It was so immense, so complex, out there working for all these companies, it was totally fascinating to me."

He had another reason to treasure the job. Hamm's first childhood girlfriend, Judith Ann Miller, had become pregnant and the couple got married, even though both were just seventeen. The move was somewhat unusual, Hamm's sister recalls, because Judy was Catholic. But getting married while still in high school was quite common at the time in the state, as was Hamm's decision to join the working world at an early age.

After a couple of years, Hamm began to notice his boss, Geer, caring more about his bottle of bourbon than their work. By then, Hamm was running the operations of a business that had grown to ten trucks from just three. Geer's drinking began to cause drivers to quit, making it hard for Hamm to run the business.

Hamm was friendly to Geer but intense about his job. Soon he was fuming about his boss. One day, he pulled Geer aside for a talk. "The next time you show up here drunk, you better bring my last check," he told him.

A few days later, when Hamm saw Geer inebriated, he quit. Hamm spent a few months working at one of Champlin's big refineries but grew unhappy. He couldn't understand why union rules prevented him from helping others at the refinery, or allowed employees to sleep on the job and then complain at union meetings. He also missed the oil fields.

He heard that a local man was having trouble making payments on a "bobtail truck," or a truck without a trailer, that he used to service those drilling for oil in the area. The man agreed to let Hamm take over payments on the truck and Hamm borrowed $1,000 and found a cosigner for the note. He was twenty and had his own business, Harold Hamm Tank Trucks, which later was renamed Hamm & Phillips. Hamm and his young wife rented a small house twenty minutes away, had a second daughter, and tried to make a success of their new life.

Competition was fierce in 1966, but Hamm outworked rivals. He took the dirtiest, nastiest jobs that no one else would touch, sometimes getting up in the middle of the night to clean tank bottoms or haul water to drilling sites. He'd pull on a pair of "high wader" boots, take a long mop, and crawl into oil tanks to rake sediment and mud out of their bottoms. He was so tired that sometimes he fell asleep with his arm in a tank, a friend remembers. When oil was pumped in and reached his arm, he'd awaken, startled.

Friendly, upbeat, and eager to learn, Hamm continued to pick the brains of industry veterans to divine secrets of the business, hoping to one day search for oil and gas. He spoke to drilling pros, well-service experts, and well-completion veterans. They taught him how to read well logs, analyze rock, and other tricks of the trade.

Hamm had a few hours during the day that still were free, so he built a library in his home and borrowed books about geology and geophysics. Around that time, his mother died of cancer. For religious reasons, she never saw a doctor, and her death was a blow to a young man juggling a young family, self-education, and his oil-service business.

"It was like a ton of bricks," he says.

Hamm pushed himself to work even harder. He drove a beat-up car and rented his home but still didn't have much in the bank. "I wanted to pay for my own house and have a new pickup," he recalls.

Over a short time, though, his business grew and Hamm hired employees and bought more trucks. He gained the trust of clients by using quality equipment and not overcharging, like some rivals. Soon, he had the largest oil-field fluid and hauling transport business in the region and even found customers in Texas and elsewhere.

Hamm had a bit of wealth, something unique in his family, but it wasn't enough for him. His dream was to find oil, just like the men he had encountered at the service station, not to clean out tankers. He wanted serious money.

"I wanted to put myself in a position to find the ancient wealth, I had this burning desire," he says. "We all want to change the world and be bigger than our daily existence. I'm no different."

In 1967, at the age of twenty-two, Hamm incorporated an exploration company, Shelly Dean Oil Co., named for his two daughters. He kept his service company and still did some drilling for others, but he began searching for oil and gas.

Hamm looked for an ideal spot for his first well. He had heard about an old well drilled back in 1943 by Royal Dutch Shell that had blown out and burned its rig down. Shell eventually gave up on the area. Years later, the well was redrilled and managed to produce some oil.

Hamm was encouraged by the well's consistent production, which suggested to him that it might be part of a larger reservoir. He began looking at well logs and noticed that drillers had bored through a zone, or layer, of rock that seemed to have high permeability, or one in which fluids seemed to flow easily. He deduced that the rock in the layer was thick with oil that previous wildcatters and larger explorers had missed. Hamm slowly acquired more than one thousand acres from various companies, such as Getty Oil, and borrowed $100,000 to drill his first well, showing an unusual confidence for a young man with no exploration experience.

"I knew I could pay it off—I just didn't have the money just then," he says.[3]

In 1971, oil began to pour out of the well at a rate of about twenty barrels an hour. The well went on to produce a steady flow of oil for more than four decades. Hamm's second well produced seventy-five barrels an hour. It turned out to be on the edge of a huge field that would generate six million barrels. Hamm's young company was making more than $37,000 a month.

Hamm used his early proceeds to pay for classes in geology, mineralogy, and chemistry at a nearby college, though he didn't have time to earn a degree. Now twenty-five years old, he embraced early technologies, such as computer mapping and directional drilling, the precursor to horizontal drilling.

Hamm seemed to have an unusual ability to find oil and natural gas, even when others scoffed at his efforts. "I can't believe you think there's oil there," Ralph Bradley, a local farmer, told him when he proposed drilling on Bradley's property. Bradley and his wife, an older couple who hauled water from town to their home each day in an old Chevrolet, advised him to try somewhere else.

Disregarding their advice, Hamm borrowed $90,000 to cover the expense of the well and hit a gusher producing two thousand barrels a day. The Bradleys, who received royalties from the well, took their cash and moved to Idaho to live with their grandchildren.

Hamm purchased a company that drilled wells and operated rigs for other exploration companies. As oil prices soared on the heels of the oil shocks of 1973 and 1979, demand grew for his service and drilling companies.[4]

Hamm sometimes used his humble beginnings to outmaneuver rivals. In 1981, Bob Moore, a competitor in the trucking business, tried to entice Hamm to buy his operation. Moore traveled to Enid and spent over an hour discussing his business with him, according to Art Swanson, who worked for Moore at the time. Moore discussed which customers he had, where he got his employees, and more. Hamm listened quietly as Moore shared more details of his operations.

When it came time to discuss the purchase of Moore's business, Hamm suggested that he wasn't equipped to negotiate with a veteran like Moore. "All I know is driving trucks," Hamm told him. "I didn't go to college."

On the way out, Swanson realized Hamm had never had any interest in Moore's operation. He had taken the opportunity to glean crucial intelligence from his competitor, downplaying his abilities in order to get Moore to open up.

"He had a blistering intelligence, but he was playing dumb," Swanson says. Hamm ended up buying up other trucking companies while Moore's went bust, according to Swanson.

Hamm sold his drilling company for more than $30 million in 1982, just before Penn Square Bank, a sizable Oklahoma bank that had profited from high-risk energy loans in the state and in Texas, collapsed, triggering huge losses for banks around the country and a painful shakeout in the energy business.

When it came to searching for oil and gas, Hamm was a successful local operator, one of many at the time. He was so parochial that in 1983 he began drilling a series of sixteen wells under Enid itself. At that point, being a neighborhood baron was enough for him, or at least for his family members. One day, his father visited the home Hamm and Judy had purchased, ten miles east of Enid. As Hamm drove up with his father, Hamm pressed a button to open the garage.

"How'd you do that?!" Leland Hamm asked his son.

"I used my garage door opener," Hamm replied.

"But there are no wires," Leland Hamm said, looking stunned. At that moment, he knew his son had made it.

Hamm had his first taste of wealth and he wanted much more. The Penn Square Bank failure made it harder for local drillers to get financing, but it had also reduced Hamm's competition. He ran into bad luck of his own, though. In 1983, he endured seventeen straight dry holes trying to find productive wells in the area, a string of failures that cost his young company over $10 million. Hamm still had some cash from the sale of his drilling company, but it was running out and he began to feel the pinch. At the rate he was spending, he estimated that he'd only be able to drill for about a year more.

He sat Judy down in their family living room and warned her that their lifestyle was in jeopardy. In the nick of time, though, he hit enough oil to keep his company going.

"We were bailed out by oil," according to Hamm.

The strike left an impression. Others were becoming enamored with natural gas, but Hamm became a believer in crude. He ordered a study to find out which regions were "oilier," or had the potential to produce large quantities of oil, and he hired a pro to look at areas in North Dakota, Montana, and Wyoming.

Hamm wanted to check things out for himself, so he flew to Rock Springs, in the southwestern corner of Wyoming. He wore overalls and awkward "Mickey Mouse" boots that reached halfway up his calves, but he still wasn't prepared for the bitter winter cold. His guide, who ran a local company that hauled water to drilling rigs in the area, had a similar outfit but tucked a flask of whiskey into his overalls, managing a bit better than Hamm.

Hamm ended up doing some drilling in the Midfork oil field in Montana, but the results weren't very impressive. He survived the oil busts of the 1980s in reasonably good shape, though, even as major energy companies, such as Exxon and the Atlantic Richfield Company, focused abroad and some local competitors ran into trouble.

As rivals pulled out of the Oklahoma region, Hamm, who in 1991 renamed his company Continental Resources, was able to hire talented employees such as Jeff Hume, a young engineer and native of Enid who had worked at companies like Sun Oil, the parent of what would become Robert Hauptfuhrer's Oryx Energy. Hamm also managed to buy acreage discarded by companies fleeing the area amid the industry's slump. In 1985, Hamm made a deal with a larger company called Petro-Lewis Corp. to purchase more than five hundred oil and gas wells, including some in the town of Ames. "They were considered the dregs of what Petro-Lewis was selling," says a Continental executive.

Hamm and his staff became intrigued by the area around Petro-Lewis's wells, despite their feeble production. At the time, few rivals drilled very deep into the ground, afraid of the cost. Hamm told his men to give it a shot. They gathered publicly available information and fed it into a computer to map where accumulations of oil and gas might be hiding. The men collected and interpreted seismic data in three dimensions, a new technology at the time.

One day, Rex Olson, an exploration manager, showed a map to Hamm, pointing to an unusual structure two miles below the surface that looked like a giant hoofprint of a cow. "You know, that kind of looks like an astrobleme," Olson told his boss, referring to an ancient meteorite crater.

"You're kind of right, it does," Hamm replied.

They drilled down nine thousand feet and discovered an ancient crater, eight miles in diameter, created by a meteor that had pummeled the earth hundreds of millions of years earlier. The meteor, one thousand feet across, likely hit the earth at seventy thousand miles an hour, packing so much energy that surface temperatures soared five hundred degrees in a matter of seconds, scientists later estimated. More important to Hamm was that the crater, later nicknamed Ames Hole, was brimming with oil. Continental tapped a series of prolific wells that over time yielded more than eighteen million barrels of oil and a gusher of profits.[5]

By 1987, Hamm was forty-two years old and worth $16 million. For all his apparent success, though, he was dealing with family issues and deep insecurities. His single-minded focus on making Continental a bigger company began to impact his marriage. That year, he and Judy would divorce.

Soon, Hamm was dating Sue Ann Arnall, a petite and pretty thirty-one-year-old industry attorney and economist in Tulsa who reminded Hamm's friends of former Olympic skater Dorothy Hamill. In April 1988, six months after Hamm's divorce, Hamm married Arnall in a Las Vegas chapel. Hamm and Arnall would have two daughters.

Soon, Hamm began to look the part of an energy executive, not a small-town driller, friends recall. "She dressed him, polished him up, and whipped him into shape," one friend says.

Hamm's new marriage eventually would turn rocky. Eleven years after the wedding with Sue Ann, who held several senior jobs at Continental, she would file for divorce, according to state court records. The divorce petition was later withdrawn, according to the records, reported by Reuters.[6]

By the late 1980s, Hamm began to feel self-conscious about his speech, friends say. His parents were farmers, he came from a tiny town, and he hadn't received formal education beyond high school. As such,

Hamm hadn't been taught proper pronunciation and his vocabulary was lacking. When he talked to members of his team or to small groups of people, he was persuasive, enthusiastic, and confident. But when he addressed larger groups or members of the industry, he stumbled and had trouble expressing his thoughts.

For a while, his issues didn't matter very much. Hamm had cash coming in from his drilling business to finance his exploration, as well as money from his early wells. He didn't have to appeal to investors for financing, like Aubrey McClendon and Tom Ward, who were selling Chesapeake debt and shares. But as Continental grew and Hamm came into contact with more business leaders in the city and around the state, his poor elocution and awkwardness in public stood out. Friends noticed that he avoided interviews or speeches and ducked industry media.

When Hamm met people for the first time, he often came across as a local yokel. In a 1991 lawsuit with Occidental Petroleum, the energy giant's strategy was to put Hamm on the witness stand for three straight days and wait for him to stumble. In his testimony, he mumbled, stuttered, and used incorrect tenses, according to someone in the court. Occidental's attorney even got Hamm to suggest that he had improperly backdated a contract with the energy company.

Hamm's own attorney, Allan Devore, won the suit by convincing the jury that Occidental had taken advantage of an unsophisticated Oklahoma country boy, which is precisely how Hamm came across at the time.

"He just butchered the English language, it was hard to understand what he was trying to say," recalls an old friend. "He acted tough, talked with an Okie accent, and used the wrong words. . . . He didn't sound smart unless you really listened to him, and then you realized he was brilliant."

Some friends said Hamm's insecurities made him sensitive to perceived affronts and eager to fight if he felt he was wronged. They cited a series of squabbles with companies and individuals in the industry. At one point, Hamm was sure natural gas prices had been unfairly pushed down, so he levied pressure on state legislators to impose a statewide ceiling on natural gas production to help elevate prices.

"He was a bulldog, you didn't want to get in a lawsuit with him," says Roger Clement, Continental's chief financial officer at the time. "He would have gambled away the company to get even with somebody."

Hamm's friends theorized that he got into fights to send a message that he wasn't going to be taken advantage of, perhaps because he worried about being taken lightly due to his lack of pedigree and poise. "He was looked down upon, people thought he wasn't smart or savvy because he wasn't polished," says Devore, Hamm's friend and attorney. "That's why he established a reputation of 'Don't screw me.' He intended to be a big deal and needed to show people they couldn't take advantage of him."

Hamm was well aware of his glaring flaws and resolved to do something about them. Just as he taught himself how to find oil, he'd learn how to speak properly and with more confidence. He began taking classes at a local Dale Carnegie training center, where he worked on his speech and poise. He also worked on his leadership skills by reading books by management guru John Maxwell. Soon, his delivery and self-assurance improved. Even after finishing his courses, he returned to help with other Dale Carnegie classes and he required all of Continental's middle managers to attend their own Dale Carnegie courses.

Some found other reasons to snicker about Hamm. Around 1995, he was in the cockpit of his single-engine airplane in a hangar in Enid when the engine wouldn't start. He climbed a ladder to try to get the propeller moving, a friend says, hoping it would engage the starter, when the plane suddenly rolled off. Hamm barely got out of the way before it began careening down the runway, crashing into a hangar in violent fashion, tearing the hangar's door off. For years, friends tweaked Hamm about the episode, and some in Oklahoma City spoke about it behind his back. Hamm was sensitive about the incident, which elicited one more lawsuit from him, this time against the plane's manufacturer. He won that suit, a friend says, like many of his disputes.

For all his remaining insecurities, Hamm was one of the few in the business with the confidence to risk serious money exploring new formations. His dreams remained grand, despite the industry's shakeout. He told colleagues he wanted to focus on high-risk, high-return plays, like a quarterback favoring long, dangerous passes rather than safe completions.

Hamm began to attract geologists and engineers who also dreamed of big, historic finds. One of these men was Jack Stark.

J ack Stark would have been excused if he felt a dark cloud was following him as he drove to Enid in 1992 for a job interview with Hamm. A soft-spoken native of Norwalk, Ohio, Stark had attended Bowling Green State University, hoping to follow in the footsteps of his childhood idol, Jacques Cousteau, the underwater researcher. It didn't take long for him to realize that the school, in a landlocked part of Ohio, probably wasn't the best place to learn oceanography. Stark, who had also been fascinated with fossils, rocks, and earth, started taking geology courses. He found he enjoyed them and decided to make geology a career.

After becoming the first in his family to graduate from college, Stark worked for a Denver-based subsidiary of Phelps Dodge that was mining uranium, a fuel for nuclear power. The company funded Stark's graduate degree and his career seemed on strong early footing. As he took his dog with him to work in the foothills outside Denver, he thought he had found the perfect job, especially when he received high early marks from his boss.

A partial meltdown of a nuclear reactor at Three Mile Island in 1979 put a quick halt to the nation's growing interest in nuclear power and uranium, however. Stark switched to a different Phelps Dodge subsidiary searching for gold in Arizona, a move that seemed fortuitous as gold prices soared. But gold collapsed in 1981, forcing Stark again to look for a job.

Stark began to work for an energy company and received another promotion. But oil prices weakened during the 1980s and exploration slowed. Stark's bad luck began to feel familiar. "It was like I was stepping on stones, walking across a stream" while trying not to fall in, he remembers glumly.

Stark remained in the energy business, despite the tough times, and found work at various oil companies. He didn't always enjoy it, though. They might have been big, but the companies usually seemed too timid to put much effort into exploring oil and gas deposits in the United States. They were more focused on selling assets and shutting offices.

In the spring of 1992, as Stark drove to Continental's office in Enid to interview for a geologist manager position to replace Rex Olson, who had died of an aneurysm, he became concerned. All he seemed to see on the side of the road was vacant buildings and empty homes. The state, still struggling to recover from the energy downturn and the Penn Square Bank fiasco, looked a mess.

Enid hadn't exactly covered itself in glory over the years. One story attributed the city's origin to a flub by early settlers after the state's land run. Some of them set up a wagon, began cooking food, and hoisted a sign reading "dine," hoping to start an eating establishment. But the sign got turned around somehow, so it read "enid." The name stuck, according to this local legend.

As he drove, Stark became convinced that even if he got the job, his wife, who had grown up in a suburb of Columbus, Ohio, and was waiting in their home in Houston to hear about the interview, likely would veto a move to the state.

"Reality started to hit that this was small-town America," Stark recalls. "And it looked depressed."

When Stark reached Enid, which at the time had a population of about forty-five thousand, he noted that the city seemed to be holding up better than the towns he had passed along the way. But he was still concerned as he sat down for the first time with Hamm.

A few minutes with Hamm changed Stark's view. Hamm was revved up, talking about expanding his exploration and how he was embracing new production methods. He appeared genuine in his determination to get on the national map.

"I really want to be a bigger company," Hamm told Stark.

Hamm's enthusiasm got Stark pumped up. He drove home and sold his wife on all the advantages of living in Enid, which he called "one of the last *Leave It to Beaver* towns where everyone knows each other."

*   *   *

By 1993, Hamm and his team were dealing with a new problem. Major oil and gas companies had exited the state, but more independent operators—led by McClendon and Ward at Chesapeake—were targeting Oklahoma, making it more expensive for Continental to compete for land.

The company still was making money, but Hamm yearned for something grander. He wanted that ancient wealth he had dreamed of as a kid. It would some kind of historic discovery to put him on the map, a strike so large it would win him both fame and fortune, much like the celebrated wildcatters he had studied.

Hamm ran a relatively small company and had limited formal education in energy exploration, but some of the most successful wildcatters had backgrounds at least as unusual. Throughout modern times, mysterious gas had flowed from shale below the rustic western New York village of Fredonia, delighting local children by occasionally catching fire. But it took William Aaron Hart, a gunsmith with limited education who moved to town with just a rifle and a pack, to drill the first commercial natural gas well in 1825.[7]

"Colonel" Edwin L. Drake, a retired and near-penniless train conductor who had earned his distinctive nickname despite lacking any military experience, was the first to drill for oil in the United States in Titusville, Pennsylvania, in 1859.

Four decades later, Patillo Higgins, a one-armed mechanic, lumber merchant, and self-taught geologist with a violent and troubled past, became a Christian and took his Baptist Sunday school class for an outing near a hill in Beaumont, Texas. There, Higgins came across a half dozen springs bubbling with gas. He became convinced he had stumbled onto an oil field, even though experts scoffed. In 1901, Higgins and Captain Anthony Lucas, an ex-hand of the Austrian navy, tapped a geyser of oil that would fill eighty thousand barrels a day at the hill—called Spindletop—a discovery that ushered in the modern age of oil.

Hamm knew that some of the nation's largest fortunes had been created by wildcatters. Many, such as Haroldson Lafayette Hunt Jr., who was the inspiration for the J. R. Ewing character on the television show *Dallas*, were showmen with a flair for the dramatic. At thirty-two, Hunt, who went by H.L., is said to have turned his last $100 into a $10,000 windfall thanks to a successful run of five-card stud. Hunt used his winnings to buy up oil rights, eventually controlling the huge East Texas Oil Field. He became one of the world's richest men, building a fortune amounting to $5 billion. Hunt also managed to find time to sire fifteen children while juggling two wives and separate families in different cities for over a decade.[8]

In many ways, the wildcat profession is quintessentially American. It takes a heavy dose of self-assurance and comfort with risk to bet on what might or might not be far below the surface, well out of sight, as well as an unbridled optimism that Americans seem to have in abundance.

At the same time, U.S. homeowners usually own the land under their property, unlike citizens in most other countries. As a result, wildcatters can directly negotiate with landowners to buy their drilling rights, rather than go through government officials. The word "wildcat" itself comes from an early-nineteenth-century American slang for a risky business venture.

For all the acclaim and fortune bestowed on wildcatters, these individuals also have suffered among the most costly and embarrassing failures in business life. One experienced wildcatter around the turn of the twentieth century named Charles Lewis Woods earned the unfortunate nickname of "Dry Hole Charlie" for hitting a string of unproductive wells. Charlie's luck seemed to change when a project in California's San Joaquin Valley turned into a massive gusher in 1910, producing 125,000 barrels of oil on its very first day. After eighteen months, however, the oil stopped flowing and Dry Hole Charlie returned to drilling dry holes.[9]

During the 1960s, wildcatters were only successful in one of ten wells, according to the American Oil and Gas Historical Society. It soon became even tougher keeping up with multinationals sporting deep pockets and huge staffs. The plunge in oil prices in the 1980s crippled wildcatters in Texas, Oklahoma, and elsewhere and many went out of business. By the 1990s, most of the biggest exploration and production companies had left the country to people like Hamm. It seemed a booby

prize. Here you go, guys, the country's all yours. See if you can find anything.

The allure of the big strike remained for Hamm, as it had for George
Mitchell. Hamm didn't have to do much convincing to get his team on
board with his plan to try to find a lot of oil somewhere else in the country besides Oklahoma.

"We can continue here, but we're just going to tread water," Stark told
Hamm one day, echoing his boss's sentiments.

Hamm, Hume, Stark, and Continental's exploration executives met
to decide on a plan. The company didn't have boatloads of money, so
they'd have to focus on out-of-favor areas in the country where it might
be less costly to establish a foothold. Passing around a box of buttered and
salted popcorn, they eyed a series of coffee- and grease-stained maps
pinned to the walls. They agreed to take a look at acreage in North Dakota, hoping they might find oil overlooked by larger companies.

Hamm had done some prospecting with a friend in North Dakota
and was familiar with the region. Stark also had done some work in the
Williston Basin, which included parts of the state. He noted that North
Dakota had always produced evidence of oil, but the region was infamous
for dashing the dreams of wildcatters. Hamm figured the state's unpop-
ularity might enable them to buy some acreage on the cheap. He began
making five-hour flights in his single-engine Piper Cub to check out the
area and learned that another company, Burlington Resources, was already making progress in the region.

Burlington's success was intriguing. Rock in the North Dakota region
was dense and had very low permeability, making it hard to get much oil
out, at least at a reasonable price. Dozens of companies already had failed
to produce very much oil from the region by drilling vertically, making
Burlington's production surprising.

Burlington wasn't sharing much information about its success. But
Hamm knew the company would have to reveal some details when it
asked the state for drilling permits.

In April 1994, Hamm and Stark sat in the back row of a hearing in
Bismarck of the North Dakota Industrial Commission, listening quietly
as Burlington executives addressed state officials. No one noticed Hamm,

a sign of how under the radar he was at the time, at least outside his corner of Oklahoma.

At the hearing, Burlington executives described how they were producing seven hundred barrels of oil a day in the Cedar Hills field in the southern part of the state by drilling horizontally. They didn't even need to frack the rock to get the oil to flow, it seemed.

Now Hamm was excited. Not only was Burlington managing to extract a substantial amount of oil, but it was finding success with horizontal drilling, a method that was relatively new. At the time, Chesapeake was seeing its own success using horizontal drilling in Texas's Austin Chalk region, and companies like Oryx Energy had already made some inroads with it. Stark had just come back from a conference in Saskatchewan, Canada, where a Shell geologist talked about how much oil the company also was extracting by drilling in a lateral fashion. It was more expensive to drill horizontally, but Stark and others on Hamm's staff were becoming convinced they had to try it.

Stark noted that Shell was producing its oil from the same Williston Basin that Continental was examining, suggesting that Continental might have similar success in North Dakota if it aped Shell's methods.

Wow, this is a game changer, Stark thought.

Hamm and the Continental team agreed to lease acreage in Bowman County, in the southwestern corner of the state, along North Dakota's border with South Dakota and Montana. They mapped the region, saw a belowground anticline, or a fold made up of a sequence of rock layers, and examined what appeared to be evidence of oil in some of these rock layers.

Hume and Stark recommended that the company lease ten thousand or maybe even twenty thousand acres in the Cedar Hills field to test whether they could produce oil. When he heard the recommendation, Hamm scolded his team. They weren't being ambitious enough. He told them to assemble 100,000 acres, though he knew the company couldn't really afford that kind of spending.

"It was all the money we had at the time," according to Hume.

The region was so out of favor that Continental's landmen locked up their land within months, paying about twenty-five dollars an acre. They began drilling in one of the field's higher rock formations, which was called the Red River B.

Until then, Hamm had resisted borrowing money or selling shares to raise cash, unlike McClendon and Ward at Chesapeake. Instead, he relied on cash coming from existing wells and from his side businesses, such as his drilling company.

"I don't like using other people's money, it changes you," he says, citing the need to be a salesman. "You believe your own bull, it distorts you."

But now he was trying to enter the big leagues. Between all the land in North Dakota and the cost of horizontal drilling, the company's expenses dwarfed those of its wells in Oklahoma. Hamm decided to sell $150 million of debt and pay an expensive interest-rate coupon of 10.25 percent on the bonds to finance the project.

In April 1995, Continental drilled its first well in the Cedar Hills field. The first wells were immediate successes. Three years later, though, just as the field was reaching peak production of seven thousand barrels of oil a day, bad news struck as crude prices tumbled. By late 1998, oil prices were barely above eleven dollars a barrel. At one point, Continental's North Dakota oil commanded just $4.50 a barrel, making it hard for the company to make any money. That year, the company lost $18 million.

A friend tipped Hamm off to the dramatic improvements that Mitchell's team was making in the Barnett Shale. "Harold, you really have to get down here," the friend told him. But Hamm didn't have any spare money to buy acreage in the Barnett.

Adding to his pressures was that Continental hadn't protected itself like many of its rivals by selling some of its future oil production in financial markets to lock in prices before they had plunged. And Hamm had overruled key members of his team and agreed to spend about $80 million of precious cash to buy acreage in a field in Wyoming that would prove a disappointment.

Hamm had to halt almost all of his company's drilling and let most of his drilling pros go. By the end of 1998, Continental was operating just one oil rig, down from eight, and only had about fifty employees on staff. To avoid additional layoffs, every employee took a 15 percent pay cut.

Almost everyone in the business understood that low oil prices were due to excessive global supply and an economy that seemed less reliant on oil and gas. Hamm wasn't like everyone else, though. He was convinced OPEC nations were dumping oil, or selling it in the United States below its cost, to drive Continental and other American producers out of business. He took it as a personal insult.

Hamm rounded up a group of small oil and gas producers to fight OPEC. The group, which they called Save Domestic Oil Inc., petitioned the Department of Commerce to impose tariffs on crude oil coming from Saudi Arabia, Mexico, Venezuela, and Iraq.

Hamm's fellow energy producers scoffed. Exxon and other major companies were big enough to survive the price decline and they wanted to keep good relations with OPEC nations. Still others thought it was silly to blame OPEC for low oil prices.

"Everybody made fun of Harold that he would file a dumping case," says Mickey Thompson, an Oklahoma political veteran who worked on the case for Continental. "He was viewed as a champion of lost causes."

Hamm and the group managed to sign up thousands of tiny producers, mom-and-pops from Kansas, Arkansas, and elsewhere who were just as scared as Hamm of being crushed by low prices. But OPEC members such as Saudi Arabia and Venezuela locked up the top law firms in Washington, D.C., to fight the effort, leaving Hamm and his group with few options for legal representation.

The group's complaint eventually was rejected, though Hamm viewed the effort as an effective warning shot at OPEC. Either way, oil and natural gas prices moved higher in the summer of 1999, providing Hamm and his company some relief, just as rising natural gas prices were giving Aubrey McClendon and Tom Ward new confidence around the same time.

Through it all, Continental's Cedar Hills field continued pumping an impressive amount of oil. It became the seventh largest onshore field in the lower forty-eight states, and the first to be developed exclusively through so-called precision horizontal drilling.

"The majors thought it was a fluke," Hamm says.

The Cedar Hills gusher opened Hamm's eyes to the possibilities of drilling for oil in the region's dense, difficult rock. He noted how many thick, long layers of rock were in the Williston Basin—a huge, shallow underground hole filled with accumulated sediment that was named after the North Dakota city of Williston and was under eastern Montana, western North Dakota, South Dakota, and Saskatchewan. Hamm began to wonder if he could tap the region's other layers with the horizontal drilling techniques he and his team had become comfortable with.

"Where else can it work?" he demanded of Jack Stark. "I want to know where the next really big field is."

It was up to Jim Kochick, a senior geologist on Stark's team, to keep Hamm happy. Within months, Kochick became excited about a 200,000-square-mile formation, or series of rock layers, not far away from Cedar Hills. The formation was called the Bakken, and it encompassed three rock layers. A top layer of shale, called the Upper Bakken, was two miles below the surface. Another shale layer below that was called the Lower Bakken. And a long, thin rock layer sandwiched in between was made of a type of limestone called dolomite. Appropriately enough, this layer was called the Middle Bakken.

The Bakken formation was well known to any geologist in the area. Continental itself had drilled right through the Bakken rock layers on its way to Cedar Hills' Red River formation below. For years, major oil companies, midsized independent producers, and small wildcatters either had ignored Bakken rock or had failed to produce much oil or gas from it. The Bakken was so hard and compact that it looked more like tombstone, just like the Barnett Shale, unlike more traditional and permeable rock.

Kochick told his boss, Jack Stark, that there was a decent chance the Bakken might be just as successful as Cedar Hills. Stark gave Kochick the green light to make his case to Hamm and the rest of Continental's exploration group at their next meeting at the company's Enid headquarters. Facing his colleagues, Kochick presented evidence of oil "shows" throughout the Bakken region. In other words, there were clear indications that these rock layers held oil.

Kochick acknowledged that some vertically drilled wells already getting oil from the Middle Bakken layer of rock were generating a dismal amount of crude, with some wells producing as little as five barrels a day.

That shouldn't discourage us, Kochick told the gloomy executives. Some of those wells were thirty years old, so their meager production should be viewed with respect, not disdain.

Further, those weak results were similar to those Continental had seen from vertical wells in the Cedar Hills field before the company began applying horizontal drilling to make that field a winner. Maybe horizontal drilling also could be the key to extracting oil from the Bakken's rock, especially the long, shallow Middle Bakken layer, Kochick said. He proposed that Continental begin buying acreage in a Montana field called the Elm Coulee and drill down to the Bakken layers to see what they could find.

A few engineers at the meeting were less than enthusiastic. The Bakken just didn't seem like a traditional reservoir eager to give up its oil. Its compressed rock had long been a big tease. They wanted to see more evidence that the Bakken was as attractive as Kochick suggested before Continental shelled out any precious cash.

Hamm turned to Stark, asking if he agreed with Kochick's assessment. Stark said he did.

"I like it," Hamm said, becoming excited. "How many brokers do you have?" He wanted to know how quickly they could buy up land in the eastern Montana area.

For months, Hamm couldn't stop thinking about the possibilities of the new formation, prodding his team to move faster to develop a plan to target the area. He didn't tell his team the Bakken was any kind of huge reservoir, and it wasn't yet clear they had the technology to produce oil economically.

But Hamm became convinced there was a lot of oil in the Bakken that might someday be tapped. Flying on his private plane to a fishing and hunting vacation in Branson, Missouri, he kept talking about the Bakken. He wondered if there might be more oil in those rock layers than even his optimistic geologists believed. On the plane, Hamm grabbed a cocktail napkin, got the attention of an old friend, Ron Boyd, and began to draw little circles within a larger circle.

"Geologists say there's oil under each of these," Hamm said, pointing to a few isolated circles on the napkin.

He gave Boyd a mischievous look.

"But I think there's oil under the whole damn thing."

In the spring of 2000, Hamm and Continental's executive team flew to Cambridge, Massachusetts, to take part in executive business classes at Harvard University. Hamm was preoccupied, though. He had received new information about how innovative drilling and well-completion technology was making it easier to extract oil from challenging rock. He also knew two other companies already were buying up acreage in the Bakken area, also on the Montana side of the border.

One morning, he got up early and convened a meeting of his top executives. Land pros back in Enid were on the phone, listening. "We can't wait anymore," Hamm said. "Boys, go and get that acreage leased," he told his land team, referring to the Elm Coulee field in Montana.

When some of Hamm's employees expressed doubt about the likely drilling cost, Hamm turned angry. "You didn't hear me—go out there right now and get that acreage leased," he responded impatiently.

Hamm was determined to find oil in rock that had bedeviled dozens of oil companies and wildcatters for years. He had no clue how hard it would prove to be.

CHAPTER SEVEN

By 2000, Aubrey McClendon and Tom Ward were searching the country for natural gas. Harold Hamm was set to make a huge bet that he could find massive amounts of crude in overlooked U.S. fields. Even in his retirement in Houston, George Mitchell was preparing to team up with his son Todd to do some drilling in other shale fields that might be as prolific as the Barnett.

Charif Souki was just as sure America needed new energy supplies. And he was certain he had a way to get his hands on enough natural gas to change his country's fortunes, as well as his own. But Souki was a rank outsider to the oil patch who had no business imagining himself as any kind of energy power. He hadn't taken a single geology or engineering course or even worked a day in an oil or gas field.

A shaggy-haired Lebanese immigrant who spoke English with a French accent, Souki had spent a full seven years in Aspen, Colorado, where he skied and bummed around, and in Los Angeles, where he ran bars and restaurants, including one involved in perhaps the most famous murder of the century. Before the late 1990s, the closest Souki had come to oil was in the salads on his menu.

The nation desperately needed help discovering natural gas, but it didn't seem to need it from a guy like Charif Souki. Relying on Souki to find new energy supplies was like a football team, desperate for a miracle completion, asking the water boy to take over at quarterback.

Few wanted to hear it, but Souki did have an audacious plan to find energy, one only he was sure would work. It was an approach as unusual and quirky as he was.

Souki was born in Cairo in 1953, a year after a coup d'état by a group of Egyptian army officers led by Muhammad Naguib and Gamal Abdel

Nasser. After seizing power, the officers received an urgent telegram from America:

Congratulations on your successful coup

The Pentagon

The message confused the officers, who didn't speak English and weren't experienced in the ways of international diplomacy. To make sense of it, one of them called Charif's father, Samyr Souki, asking for his help. Samyr was the chief of *Newsweek* magazine's Cairo bureau and earlier in his career had reported for United Press International, so he was considered a local expert on the ways of America.

An officer had a simple question for Samyr: "Who is the Pentagon?"

After explaining the telegram's meaning and getting to know the officers, Samyr embraced the new Egyptian leadership. Soon the American administration did as well. Samyr, who had received a British education and spoke English fluently, was so optimistic about the new chapter for the nation that he took a leave of absence from *Newsweek* to work in Egypt's embassy in Washington, D.C.

Soon after he returned to Cairo in 1954, though, relations between the United States and the Egyptian leadership soured. Samyr, a Greek Orthodox Christian, realized that his writing was being censored and that he was being tracked by members of the regime, making it hard to do his job. After a while, he began to contemplate moving his family to a new country.

When Charif was four, the Souki family left for Beirut, where Samyr and his cultured and well-educated wife, Nicole, had been born. Samyr, who also went by Sam, became *Newsweek*'s chief correspondent for the Middle East. As he covered subsequent upheavals, including the 1957 war, as well as Beirut's growth into an intellectual, tourist, and banking center, Sam emerged as a star reporter. He had a sense of where American policy was headed and enjoyed unique access to senior officials in several countries.

It didn't take long for Sam to become a key adviser to political and business leaders throughout the region. The young Charif sometimes sat with his father as he entertained various rulers, including King Faisal of Saudi Arabia, top United Nations officials, and a range of diplomats and businessmen, including chief executives of U.S. military contractors. Whenever they passed through Beirut, dignitaries made sure to stop by the Souki home to trade gossip, share intelligence, and pick Sam's brain.

"Back then, there was no CNN," Charif recalls. "You got your news from people you knew."

As a youth, Charif had little interest in politics and business, despite his father's unique vantage point. Tensions sometimes arose between Lebanon's Muslim and Christian political leaders, but Charif moved easily between both communities. He had Christian and Muslim friends and sometimes played with boys from the smaller Jewish community, all of whom lived near the Souki home.

The country's phones didn't work well, there were no video games to play, and the Souki family didn't own a television until Charif turned eleven. He didn't miss much, since entertainment programming in the country amounted to little more than a couple of hours each day of imported American fare, such as *Bonanza* and *Peyton Place*.

Souki discovered other kinds of local fun. Beirut was a cosmopolitan city with spectacular beaches, and the city was just an hour from prime skiing in the Lebanese mountains. Young instructors from Austria and France came to the country to fulfill their military service, working with Charif and his friends as they became expert downhill racers.

During his teenage years, Charif began to notice street demonstrations involving various religious factions, but his family and neighbors generally shrugged off the early signs of sectarian unrest. "I was mostly interested in skiing, surfing, the beach, and girls," Souki says.

Teenagers in Beirut didn't have a great deal of supervision and Charif enjoyed his freedom. His grades at the International School, which was affiliated with the American University of Beirut, usually were quite good. That sometimes earned him an even longer leash from his parents.

Some days, Charif skipped out early to surf or sail. Other times, he and his friends avoided school altogether to drive to a resort in the mountains for a day of skiing. They tried to get home around the time classes let out, to avoid arousing suspicion. It didn't always work, though.

"My parents had some nervous moments," Charif recalls. "I didn't have much affinity for authority."

Charif's father encouraged his son to attend college in America. As a teenager, Sam Souki's own dream had been to leave Lebanon to go to a U.S. university. World War II had intervened, however, and he became a war correspondent. Sam encouraged his son to apply to American schools and was thrilled when Charif was accepted at Colgate University in Hamilton, New York.

"I was fairly restless and rambunctious, so my father thought it was a good idea" for him to go away to school, he says.

Charif had never been to the United States and he wasn't particularly eager to leave Lebanon, where he was having so much fun. When he heard Colgate was in New York, though, he became more enthusiastic, anticipating four years of excitement.

When he arrived on Colgate's campus in 1971, Charif was stunned. It turned out that the university was in a little hamlet over two hundred miles from New York City. Racial tensions on campus were another surprise, and even the skiing was horrible. Worst of all, there weren't any female students to be found.

"The notion of an all-male school never even entered my mind," Souki says. "I was in cultural shock for a few years."

There wasn't much point complaining to his parents back in Beirut since it was almost impossible to contact them by phone, so Souki decided to try to complete his studies as quickly as possible. He enjoyed the classes, at least, so he took extra courses in economics, history, and literature and received a bachelor's degree in three years.

By then, Sam Souki had left *Newsweek* to start his own publishing, consulting, and public relations enterprise and to do his own investing. The shift inspired Charif to consider his own career in business. He figured a master's in business administration might help him find a job when he returned to Lebanon, so he enrolled at Columbia University's business school.

Columbia, in New York City, was a breath of fresh air. Friends from Beirut lived nearby and there always seemed to be something fun going on. The city was in the midst of a financial crisis, adding drama to the times. Souki even fell in love with a fellow student, Deborah Taylor, and they soon wed.

By the time Souki graduated from Columbia, a brutal civil war had broken out in Lebanon and Christians felt under siege. Souki's parents, who had been on a visit to Paris when the fighting began, decided to stay in France.

Unable to return to Beirut, Souki began interviewing for jobs in the United States, quickly realizing he was a hot commodity. Banks were racing to various Gulf states to court wealthy businessmen and emerging oil barons looking to invest their growing piles of petrodollars. There was intense demand for young men like Souki who spoke Arabic and held a business degree from a top American school.

Souki was hired by an investment banking firm called Blyth, Eastman Dillon & Co. and began wooing Middle Eastern moneybags, trying to entice them to diversify their holdings into real estate, hotels, and other projects championed by the bank in various Western cities.

Souki had clear advantages over his rivals. For one thing, his father remained well known in the region. Many of the petroleum potentates had trekked to the Souki home to seek advice, and they proved eager to meet with Charif. Because he had been exposed to power at a young age, he wasn't impressed or intimidated by the wealthy men he was dealing with.

Souki had grown up traveling the region, so he enjoyed jetting between New York, Paris, Athens, and various Middle East capitals. He also realized that he had an ability to differentiate between attractive deals and potential money losers, and quickly gained a following by steering clients to the winners.

"At first they received me because of my dad," Souki says. "Later it was because of me."

Cutting deals in the region had its fair share of challenges, however, even for Souki. For one thing, phone communication wasn't reliable in Saudi Arabia and other countries in the Middle East, so he couldn't even book a hotel room before getting on a flight to the area. He tried sending telex messages, but the hotels never seemed to receive them. He typically showed up in a lobby, hoping for the best. To improve his odds with hotel personnel, he became a big tipper. Soon, hotel staff members were racing to grab his suitcase and whisk him to a comfortable room.

Secretaries were rare in that part of the world, making it just as difficult to schedule an appointment with anyone of importance, and tips
didn't go nearly as far in the world of oil riches. Instead, Souki would
show up at the home or business of a local baron or family patriarch and
take a seat in a large waiting room next to dozens of bankers, businessmen, and well-wishers.

Souki and his competitors lounged and chatted for hours, waiting for
the tycoon to arise from his midafternoon nap and convene his late-
afternoon *majlis,* or daily social gathering. There, those vying for the titan's attention would try to get invited to another, more private
get-together later that evening. Souki usually was among the lucky few
asked to return for these one-on-one meetings, which usually featured
exotic food and drink and took place around midnight. That's when
hands were shaken, deals were cut, and checks were written.

During that period, Souki met billionaire arms dealer Adnan
Khashoggi, members of the Saudi royal family, up-and-coming energy
power brokers, and others. Personable and smooth, Souki managed to
broker huge investments. Once, an oil magnate wrote a check for $300
million, or the equivalent of over a billion dollars in today's money. Souki's employer usually received a healthy commission of 1 to 5 percent of
each investment, he says.

Souki spent three years wooing various moguls, emerging as a star
at his bank. The job began to grate on him, though. For one thing, the
lifestyle was hard on his marriage and family, which then included two
small children. It wasn't just that texts and e-mails hadn't been invented,
making it hard for him to stay in touch as he spent weeks traveling; sometimes he waited in line for three or four hours just to get access to a phone
to call home. The Middle East was seven hours ahead of the U.S. East
Coast, adding to the challenge. If Deborah wasn't home, he couldn't even
leave a message for her because they didn't have an answering machine.

Souki was meeting interesting and powerful people, and he was dealing with big sums of money, but he wasn't doing any actual investing. He
chafed at raising boatloads of money for deals he wasn't participating in
and felt he was little more than a well-paid matchmaker introducing
wealthy oil barons to his bosses.

"It was 'just go to the Middle East, bring money back, and we'll decide what to do with it,'" he says. "The job sounded more interesting than it actually was."

Souki quit in 1978 and began raising money for deals that he himself found. There was a refinery in Long Beach, California, a hotel in Hawaii, an office building in Paris, and more. He continued to crisscross the Middle East, but he also made inroads with wealthy investors in various European capitals.

Souki became an even more effective salesman and was thrilled to work on his own transactions. He kept a small percentage of each deal for himself, hired bankers to work for him, and began investing his own money in the most attractive opportunities. His business thrived, but the stepped-up pace put even more pressure on his marriage. Soon he and Deborah divorced.

As his reputation grew, top businessmen around the globe sought him out, asking for help raising money or requesting Souki's advice on tricky transactions. A few of the deals were in the energy world, but Souki mostly worked on retail, apparel, and real estate transactions, making several million dollars a year for his work. In 1986, Marvin Traub, the impresario who ran Bloomingdale's, turned to Souki, thirty-three years old at the time, when his upscale department store attracted takeover interest. Souki also spent time working with Calvin Klein, Ralph Lauren, and others in the fashion business while advising the chairman of Bidermann Industries, the largest apparel manufacturer in North America.

By 1986, Souki had had enough of the globetrotting and dealmaking. On a visit to Paris, he had met a New York model named Rita Tellone, who had appeared in magazines including *Vogue, Elle,* and *Sports Illustrated*'s annual swimsuit edition. He and Rita became close and soon married atop Aspen Mountain, outside the city of Aspen, in a daytime ceremony before fifty friends and family members.

Souki told his new wife that he was done with banking and traveling. He just wanted to ski, relax, and spend time with her. His frenetic pace had ruined one marriage, and he didn't want it to crater another one. The couple decided to stay in Aspen and make it their new home.

Souki built an entirely different lifestyle. He had saved enough money to retire so that he could fully enjoy the city that singer John Denver described as a "sweet Rocky Mountain paradise." Souki hiked, biked, and dined with locals. Racing a local ski instructor down various mountains was as close as he came to competition.

For several years, Souki led a relaxed routine. His two children spent extended vacations and all summer with him and Rita. Soon they had two children of their own, and Souki focused on them, trying to be a better parent than the first time around. "There were many years when I traveled so much I couldn't see them grow up," he recalls. "This was a counterweight."

He figured he had enough money stashed away, but he still needed a hobby. The Aspen slopes seemed to lack an upscale family eatery, which got Souki thinking about starting one. He didn't know much about the restaurant business, but he had confidence he could figure it out.

Eventually, Souki launched a restaurant called Mezzaluna in the center of town, a block from the famed Silver Queen Gondola. The restaurant, featuring pizza, pasta, and salads, was an instant hit. Movie stars who came to Aspen to ski, including Jack Nicholson, Michael Douglas, and Kevin Costner, were regulars, as was oil and movie mogul Marvin Davis. Gonzo journalist Hunter S. Thompson entertained guests at the restaurant's bar, and lines formed around the block to get in.

The experience was fun and lucrative, so Souki and his brother decided to open a few more restaurants in Los Angeles with the goal of providing income during Aspen's off-season. They also invested in Los Angeles bars. One of their hot spots, the Monkey Bar, attracted the brightest stars of the day, including Nick Nolte, Dolly Parton, and Penny Marshall.

In 1990, Souki and his brother opened a new Mezzaluna restaurant in the Brentwood district of Los Angeles. Souki began spending morning until nighttime at the new restaurant, trying to put it on firm footing and getting to know his staff and regular customers. Over time, he became friendly with a young waiter named Ron Goldman, as well as a regular patron, Nicole Brown Simpson. He also got to know her ex-husband, football legend O. J. Simpson.

One evening in June 1994, Nicole's mother called the restaurant looking for Nicole's missing eyeglasses. The manager found the glasses and placed them in a white envelope. After his evening shift was over, Goldman dropped them off at Nicole's home around 10 p.m. Shortly thereafter, Brown was found lying facedown on the house's walkway in a puddle of blood; nearby was Goldman's own lifeless, bloodied body. A Jeep-like vehicle was seen racing from the scene.

O. J. Simpson was arrested and tried for double murder. His acquittal, in one of the most controversial trials in American history, divided the nation, though Simpson later lost a civil case related to the murders.

The trial placed Souki's restaurant in the national headlines. Within days, hordes of tourists lined up to get into Mezzaluna, badgering Souki and his staff about Ron and Nicole, asking for details of her last meal, and stealing dishes embossed with the restaurant's logo. Business soared, but Souki was sickened by the experience.

"It was disgusting, tourists got off of buses from Indiana and wanted to eat in a place connected to the crime," Souki says. "I got to see a side of human nature that wasn't very pretty." He and his brother sold the restaurant and got rid of the rest of their eating and drinking establishments. What started out as a hobby had become a monumental headache for him.

Souki received a new shock when he realized he had drained most of his savings. A nest egg of several million dollars had been whittled away by years of fun in Aspen and mixed results from his Los Angeles ventures. It didn't help that he led an expensive lifestyle with four children, a wife, and an ex-wife.

"I had literally lost all my money," Souki recalls. It was a "big miscalculation."

It was a bit of an exaggeration—Souki still had about $300,000 left in savings. But he was close enough to broke to turn quite nervous and search for a new career. Restaurants and bars hadn't worked out very well. He had some experience working on deals in fashion and retail, but those businesses held even less appeal.

"You're dealing with large egos and difficult people, it's not a lot of fun when you get to know them," says Souki, who acknowledges that he wasn't a perfect fit for the retail or fashion business. "When you have to re-create yourself every six months it's a lot of pressure."

By 1996, American businesses were beginning to embrace technology, the Internet was coming into vogue, and Souki was convinced something big was afoot. He decided to try to find an industry that hadn't been fully impacted by technological change. He thought he might have an advantage if he could be the first to embrace some new advance.

By then, he and his family had moved to Los Angeles, where his two older children lived, so he looked into the entertainment industry, learning about digitization of music and movies. He couldn't figure out how to make money from it, though, which frustrated him.

His interest began to shift to energy, a sector few cared about. Oil and gas prices had been limp for nearly two decades, making it hard to score profits, even as sexy Silicon Valley technology companies were raking in investor money and attracting the best and brightest young minds. It was hard to persuade top students to join sleepy oil companies when tech start-ups were handing out stock options, installing foosball tables, and feting employees with catered lunches.

Because the energy industry was so unloved, Souki thought there might be opportunity, if only because it likely was starved for new capital. He knew a bit about the business from his trips to the Middle East and from a few early deals, but that was about it. At that point in his life, he knew more about fajitas than fracking.

So he went to an industry conference to meet geoscientists and learn. There, a Chevron geologist discussed how his team was doing more with desktop and even laptop computers than they had a decade earlier with a huge supercomputer. Not only that, but three-dimensional seismic imaging and other advances were improving the odds of finding energy deposits. Just as MRIs are better at scanning than mere X-ray machines, the pros were using the latest advances to do a better job detecting oil and gas.

Technology was a great equalizer, Souki concluded, one that might even allow someone who had been skiing and pouring drinks for seven years to discover oil or gas. All he had to do was find smart geophysicists, raise enough capital—something he always had been good at—and invest in the right exploration technology.

"It was a numbers game now," he says. "I was never going to be an expert in putting geology together, but I understand probability and statistics."

Souki searched for energy exploration and production pros employing the latest technology and in need of cash. Then he tapped friends and previous investors to raise the necessary financing. He received a commission for bringing money to the exploration executives, cash that gave him and his family some breathing room after his financial scare. Soon he was kicking his own money into the deals.

These were small-fry deals, each less than $20 million. One early investment was in a California company that saw a well burst into flames, sparking a fire that took ninety days to put out. After two years, Souki's group realized that gas just couldn't be produced economically from the company's wells.

Souki managed to have more winning investments than failures, however, and by early 1996 he was worth several million dollars. At that point, he decided to try to do some exploration himself, figuring there was more upside to that approach.

Souki's entrance into the wildcatter game had small-time written all over it. He took over an inactive Hollywood film-colorization company that happened to still have publicly traded shares. Assuming control over a "shell" company, or one without any assets or operations, is a decidedly backdoor means of accessing public markets and one that often raises eyebrows among experienced investors.

Souki and his geologists decided to focus on drilling in the Louisiana Gulf Coast area after their mapping data suggested the region held promise. Souki renamed his company Cheniere Energy after the Cajun word for the elevated land above a swamp. The name seemed sturdy and safe, and had some local flavor.

In 1998, Souki and his family moved to Houston so he could run Cheniere. He began raising money for his new company, telling investors that his team would use the latest and most expensive seismic data to find oil and gas in the Gulf of Mexico. Tapping old connections, he managed to raise cash from a former Lehman Brothers energy banker, a Cayman affiliate of a company headed by Lebanese financier Michel el-Khoury, and several high-profile executives.

Cheniere spent about $20 million for three-dimensional seismic data and leases along the Louisiana coast and began drilling. By the summer of 1999, however, Cheniere had no booked reserves, no revenues, and no production.[1]

Souki was a forty-five-year-old ex-banker who seemed stuck in the energy industry's ugly underbelly. Cheniere was worth a measly $40 million, and energy seemed mired in a never-ending bear market, making life hard for the puny company. Cheniere had drilled four exploration wells along the swampy Louisiana coastline, but just two were modest successes, and most experts viewed their shallow-water Gulf of Mexico locations as having been fully picked over and played out. The region also had environmental risks that added to the company's obstacles.

Souki found ways to attract blue-chip investors, like he usually did, by focusing on Cheniere's technology-heavy approach to squeezing gas out of old wells. In 2000, private equity giant Warburg Pincus and industry heavyweight Samson Investment Co. both agreed to kick in cash and partner with Cheniere to drill some prospects.

These investors viewed Souki as supersmart, and more ambitious and creative than executives at even the largest oil companies. His new investors had deep reservations about him, though, even as they wrote their checks. Souki was brilliant and hard-charging, but what did he really know about energy?

"Reservations" might actually be an understatement. Jack Schanck, co-CEO of Samson at the time, doubted Souki understood "an iota" about the drilling Cheniere said it would do, and he was turned off by Souki's smooth pitch. Samson only agreed to invest money in the project because one of Souki's lieutenants was a respected energy veteran, Schanck now says.

"Charif was forceful and charismatic, you really had to calm him down," Schanck recalls. "He was a little too slick for me."

Souki stood out in the energy patch, which may have added to the concerns. He had olive skin, wore his hair a bit long, favored custom-made double-breasted suits, and came off as almost too self-assured. He oozed confidence but hadn't accomplished very much.

Some assumed Souki was out to sell his company for a quick score and then would move on to his next business, maybe in an entirely different sector. Even this plan seemed a long shot, however. Throughout most of 1999 and the start of 2000, natural gas traded for less than three dollars per thousand cubic feet. But it cost Cheniere about three dollars per thousand cubic feet to get gas out of the ground. In other words, it cost more to produce gas than the company could get for it.

Souki racked his brain to understand what was happening. *I'm using the latest technology, I have great guys, and we're in a promising area, but I still can't get my costs low enough to make any money,* he thought.

All the cheap natural gas in the Gulf, and perhaps the country, had been sucked up, it seemed, and Souki didn't know how his company was going to make any money at those prices. He also couldn't figure out how other producers could be profitable if it cost more to find and produce natural gas than they could get selling it. The math just didn't add up.

He came to a simple and yet unconventional conclusion: Natural gas prices were bound to climb—the same conclusion Aubrey McClendon and Tom Ward were coming to around the same time. If prices didn't rise, the entire energy industry would go out of business, he decided. And there was no way that was going to happen, since the country still depended on natural gas. In fact, Souki saw indications of rising demand.

If natural gas prices were set to skyrocket, Souki figured he'd strike it rich if he could just find a lot of gas, somewhere. It was getting harder for him and others to produce natural gas in the Gulf of Mexico, but maybe Cheniere could locate it somewhere else.

McClendon and Ward had decided to search for gas in the United States, but Souki couldn't imagine the country held enough natural gas to get very excited about. This was America, after all. He and his team did some research and decided the nation didn't hold enough new energy supplies.

"I knew about the shale reserves," he recalls, "but it wasn't relevant, nobody believed they could be produced."

He had another, more unorthodox idea. Maybe he could get his hands on cheap natural gas being produced elsewhere in the world and ship it to the United States? He continued to travel to the Middle East and other countries to visit family and investors. He knew that a number of countries, including Qatar, Algeria, Australia, and Russia, had massive gas reserves and needed new customers. Perhaps the way to locate huge amounts of gas wasn't to dig it up in America, but to turn foreign supplies into a liquefied form of gas that could be shipped to the United States.

Liquefied natural gas is simply natural gas subjected to the intense cold of minus 260 degrees Fahrenheit. The supercooling process converts the gas to a liquid form that's one six-hundredth its original volume. As long as it remains at this temperature, liquefied natural gas can be shipped for use elsewhere. Producers transport it in a cryogenic container, which isn't much more than a very, very large thermos. At its destination, the liquefied natural gas, or LNG, is "regasified," or heated until it turns back into its natural gaseous state. Then it can be distributed through traditional natural gas pipelines for heating, electricity, and other purposes in homes and businesses.

Souki knew the LNG business was growing abroad. But the amounts being shipped weren't huge and few considered sending it to the United States. There were just four facilities in the country capable of accepting LNG and converting it into gas. All of these had been built nearly two decades earlier and two of them had been mothballed, a sign of how unpopular and unprofitable the idea had become. Most in the industry saw the low price of natural gas as a sign that the country had more than enough of it, so shipping more of it to America was the last thing on their minds.

Souki came up with a strategy that seemed unrealistic, if not harebrained, especially for a tiny company run by a relative energy novice. Cheniere would raise enough cash to secure land for an enormous gas import terminal, or maybe even a few terminals. Then the company would convince regulators to give it permission to build the facilities. After that, Cheniere somehow would raise billions of additional dollars to build the facilities. Oh, and Cheniere also had to persuade foreign companies to ship LNG to the United States and pay Cheniere to regasify it so the gas could be sold domestically.

The idea sounded crazy, but Souki figured he could get a meeting or two with foreign businessmen and government officials to pitch it, if only because some of them still had fond memories of his father. Then he'd sell them on the idea, like he had sold so many of his previous deals.

Others know about the subsurface, but I know about traveling internation-ally and moving in different cultures, he thought.

Souki quickly realized that not everyone was as convinced of the value of importing gas to America as he was. Warburg Pincus, the white-shoe Wall Street firm he had struck the earlier deal with, turned down a chance to invest in the project. Other large investment firms also rejected Souki's offer.

"I definitely lacked credibility," he acknowledges.

Souki knew he'd have to recruit experts who knew something about importing LNG, so he approached Charles Reimer, who had spent twelve years on the board of directors of the world's biggest LNG liquefaction plant in Indonesia, asking him to join Cheniere.

Reimer was skeptical. "Charif was a minor-league player," he recalls. "He was doing small things."

Souki took Reimer to lunch in the summer of 2000 at Houston's up-scale Coronado Club to outline his plan and convince him that it could succeed. Reimer shared Souki's belief that energy prices were headed higher, but Souki's scheme didn't seem likely to succeed.

"Let me put some numbers together," to see if the idea even makes sense, Reimer finally told him.

That summer, Reimer went over the math, assessing what it would cost to freeze, ship, and warm natural gas. He determined that the ex-pense would be huge—but the endeavor could be profitable if gas prices rose as much as he and Souki expected. After Souki convinced him that the financing for the huge project would come, Reimer signed on, giving Souki encouragement that his plan might actually work.

Reimer wasn't entirely convinced that importing gas to America was the brightest idea, even after agreeing to come onboard. But he had just spent a decade in Indonesia and was eager to try something new. Reimer figured he'd roll the dice on Souki and his radical idea; even if it didn't work he'd likely still have fun and learn a thing or two.

"The idea was a little bit out there, but I was at the point in my life where I could take a chance," he recalls. "You could see he was extremely smart and thought outside the box."

In the summer of 2000, Souki and Reimer began working on importing boatloads of natural gas from foreign countries. If he could get his hands on enough gas to quench the nation's growing thirst, Souki was sure he could make a fortune.

That same year, Aubrey McClendon and Tom Ward launched their own push to discover a historic amount of natural gas in the United States, while Harold Hamm at Continental Resources set out to tap rock in the Bakken formation in Montana.

McClendon, Ward, and Hamm didn't seem like the most likely candidates to transform the nation and revolutionize global energy markets. Souki was an even longer shot. Soon, all four underdogs would get a chance to pull off their audacious and unlikely schemes.

CHAPTER EIGHT

Something big was coming. Tom Ward repeated it, again and again.

It was late 1999 and Ward was having breakfast with Dan Jordan, a fellow energy executive, at Oklahoma City's Classen Grill, a well-known restaurant frequented by hometown sports legends like Keith Jackson and Barry Switzer.

Over raisin-bran toast, poached eggs, home fries, ham, and coffee, Ward relayed what he and his partner, Aubrey McClendon, were telling investors: Natural gas prices were about to soar and their company, Chesapeake Energy, was going to take advantage.

Ward and McClendon were just trying to get investors excited about Chesapeake shares, Jordan figured. His friend was like a salesman reciting a pitch falling on mostly deaf ears. After all, natural gas prices were barely above two dollars per thousand cubic feet, about the same level as a full decade earlier.

Ward went on and on, though, explaining why anemic production and rising demand would send prices higher. Jordan began to realize his friend wasn't just mouthing a favorite line.

He's really starting to believe his own bullshit, Jordan thought.

Ward took his pitch up a notch. "Natural gas will be at double digits," he said confidently, predicting that prices would soar fivefold, from two dollars per thousand cubic feet well past ten dollars.

Jordan almost choked on his coffee. "What?!" he stammered. "Where are you getting that from?"

"Just look at rig counts," Ward responded. They were dropping and gas supply was drying up, so prices were bound to skyrocket, he argued.

Jordan shook his head. "Yeah, maybe when we're seventy-five years old," he told Ward dismissively. "You're crazy."

Ward smiled. Just watch, he seemed to be saying.

Natural gas prices drifted over the following few months and Jordan's skepticism looked well placed. By the spring of 2000, however, the market perked up and prices moved past four dollars per thousand cubic feet. By the winter, gas was above five dollars.

Many experts viewed the rise as a temporary blip. But McClendon and Ward insisted to colleagues, investors, bankers, and pretty much anyone else who would listen that gas prices were heading still higher. The Chesapeake executives had heard that Charif Souki was crafting a plan to import natural gas to the country, a step that might add to supplies. But Souki's company was tiny and insignificant; it was doubtful he or anyone else could raise the billions necessary to bring much gas to the country. The biggest energy companies still weren't wasting much time drilling for new deposits in America, another reason the Chesapeake cofounders were so sure demand would outstrip supply.

The company began buying wells, trying to get its hands on natural gas. McClendon told investors that Chesapeake had a chance to improve the nation's well-being by leading a shift to natural gas, which was both homegrown and cleaner than oil and coal. McClendon repeated the message to colleagues, a sign he truly believed in his pronouncements, no matter how naïve, corny, or self-serving they sounded.

"This can change our *country*," he told Marc Rowland, the company's chief financial officer.

Rowland, who had the job of figuring out how to pay for McClendon's grand vision, rolled his eyes. "Okay, but in how many years, Aubrey?" he responded.

"I wondered if I'd still be alive," Rowland recalls.

McClendon wore Rowland down, though, as he did others at Chesapeake, convincing them that consumers and businesses would shift to natural gas from other energy sources.

"This is the fuel of the future, it will be in demand," McClendon said in another internal meeting. "Everything will change."

McClendon and Ward had embraced horizontal drilling before most competitors. Now they closely monitored how specialists were improving hydraulic fracturing techniques, enabling drillers to shatter gas-soaked rock to free natural gas.

It dawned on the Chesapeake executives that a unique opportunity—maybe even a historic one—might be within reach. Huge oil companies with talented geologists and engineers once held an insurmountable lead over small, "independent" companies like Chesapeake because big organizations were best equipped to pinpoint new reservoirs and then extract oil and gas. But now, thanks to improved fracking and drilling techniques, it seemed that finding and pumping meaningful energy deposits was the easiest part of the equation.

The new challenge, McClendon and Ward decided, was to grab prime acreage before their rivals figured out that energy prices were headed higher. Chesapeake had to get its hands on gas-producing wells as quickly as possible. McClendon and Ward seemed the perfect men for the task. After all, who knew how to quickly lock up land better than two land experts?

The Chesapeake cofounders began testing their theory with friends and others in the business. "We'll never be able to compete with" large exploration companies and their huge staffs, McClendon told his old college friend Ralph Eads, who had joined pipeline company El Paso Corp. after a stint as an energy investment banker. "But I've got the skills to get land."

Eads and others lent encouragement, agreeing that a new age had begun, one in which even a small company like Chesapeake might hold a unique advantage. Who knew—maybe the company could even become a true energy power.

McClendon began telling bankers and rivals that Chesapeake was eager to bid on almost any gas well in the country up for sale. The company still wasn't very interested in acreage that held shale and other challenging rock. It didn't seem likely to McClendon and Ward that George Mitchell's extraction methods could work outside Texas's Barnett Shale formation, at least not in an economical way. But there were so many "conventional" wells available that they didn't need to look at shale formations.

The deals started small: In July 2000, Chesapeake paid $28 million to buy a competitor, Gothic Energy Corp., which controlled some of the best natural gas assets in Oklahoma. Chesapeake was willing to assume Gothic's $316 million of debt as part of the deal. Other purchases followed, most just $200 million or so in size. Chesapeake slowly expanded its gas production and investors sent the company's stock price climbing above ten dollars a share in March 2001, a move that encouraged the executives to look for more acquisitions.

Around that time, McClendon placed a call to John Penton, a local rival who helped run Canaan Energy, which owned wells in Oklahoma, Texas, Arkansas, Nebraska, and elsewhere.

"Why don't you sell your shares to us?" McClendon asked Penton. Chesapeake was willing to buy Canaan at a price that was more than 20 percent above its share price, he said.

Penton and his longtime business partner told McClendon to get lost. They had spent a decade assembling a collection of natural gas wells that was enjoying strong and steady production. Penton later told a local reporter that he felt a "strong pride in ownership" and wasn't interested in any kind of sale.

That fall, the September 11 terrorist attacks destroyed the World Trade Center, a terrible blow to the nation that sparked immediate fears of an economic recession. Rather than halt their acquisition spree, McClendon and Ward decided that only a temporary drop in energy demand would result from the attacks, so they should continue to search for deals.

Throughout the dark period, McClendon couldn't let go of his pursuit of Canaan. Chesapeake managed to buy 10 percent of Canaan's shares as a foothold investment. After one more failed attempt to sweet-talk Penton, McClendon called him with a threat: Agree to a sale or Chesapeake would announce an offer to buy the company.

"I'm going to take it public," McClendon promised him.

Penton and his team told McClendon to go away. His offer of twelve dollars a share still seemed insufficient, even though Canaan shares were trading at just above nine dollars. Couldn't McClendon just leave them alone already?

McClendon and his fellow Chesapeake executives grew more frustrated. In November, they followed through on their threats and went public with a hostile takeover offer for Canaan. To the relief of Penton and his partner, Canaan's shareholders didn't seem interested in the bid.

In early 2002, McClendon and Chesapeake boosted their offer to eighteen dollars a share, a price so high that Canaan's brass felt it had no choice but to hire a banker to conduct a formal sale. By then, McClendon had changed his tactics and had turned on the charm, forging a friendship with Penton, who came to respect McClendon's persistence. Canaan even included seven thousand acres in the Barnett Shale formation in Texas in the sale as a "sweetener" to help McClendon justify his higher price to his board of directors and shareholders. It was a nice gesture, but McClendon and Ward had little interest in costly shale wells and decided to abandon them if they succeeded in buying Canaan.

When executives couldn't find a better offer, Canaan agreed to sell itself to Chesapeake for eighteen dollars a share, or $118 million, in an April 2002 deal.

"Aubrey was tenacious," Penton recalls. "He knew he wanted the company, it took him a year and a half, and he got it."

"It was a key first component in my plan to build our presence in the Anadarko Basin," McClendon says, referring to a formation in western Oklahoma and the Texas Panhandle and into Kansas and parts of Colorado.

Ward helped develop Chesapeake's deal strategy and he endorsed the acquisition push, as did Rowland, the CFO. But rivals whispered that the deals were reckless. Some Chesapeake senior executives and members of the board of directors began to question why the small company was spending so much money. "They're giving me grief," McClendon confided to an industry executive during the Canaan negotiations.

Bowing to the criticism, in the summer of 2002 McClendon announced plans to focus on low-cost prospects in the Midwest and to sell wells in the Permian Basin in Texas. Chesapeake didn't receive much interest in the Texas acreage, though. Soon, McClendon and Ward were buying even more wells in the area, shelling out $420 million. It was as if they couldn't help themselves.

Over time, McClendon's arguments won over Chesapeake's board of directors, which gave him the green light for still more acquisitions. It likely helped that some on the board had close relationships with the Chesapeake cofounder. One director, Wall Street executive Frederick Whittemore, had even lent money to McClendon in the late 1990s, despite the fact that Whittemore served on the board's committee that determined how much McClendon should be paid.[1]

The Chesapeake cofounders crafted a division of labor that allowed each to excel at what they did best—and stay out of each other's way. Ward and his team vetted all the assets up for sale and determined how much gas the wells could produce. McClendon and his own group, located in a separate building, decided how much to pay for the deals and how to finance them. McClendon did the vision thing, while Ward made it all happen.

"Aubrey would get the first call from bankers for every package" of natural gas assets up for sale, Ward recalls.

Soon, Chesapeake was outbidding rivals for the nation's best wells. Because McClendon and Ward were convinced gas prices would continue to rise, they were willing to enter hefty bids. In the industry's lingo, Chesapeake used a higher "price deck," or estimate of where prices were going, than rivals. In 2003, Chesapeake spent $530 million to buy natural gas assets, and the company emerged as the eighth largest natural gas producer in the country.

McClendon and Ward's plan was to establish a foothold with a group of newly purchased wells. Then they'd quickly assemble a team of landmen to buy additional acreage near the wells, extending their conquest and conquering new territory, as if they were playing the board game Risk. They would drill new, productive add-on wells in this virgin territory before natural gas prices climbed further and rivals caught on to their "acquire and exploit" strategy.

Since they were land pros, McClendon and Ward figured they had a good shot at hiring a battalion of the best landmen around the country, especially since rivals remained scared to spend a lot of money on U.S. drilling. Landmen would be the key to the company's future. It was among the first times in modern history that a company bet so much on these pros, the Rodney Dangerfields of the energy patch, who usually command little respect.

The landman profession is uniquely American. That's because the United States is among the few countries in which citizens, rather than the government, own the mineral rights under their own properties. If an energy company wants to produce oil or gas in a certain area it generally has to persuade landowners to sell or lease their mineral rights before drilling can begin.

There are two types of landmen—financial and legal specialists who sit in a company's office, and "field brokers" who negotiate with landowners. The field brokers are the industry's extroverts and natural salesmen, the backslappers and handshakers who can knock on a door, find a topic in common with a landowner, and cut a deal. To be a good field broker, "you've got to be able to sell a refrigerator to an Eskimo in the middle of January and make them think they need it," says Ted Jacobs, who runs the energy management program at the University of Tulsa, which trains landmen.

During the industry's good times, landmen find adventure and wealth. Starting salaries have reached as high as $85,000 in recent years and work has been easy to find. When things slow down, though, landmen scrape by. Those in the field usually are contract workers who have no salary or benefits to fall back on. It's also not uncommon for a landman to find a door slammed—or a gun cocked—by a suspicious landowner. These "lease hounds" are always on the move. It's a lifestyle that can wreak havoc with family life.

"You have to go where the action is and where you can find work," Morris Creighton, the veteran landman who was part of Tom Ward's team in the late 1990s, says. "Lots of people see it as romantic, and it is, but it's very, very tough on the young."

Wildcatters who risk it all to find oil and gas usually get notoriety, wealth, and even women. Landmen are viewed as little more than used-car salesmen. Movies are made about wildcatters. Jokes are made about landmen.

But McClendon and Ward appreciated the talents of these men and women and set out to build a land machine to profit from the natural advantage they had in this part of the business.

"Engineers look at landmen the way some people look at accountants, as necessary evils," says Ted Jacobs, who was a landman in Oklahoma while Chesapeake expanded. "If you're a landman, you wanted to work for someone like Aubrey McClendon. He values landmen more than engineers."

To be sure, McClendon and Ward weren't the first to use aggressive tactics when it came to securing land and title rights. There's a long history of fortunes created by those focused on these kinds of deals, rather than on new oil or gas discoveries. In 1930, for example, H. L. Hunt, founder of the Hunt Oil Company, spent two days in a Dallas hotel convincing a beleaguered peer, Columbus Marion "Dad" Joiner, to sell him the title to what is now the East Texas Oil Field. After all the wooing and negotiating was over, Hunt wrote Joiner a check for just over $1 million to secure the title. The field turned out to be the biggest and best-producing oil reservoir in the United States, eventually producing over five billion barrels.

Chesapeake would lean on these hustlers as few companies ever have. Ward began searching for the best and most aggressive landmen, directing them to buy up drilling rights as quietly and efficiently as possible, so as not to tip off competitors. The team traveled to county courthouses and "abstract offices"—privately owned businesses that own summaries of ownership documents—to figure out who owned the mineral rights of the most attractive land. These field brokers then visited individuals holding these rights and pitched them about selling to Chesapeake.

The strategy led to huge growth in acreage. Soon, however, it became clear—McClendon and Ward had made a huge mistake.

After purchasing Mitchell Energy in 2002, Devon Energy extracted impressive amounts of gas from the Barnett geological formation. Skeptics who had doubted the wisdom of paying over $3 billion for George Mitchell's company saw the success Devon was having in parts of the Barnett, including Wise and Denton counties, and acknowledged that Mitchell in fact had figured out how to extract gas from shale at a reasonable cost, the holy grail of the energy business.

Getting gas from shale in nearby Johnson County was an entirely different story, however. When the Devon team drilled down and hydraulically fractured the shale in this area, they got too much salt water from the ground, and too little natural gas. The bottom of the shale layer couldn't contain the fracking liquid; it kept leaking into rock layers below.

Devon's fracking fluid didn't seem to be the problem. A growing number of energy producers were adopting versions of the radical, water-based liquid that Nick Steinsberger, the Mitchell Energy engineer, had discovered almost accidentally a few years earlier, and they were meeting with great success.

Before giving up, the Devon engineers figured they should try to change the way they were drilling. Maybe by going laterally, rather than vertically, they wouldn't rupture that layer below the shale, the one that was causing all the water to rush into their wells. Devon already had experience with horizontal drilling and was doing it elsewhere in the Barnett.

In 1991, Mitchell Energy, together with an arm of the U.S. government, had tried combining horizontal drilling and fracking in certain Barnett fields, but it didn't work. The Mitchell team revisited the idea a few years later but gave up once again. The company's president, Bill Stevens, was no fan of horizontal drilling in the Barnett, and it just cost too much to combine both approaches.

By 2002, though, a number of technologies, such as three-dimensional seismic imaging, made it easier to understand the faults and other complex parts of the Barnett formation, while the advent of the diamond-studded drill bit allowed for faster drilling through the rock layer above the shale. Meanwhile, natural gas prices had climbed above four dollars per thousand cubic feet, making the added expense of combining horizontal drilling with hydraulic fracturing more palatable.

What came next was another giant leap for American drillers, who were getting more comfortable working with shale even as the rest of the world barely experimented with such challenging rock. Devon mixed the two methods—horizontal drilling and fracking—and began to see a surge of gas production in its Barnett acreage. A company called Hallwood Energy was also seeing success with the same integrated approach.

News about both companies' activities spread throughout the industry. Drilling horizontally, and then completing the wells with hydraulic fracturing, seemed a fresh breakthrough, one that turned the Barnett into a truly world-class reservoir that was a model for shale formations around the country. In April 2004, a paper presented at a convention of the American Association of Petroleum Geologists in Dallas determined that the Barnett formation held gas that was two and a half times what the geological group thought in 1998, when only George Mitchell and his team believed in the region's potential.

The news stunned Ward and McClendon. "What Hallwood did opened our eyes," Ward says.

They quickly realized they were late to the shale game. Really late. The only land Chesapeake had in the Barnett was the seven thousand acres that came from buying Canaan—the land McClendon and Ward had expected to walk away from. If it was truly possible to get serious gas from shale formations, they knew they'd have to do something drastic to catch up to their rivals.

Trying to make up ground, Chesapeake in late 2004 shelled out nearly $300 million to buy Hallwood's eighteen thousand acres in the Barnett. It wasn't nearly enough, though. With each month, McClendon and Ward seemed to fall further behind competitors like Bob Simpson. Simpson was a hard-charging Texan who had started a Fort Worth, Texas, company, eventually called XTO Energy, with help from Robert Rubin, at one time a senior trader at Goldman Sachs and later the nation's secretary of the treasury.

Simpson favored custom-made cowboy boots and searched for a certain type of employee. He wanted to hire those who "grew up poor and were disciplined by their parents, including corporal punishment," he told a reporter. "The most successful people I know were disciplined as children."[2]

Simpson was just as convinced as the Chesapeake executives that remarkable wealth could result from natural gas wells. In January 2005, XTO outbid Chesapeake for Antero Resources, a prominent producer in the Barnett. That year, the Energy Department determined that the Barnett held as much as thirty-nine trillion cubic feet of recoverable gas, enough to supply the entire country for almost two years, just as Kent Bowker and the Mitchell Energy crew had predicted. Despite that, McClendon and Ward couldn't bring themselves to spend the $685 million that XTO paid for Antero, a loss that stung the Chesapeake team.

Meanwhile, Devon's field brokers were rushing to abstract offices and county courthouses in the Barnett region, pulling files on land ownership and then racing to buy or lease mineral rights from owners. Everyone seemed a step ahead of Chesapeake.

Larry Coshow, a veteran Chesapeake landman, was eager to help McClendon and Ward make up lost ground in the Barnett Shale. Coshow knew he had his work cut out for him, so he went on a hiring spree. Soon he was directing over seven hundred brokers to blanket six counties in East Texas, checking records and buying leases, directing twice as many men in local courthouses as any other company.

Coshow told his team they were in for a brutal fight. "If you guys don't want to do it, I'll find someone who will do a better job," he told one group of landmen.

Chesapeake resorted to creative methods to lock up remaining land in the region. Many landowners didn't understand the process of selling and leasing their mineral rights, so Chesapeake began hosting town hall meetings throughout Texas. The meetings sparked a buzz and sometimes media coverage, publicity that helped send a message that Chesapeake was on its way to town and that it wanted as much land as it could get its hands on. The public efforts caused rivals to whisper to landowners that Chesapeake wasn't to be trusted.

One day, Coshow arranged for a town hall meeting in Centerville, a town of about one thousand citizens in East Texas, midway between Dallas and Houston. Coshow found a local man who agreed to prepare a barbecue lunch for all the attendees. Coshow had his team of landmen haul computers and printers to the meeting so they could immediately sign contracts with landowners.

Soon after Coshow delivered some introductory remarks, he began to field tough questions, some of which sounded a bit rehearsed. Coshow thought he recognized a few of the faces of his new critics. He realized

Chesapeake's rivals had planted people in the audience to try to spark an ugly confrontation.

Coshow tried a friendly approach. "You guys are probably getting a lot of comments" trashing Chesapeake, he told the audience. "But I don't have anything negative to say about our competition, I really don't. We must be doing something right for them to talk about us, though."

That night, Chesapeake cut a dozen or so deals, paying about $250 an acre to lease most of the land, though some got as much as $800 an acre.

Chesapeake's landmen became frustrated by how long it took to check land ownership records to get a deal done. Many small abstract companies in the area lacked modern computers to accelerate the process. They relied on paper records that could be cumbersome to leaf through.

"What if we put all your records on computers?" Coshow asked some of the abstract specialists. "We could digitize all your records" and leave workstations for the abstract companies to keep.

The companies loved the idea, and the gift of free technology wasn't easily forgotten. "You betcha we got the first shot at those workstations if I had landmen in those towns," Coshow says. "The competition had to go to courthouses [for the records] and it took them two or three times as long."

Within a year, Chesapeake had overtaken its competitors in the Barnett. The company had purchased over a million acres of land around the country, thanks to the work of over a thousand landmen crisscrossing the nation's fields. Natural gas production was soaring. And the wells in the Barnett Shale formation that Chesapeake had received as a throw-in from Canaan—once viewed as a liability—now were Chesapeake's greatest asset.

McClendon, smooth and comfortable with investors, emerged as the company's public face. He courted Wall Street analysts and top investors, persuading them of the wisdom of Chesapeake's evolving strategy. The salesmanship was crucial because the company relied on selling new shares and bonds to raise cash for each new purchase.

Tall and rangy with unruly blond hair, McClendon was relaxed around the office, a friendly face with rimless glasses whom employees called by his first name. He usually wore dress slacks and a crisp shirt with rolled-up sleeves. He came to favor ties patterned with drilling rigs and hard hats.

With investors and bankers, McClendon really turned on the charm, donning a Hermès tie and a custom-made suit as he wowed Wall Street. He had a good grasp of the geology and technology of the business, not to mention the latest information on land buying. But he also had a sense of the industry's history and made a compelling argument that natural gas would be the nation's most important energy source for the next few decades. It was clean and plentiful, no matter what the skeptics said.

McClendon had a unique ability to connect with employees, investors, and industry members. When he met them, he asked at least as many questions as he answered. How's your family? What are you working on? Where do you see the industry going? Even the most cynical, hard-bitten investors were smitten. Many compared McClendon to former president Bill Clinton for his ability to make everyone he met feel like the most special person in the room. Some said he was the most upbeat person they had ever met.

"Aubrey is larger than life, he's energizing," says a banker who lent Chesapeake money. "I enjoy spending time with him, he makes you feel good, he's like the best salesman ever."

Chatter grew on Wall Street that McClendon and Ward were at the vanguard of a new generation of energy explorers and that Chesapeake had mastered cutting-edge drilling technology that even the largest oil powers hadn't fully embraced.

Ward, the company's chief operating officer, remained something of an introvert. In company meetings, McClendon usually spoke and Ward took notes. He kept his own office in a separate building hundreds of yards from McClendon's office in Chesapeake's headquarters. They had once run separate companies and didn't see any reason to change the way they operated. Each kept his own assistants and teams. Their office buildings were separated by a parking lot and retail offices, including an insurance company, a small law firm, and a hairdresser.

Ward and McClendon rarely even met. Instead, they spoke on the phone, faxed, or e-mailed each other hundreds of times a day, from before six in the morning until past midnight. Ward spent more time outside the office, quarterbacking the company's efforts in the field and coordinating between Chesapeake's geology, engineering, and land personnel.

Ward began to learn from McClendon and to become more comfortable dealing with investors and others. "I was a hick," he recalls. "I stayed in nice hotels for the first time. Aubrey taught me how to dress, wear Hermès ties, and what type of suit to wear" to meet bankers or industry executives.

As the acquisitions multiplied and the company expanded, Ward's workload swelled. He and his team were analyzing hundreds of wells. Most companies exaggerated their reserve numbers, so he had to examine the history of the area, the type of rock in the fields, and other data to independently verify the quality of the wells. He hardly slept and found he had less time for his three children, relying instead on his wife, Sch'ree, to run the household. He never took a vacation. The closest Ward got was bringing his family to a drilling location. He didn't listen to much music, preferring quiet in the car, nor did he go to movies.

To let off steam, Ward cut out around lunchtime each day to play basketball for an hour or so with a group of local young men at the YMCA. A quick and expert ball handler, he developed a bond with Mike Harrison, an African American man in his thirties who had grown up in Oklahoma City's poorer Eastside. Harrison had given up a potential college scholarship to work in a convenience store and help his family get by. Later, after he lost his job, Harrison would get a position working for Ward.

On the court, Ward was just as measured as in Chesapeake's offices. The players had no clue he was one-half of the hottest drilling duo in the industry. Others talked trash, but Ward had only positive things to say, or kept quiet. One day, though, after he was teased for the umpteenth time about how rarely he drove to the basket with his left hand, he had had enough. "I'm going right and you're not going to stop me," he said, to the surprise of his court mates. With that, he drove to the hole for an easy layup, an indication of the fierce, underappreciated drive propelling Ward to prove the skeptics wrong, off the court and on.

★   ★   ★

The Barnett was proving such a rousing success that McClendon and Ward became converts to the idea that drilling in these once overlooked rock formations would produce a gusher of gas. Ward began reviewing other shale formations around the country. Sitting at a broker's office in late 2004, he heard that a rival called Southwestern Energy had made a discovery in Arkansas's Fayetteville Shale formation. Ward immediately sent hundreds of landmen across Arkansas to grab a position for Chesapeake. "If we can control the land, we can control the resource," he told a colleague.

The approach, which became known as the "resource thesis," won a following among investors, who cheered the land grab. They sent Chesapeake shares climbing from five dollars in the summer of 2002 past thirty dollars a share in the summer of 2005.

Most environmental groups, including the Sierra Club, also applauded the push into shale drilling, hoping natural gas might take market share away from dirtier coal and oil sources. Gas also could serve as a "bridge" fuel until the cost of wind, solar, and other renewable sources were ready to power the nation.

Throughout 2005, natural gas prices soared, moving past ten dollars per thousand cubic feet late in the year, just as McClendon and Ward had predicted. Investors, politicians, and industry experts fretted over how the nation would find enough energy supplies to meet its seemingly inexorable demand. McClendon and Ward, sitting pretty on vast amounts of natural gas, seemed on their way to becoming modern-day tycoons.

Below the surface, however, trouble was brewing.

Aubrey McClendon and Tom Ward weren't the only ones excited about new shale formations in 2004. Even George Mitchell was betting on a new field. By then, Mitchell was eighty-five years old and his pace had slowed. It had been two years since he pocketed two billion dollars selling his company to Devon Energy.

The wildcatter was hungry for another home run, however. As he spoke with his son Todd about acreage in Arkansas's Fayetteville Shale that he and his business partner were considering buying, Mitchell became convinced the new area might yield huge amounts of gas.

Thanks in part to financing from Mitchell, Alta Resources—the company Joe Greenberg and Todd Mitchell had formed—began spending nearly one hundred million dollars, emerging as the second largest land holder in the Fayetteville region. They even hired Nick Steinsberger, the geologist who had helped Mitchell Energy extract gas from the Barnett, to develop the best techniques to fracture rock in the Fayetteville. George Mitchell's new enthusiasm was a clear sign the shale revolution was picking up pace.

M cClendon, Ward, and Mitchell chased natural gas, but Harold Hamm was convinced the Bakken rock formation in Montana was packed with oil, making it a more promising field than the shale gas areas. But crude prices averaged less than thirty dollars a barrel in 2000, and were a mere twenty-five dollars a barrel in 2001. Unlike a toothpaste or cereal maker, oil and gas producers are forever under the thumb of market prices that can fluctuate wildly. For Continental, which rang up losses of $20 million in 2002, weak prices for its main product meant it couldn't afford the expensive fracking and horizontal drilling necessary to tap oil in the Bakken's difficult rock.

Hamm began to hear rumors that two small competitors from Dallas—Tim Headington's Headington Oil, and Lyco Energy Corp., run by Bobby Lyle, an engineer and former dean of Southern Methodist University's business school—were scooping up acreage in Montana. When Hamm learned that Headington and Lyco somehow were producing about fifteen hundred barrels of crude a day from the region, he began to worry about falling behind.

Oil prices moved a bit higher in 2002. OPEC was reining in production, the U.S. economy was recovering from the devastating attacks of September 11, 2001, and Asian economies were rebounding from their own late-1990s slowdown. Rising global oil demand gave Hamm even more impetus to get going in the Bakken.

In late 2002 and early 2003, the company began leasing land in eastern Montana, focusing on the Elm Coulee field in the Williston Basin in Richland County. The Bakken rock layer was about ten thousand feet below the surface and just forty-five feet thick, adding to the challenge of finding oil.

Because Hamm's competitors had already drilled wells in the heart of the formation, Continental would have to settle for surrounding positions. "They'll have the yolk," a geologist mapping the field told one of Continental's landmen. "We'll get the white of the egg."

The Lyco team dismissed Continental's efforts as a waste of time. "Let them have the edge," Richard Findley, a local wildcatter advising Lyco, told the company's geologists. "We'll do the low-risk" portions of the Elm Coulee. "No one knew and no one cared" what Continental was up to, recalls Findley, a Bakken expert.

By August 2003, it was clear that Continental's first well was a winner, producing about thirteen hundred barrels of crude a day. Headington and Lyco were pumping more crude from their own wells in the yolk of the field, but Continental's production was enough to score profits. The Bakken formation was so rich with oil that even fields on its edges could generate substantial amounts of crude, it turned out.

More important, the three companies had proved that oil, not just natural gas, could be extracted from rocks with extremely low permeability. Oil molecules, which move less easily through tight rock than gas molecules, were flowing, thanks to improved horizontal drilling and hydraulic fracturing techniques.

"We need to get more rigs in here!" Hamm told members of his team, encouraging them to drill wells elsewhere in Montana.

Continental's land brokers persuaded farmers to lease 100,000 acres in the western part of the state, paying about seventy dollars an acre, or a total of $7 million.[3] The progress in the Bakken garnered no headlines. Continental, Headington, and Lyco were small and privately held, so they had no shares to hype and no press releases to send. Besides, most major oil companies were busy crafting a series of corporate mega-mergers, and they still thought the high cost of producing oil from the Bakken would prevent any gusher of profits.

Hamm was jovial around the office. Crude prices continued to edge higher, and his Montana wells were showing impressive early production. "You guys are doing a great job, this thing is looking good," he told them one day. "I think we've got a major discovery here."

The early returns in Montana were impressive, but Brian Hoffman wondered if they could do better. At thirty-three years old, Hoffman was a junior member of the exploration team. He already had dealt with his share of troubles from the business, however.

After receiving an undergraduate degree in geology in 1994 from Oklahoma State, Hoffman couldn't attract a single job offer, as the energy business suffered from low prices. While he waited for the industry's fortunes to improve, he went back to school to get a master's degree, just to do something productive with his time.

A few years later, Hoffman was thrilled to find work at Continental Resources. He was especially happy to learn that he'd be joining the effort in the Williston Basin. It wasn't that he was inspired by Harold Hamm or convinced that any resurgence in American oil production was in the offing. Instead, he hoped his new job might give him a chance to move to Denver, the picturesque city he recalled fondly from his youth. He figured that if he worked for Hamm in the area for a few years, he might get a transfer to Colorado, at least someday.

Sure, Denver was a full seven hundred miles away from the heart of oil country in Williston, North Dakota, which was Hoffman's area of responsibility. And Hoffman actually was based far away, in little Enid, Oklahoma. Still, a guy has to have a dream.

"I really figured I'd work for a few years and then go to Denver," Hoffman says.

Hoffman spent the early 2000s helping Continental extract oil from a layer of rock in North Dakota called the Mission Canyon. They *tried* getting crude from this layer, anyway, which was relatively close to the surface. In reality, Continental got too little oil and too much water.

During his work on this exasperating rock, Hoffman became intrigued by the thin Bakken layers lower in the ground. Rock in the Williston Basin can be imagined as a layer cake; the Mission Canyon layer was two levels above the three Bakken layers smack in the middle of the cake.

By 2003, Hamm and Continental had their hands full in the Bakken formation in Montana. But Hoffman couldn't stop thinking about his time in North Dakota. He remembered how competitors had drilled through the three Bakken rock layers in the state and saw evidence that it held oil, though few managed to get much of it out.

Now that Continental was managing to use horizontal drilling to get oil out of Bakken rock in Montana, Hoffman thought he'd go back and investigate the same rock over in North Dakota. "I wasn't told to look at North Dakota. We were all focused on Montana at the time," he recalls. "But you want to squeeze every drop of oil" out of an area.

Hoffman and his colleagues returned to North Dakota and tested the Bakken rock there. The Upper Bakken layer was composed of black shale, as was the Lower Bakken. Sandwiched in between, like the vanilla of an Oreo cookie, was the Middle Bakken. It was made of a different type of rock called dolomite that seemed easier to drill than the shale.

Hoffman visited the state's geological library to examine its collection of cores, or cylindrical samples of rock from old wells in the region. He read how others periodically had been enticed by this rock but later gave up.

Most experts viewed the Bakken's upper and lower tiers as "source" rock, or spots where oil originated before migrating to other layers until it reached an impermeable layer where it was trapped, preventing it from migrating farther. It was very much like how industry pros had considered the Barnett's shale to be source rock, and not a reservoir worth drilling, until George Mitchell's men proved the rock could produce natural gas. "In school they taught us these layers aren't reservoirs," Jack Stark says, referring to the Bakken's shale.

As Hoffman and his colleagues researched the Bakken layers in North Dakota, Aubrey McClendon and Tom Ward discussed drilling their own Bakken wells. Ward eventually decided it would be too costly to get crude out from the formation, and that its fans were hyping its potential, so Chesapeake passed on North Dakota.

But the more Hoffman examined the Bakken rock, the more he thought it might hold larger amounts of energy than generally believed, especially that middle level, the vanilla in the Bakken Oreo, which seemed really thick with oil.

Maybe the experts were wrong after all. *This really seems like a reservoir,* he thought.

Hoffman approached his boss, Jack Stark, as well as Hamm, to suggest that the company extend its drilling from Montana back into North Dakota. "We need to go up and take a look" at the acreage in North Dakota's northwest corner, he told them.

Stark was an instant convert to the idea, rejecting his earlier education. "Heck, it can't help but get better," he concluded after doing his own research and noting that the Middle Bakken layer seemed to be thicker in North Dakota than in Montana, and presumably more oil-rich.

The idea also made a lot of sense to Hamm. If Hoffman and his colleagues were right, rock across the state's border might even have more oil, Hamm decided. If the Bakken proved more fruitful in North Dakota, the country itself might be impacted, Hamm insisted to Mike Armstrong, an old friend. Even energy self-sufficiency was a distinct possibility.

"He absolutely believed we could be energy independent," Armstrong says. "I thought he was full of shit. . . . I thought he was nuts."

Hamm directed his landmen to lease acreage in western North Dakota under an assumed name with a vaguely Canadian ring to it—Jolette Oil LLC. He didn't want competitors to know they were on to something. That way they could lease the land cheaply.

"We wanted people to think it was a dumb Canadian company—they used limited liability companies all the time, so we thought they'd assume Jolette was Canadian," Hamm recalls. "We hoped they'd say, 'Oh, those crazy Canadians, there they go again.'"

He probably shouldn't have bothered. At that point, few outside Oklahoma knew much about Hamm or his company. The odd name they chose actually piqued some interest among locals; they realized it was made up and began asking who really was leasing the land. So much for Hamm's cloak-and-dagger tactics.

It didn't bother the rivals too much when they heard it was Continental, though. "We were seen as a very small Oklahoma producer that probably wasn't going to be up there very long," says Jeff Hume, Hamm's right-hand man.

Hamm's landmen assembled 300,000 acres in the area in 2003 and the company began drilling. Hamm and his men were sure a new, huge discovery awaited them.

They had no idea how wrong they would be.

Charif Souki was getting closer to his destination.

It was 2000 and Souki was driving a rental car from the city of Corpus Christi, Texas, to a nearby tract of land that seemed perfect for his company's first terminal for imported liquefied natural gas.

Souki was still a novice when it came to turning supercooled LNG into natural gas for American consumers and businesses. And Souki's company, Cheniere Energy, didn't have nearly enough money to build a plant to convert any gas. But in the front seat next to Souki was Charles Reimer, a veteran of the LNG business and Cheniere's new senior executive. Reimer was lending Souki crucial encouragement and guidance.

Souki had spent a year scouring maps for possible sites for a facility, examining nearly a dozen possible locations on both coasts of North America, from Canada down to Mexico. He first considered spots near major cities on the West Coast. But Cheniere needed some serious space. Any site would need forty-foot-deep channels and harbors wide enough to allow a 1,000-foot-long, 160,000-ton LNG tanker to turn around.

Souki worried that environmental activists in that part of the country would block any huge terminal. Working with an engineering professor at the University of Houston, he examined areas on the East Coast, where natural gas prices were higher and population centers within close reach. The land seemed too expensive for a tiny company like Cheniere, however.

Eventually, Souki and his team of five newly hired LNG specialists settled on the Gulf Coast, where local leaders welcomed oil and gas jobs. The region also boasted an extensive network of natural gas pipelines, thanks to the multitude of industrial facilities, which figured to make it easier for Souki's company to move and store gas from its proposed plant.

The choice was met with derision. There was already so much natural gas coursing through the Gulf Coast that there seemed little need for additional supplies from Souki's company, industry experts said. Import-

ing gas to America seemed silly; importing it to the Gulf Coast seemed
even more absurd. When Souki shared his idea at a major industry conference there was disbelief. "You're going to build a plant on the Gulf
Coast with all the gas there?" an audience member asked him. "Isn't that
like bringing coal to Newcastle?"

The audience cracked up like they hadn't heard anything so clever in
weeks.

Souki shrugged off the ridicule. He wasn't prepared to share details
of his prospective terminal or defend it, so he turned the other cheek.
Besides, he had gotten used to the putdowns. "They could think I was an
idiot, I didn't mind," he recalls.

Eventually, Souki found a perfect site near Corpus Christi, the nation's fifth largest port, and agreed to an option to lease a huge tract of
land in that coastal South Texas city.

As Souki and Reimer drove their rental car to the site, ready to sign
the deal, their spirits were sky high. Just as the two executives crossed a
nearby bridge, though, Souki noticed something that gave him pause.

"Charles, how tall are the ships?" he asked Reimer, referring to the
size of the tankers that would haul LNG to their prospective plant.

"One hundred and sixty-five feet," Reimer told him.

"How tall do you think this bridge is?" Souki wondered aloud.

Reimer didn't have a clue. Souki circled back to the bridge's entrance
and read a marker listing its height: 135 feet. Souki and Reimer looked at
each other, mouths agape. They were about to spend $200,000 of their
company's precious cash to lease acres of land near a bridge so low that
LNG ships wouldn't even be able to reach their terminal. Souki wiggled
his way out of the lease deal, barely averting a monumental blunder.

A few months later, he found a better spot, this one in Freeport,
Texas, next to one of the largest chemical plants in the world. If his strategy was going to work and demand emerged for imported gas, he figured
he might as well build more than one terminal, to try to reap bigger
gains. So Cheniere also leased acreage in Sabine Pass, Louisiana, also on
the Gulf Coast. This spot held little more than a hunting trailer and an
assortment of animals, including bobcats, alligators, and free-range
cattle. Souki even found a site for a third LNG conversion facility; it was

in Corpus Christi, but the site was far from that troublesome bridge. Cheniere purchased options to lease hundreds of acres of land in the three locations and began designing gas import terminals, going full speed ahead.

Souki's company couldn't really afford all its heavy spending, though. Cheniere was selling for less than fifty cents a share, or a puny $25 million. Most investors were sure the company would go under. Cheniere's own landlord in Sabine Pass figured he'd pocket a few hundred thousand dollars to lease the land for a couple of years, and when Souki's company inevitably collapsed the landlord would get his land back.

Souki hadn't even begun raising money to pay for his first plant and pipeline, something Reimer estimated could cost more than a billion dollars. It's one thing to be ambitious; at that point Souki seemed downright delusional.

By early 2002, Cheniere had run out of money, just as the skeptics had anticipated. Souki couldn't afford a few million dollars to obtain permits with various government agencies, such as the Federal Energy Regulatory Commission, to build his first plant. He was even forced to borrow $30,000 from Reimer just to make payroll.

He reached out to Wall Street investors to try to raise $30 million to pay for the application process and obtain government approval for his three gas importing terminals. By then, natural gas and oil prices were edging higher again and the country was rebounding from the September 11 terror attacks, buoying Souki.

If I can just convince 3 percent of the people I meet with, that will be enough, he thought.

He traveled up and down the streets of Manhattan, sharing his vision with big-name investors like Kohlberg Kravis Roberts & Co., the Blackstone Group, and Apollo Advisors. Souki was confident they could score big profits by investing a few billion dollars in his import facilities, or simply by helping him flip his prime real estate to a party with deeper pockets who could afford the billions necessary to actually build the terminals.

Everyone wanted to hear Souki's story. They all liked the idea and wanted more details. He still couldn't raise enough money to make his plan happen, however. "When you're talking about two billion dollars you don't have a lot of friends," he recalls.

Souki wasn't deterred. He always had been good at raising big bucks in the Middle East, and he still had prime contacts in the area. He began flying to the region, and to other foreign locales, pitching his radical idea and trying to sign customers looking to sell gas to America. Souki and his wife would first fly to Paris or London, where Rita would stay and spend time with friends, while Souki went on to Algeria, Saudi Arabia, Nigeria, Equatorial Guinea, and elsewhere.

In Doha, the capital city of Qatar, Souki managed to get a dinner meeting with Abdullah bin Hamad Al-Attiyah, the minister of energy and industry. "You're going to have a new market in America," Souki told Attiyah, trying to entice him to direct a big investment Cheniere's way so Qatar could sell gas in the United States.

The prospect of targeting the huge American market held instant appeal. Qatar had built enormous production facilities and needed to find demand. Attiyah and other local executives and officials even seemed to root for Souki, who shared a Middle Eastern descent, spoke Arabic, and was a David up against the Goliaths of the energy world. Local technocrats did their research and received promising reports about Souki's sites on the Texas and Louisiana coasts.

Souki couldn't get Attiyah or anyone else to sign on the dotted line, however. Cheniere just seemed too tiny and risky. Souki had made a career out of raising cash, but he couldn't raise a measly $20 or $30 million. There already were four LNG terminals in the United States that had been built in the 1970s, some potential investors noted. Two had been mothballed, and the other two hardly were in use. Who needed another costly LNG conversion facility?

Souki, still pitching his crazy idea, flew to a breakfast meeting with Michael Smith at Denver's Brown Palace Hotel. The two men seemed a perfect match. A native of the New York borough of Queens, Smith had become wealthy in real estate, recently sold his energy company for $400 million, and was looking for investment projects. Smith's company had drilled in the Gulf Coast. Like Souki, Smith had become convinced that energy prices were heading higher as gas discoveries became more infrequent and existing wells were depleted. Smith even had taken a year off at one point to ski in Colorado, just like Souki.

Souki described his three-step strategy: Once Cheniere received a permit to build its first plant, Souki planned to find customers willing to give Cheniere lucrative contracts to convert their LNG to natural gas. After Cheniere signed a client or two, Souki figured it would be easy to find backers willing to finance the billion-dollar construction of an expensive terminal and connecting pipeline. After he built one terminal, he'd build a few more. By then, his plan included building four LNG terminals.

When he heard Souki's idea, Smith gave an immediate and blunt reaction. "You're out of your mind," he said. "No one's built an LNG facility in this country in over twenty years. . . . One terminal is a massive undertaking, four at the same time makes no sense."

Just read our plan over, Souki insisted. I really think you'll like it.

Over the next few weeks, Smith digested the materials, did his own research, and became intrigued despite himself. Additional conversations with Souki piqued Smith's interest.

"He's such a charismatic, interesting guy," Smith says.

But Smith soon learned of several lawsuits in Cheniere's past that raised some concern for him. "This is an interesting idea but I'm going to pass," he finally told Souki.

Souki spent another six months trying to raise his money, with no luck. The company was spending about $200,000 a month and time was running out. When an investment banker working with Souki invited Smith to hear a speech Souki was giving at an energy exploration-and-production conference, Smith agreed to join the audience.

Souki usually was a smooth speaker, but he was unusually awkward that day, perhaps a sign that even he was tiring of the never-ending sales process. Halfway through his speech, he stopped to apologize to the audience of analysts and investors. He was discussing importing natural gas, but it was a subject most had little interest in, he realized. There were only about twenty-five people in a room that held over one hundred, a further sign of how few wanted to hear from him.

After the speech, Souki joined Smith in a private room. Souki told

Smith that Cheniere was going to focus on just one project, the prospective facility in Freeport, Texas, not on trying to build three others. The two men began an intense, three-month negotiation, settling on a deal for Smith to invest several million dollars with Souki.

A few days later, however, Souki's banker called Smith with news that Souki had inked a deal with Dow Chemical to both finance the terminal and become its first customer, Smith recalls. (Souki says he never told Smith he had a deal with Dow. At most, he was "posturing," he says.)

Smith became upset, believing Souki had used him to get Dow to agree to its investment. "When your deal falls apart, call me," Smith says he told Souki's banker. "But our deal will be on new terms."

"I was fucking angry," Smith remembers.

Smith was sure Dow would walk away from Souki and Cheniere. "They were a fly-by-night company with just an idea . . . it wasn't hard to see they had no money" and that Dow would have second thoughts.

Sure enough, Dow never completed a deal with Cheniere. Souki's banker got in touch with Smith to ask if he remained interested in putting up the money. Souki and the banker hadn't made it a secret that they were talking with other parties, including Dow, Souki says, so they never understood why Smith would be upset by what Souki calls a simple "break in the negotiation."

Late on a Friday afternoon in early August 2002, Souki flew to New York's Palace hotel to meet Smith, who drove in from a summer resort in the Hamptons.

This time I'm not going to get fucked, Smith thought as he approached the hotel.

In September 2002, they agreed to a new deal. Smith would pay Cheniere $5 million—$1 million up front and the rest down the road. In return, Smith would get full control over Souki's most advanced project in Freeport, Texas, with Cheniere reduced to a 40 percent minority investment. Smith wanted ownership of any company he was going to invest a lot of money in.

Smith also insisted that Souki agree not to move forward on his other LNG projects for about a year, to give Smith's Freeport facility a big head start. Oh, and one other thing: Smith demanded that Charles Reimer,

Souki's in-house LNG expert, leave Cheniere and join Smith's new company.

"I liked Charif but Cheniere was technically insolvent," says Smith. "I just wanted my new company to be fully protected and isolated from any problems Cheniere could have."

Souki didn't care about Smith's concerns. He was actually thrilled someone else agreed with his argument about the value of importing gas. "Now there were two maniacs, not just one," he quips.

His company had a million-dollar lifeline and a commitment of $4 million to help pay for preliminary work on an import facility at Sabine Pass, Louisiana, the terminal Souki would now focus on. It wasn't much, and he'd still have to raise a billion dollars somehow to build the thing. Cheniere was moving ahead, though, even if its stock still traded for about a buck a share. Souki's dream was alive, if only just barely.

CHAPTER NINE

Henry Harmon had a hunch why Aubrey McClendon was calling.

It was the summer of 2005 and McClendon was on the phone asking if he could fly to Charleston, West Virginia, to meet Harmon.

Harmon, who ran a natural gas company in the Appalachian Basin called Columbia Natural Resources, was intrigued by the call. McClendon was the industry's rising star, and Harmon was thrilled that he wanted to visit.

Days later, McClendon took a private jet to Charleston to take Harmon to dinner. Aware that McClendon was a big steak fan, Harmon took his guest to the Chop House, one of the city's most upscale restaurants, eager to hear why McClendon had come all the way to see him.

McClendon and Harmon sat in a private dining room featuring rich oak floors and comfortable leather chairs. McClendon had a juicy cowboy steak on his plate, but he couldn't focus on his food. He fired question after question at Harmon: What was it like operating on the East Coast? Are there environmental concerns that Chesapeake didn't have to worry about in Texas, Oklahoma, and elsewhere? What about the company's unions?

Harmon quickly realized why McClendon had flown to see him—McClendon was coming after his company. For three years, McClendon had privately lusted after Columbia's natural gas wells, which stretched from the Finger Lakes region in upstate New York all the way south to central Alabama, with the heart of its operations in West Virginia. By then, McClendon and Ward had made a decision that it was time for Chesapeake to head to the East Coast, after the company had gobbled up so much land elsewhere in the country. Now it was time to act.

Columbia, the second largest producer on the East Coast, hadn't given any indication that it wanted to sell itself. But McClendon knew the company was owned by a number of parties—including investment bank Morgan Stanley—that likely would pursue a sale someday. He wanted to be ready to pounce if they did.

Harmon cautioned McClendon about an acquisition. The East Coast is a more densely populated region, he noted, making it harder to drill new wells near people's backyards without sparking a backlash from some landowners and environmentalists. It's very different here, Harmon told McClendon.

McClendon shrugged off the warning, exuding optimism and self-confidence. He told Harmon that natural gas prices were heading still higher and his company would be a natural fit within Chesapeake's growing empire. "We're the best buyer" of Columbia if it's going to be sold, he said.

A few months later, Columbia's owners put the company up for sale, just as McClendon had hoped. Some of the largest oil and gas companies in the country showed immediate interest. McClendon insisted to Ward that Chesapeake needed to beat their rivals in the quest for Columbia.

One day, McClendon showed up, unannounced, in the New York City lobby of Morgan Stanley's headquarters. Chief executives of major companies normally don't fly anywhere without a handler or two, and they usually don't show up anywhere unannounced. Aubrey McClendon wasn't a typical CEO, though. He was alone in the lobby, carrying a small bag, and he wanted to speak with the banker working on the sale of Columbia.

"Can I come up and tell you why we're the best buyer?" McClendon asked a stunned Morgan Stanley banker. "Can we get a deal done?"

In October, McClendon got his way. Chesapeake announced a deal to pay $2.2 billion in cash for Columbia, while also assuming $800 million of the company's debt. It was more than Chesapeake had ever paid for an acquisition and a sign that the company's land acquisition machine was now in overdrive.

Aubrey McClendon and Tom Ward had truly arrived. Their company now controlled eight million acres, almost the size of New Jersey and Connecticut, throughout seven states. Chesapeake sat on nearly seven trillion feet of gas, making it the sixth largest natural gas producer in the nation.

Adoration and accolades quickly followed. That year, *Forbes* magazine named McClendon one of America's top-performing executives. The company is "the closest thing you're going to find to a Bill Gates story in the energy industry," an energy industry analyst, Stephen Smith, told a reporter.[1]

Those who challenged McClendon and Ward's strategy usually were dismissed as naïve or excessively cautious. On a quarterly conference call with investors and others, Duane Grubert, a stock analyst at Fulcrum Global Partners LLC, asked when Chesapeake was going to stop borrowing and buying companies at expensive prices. "When is enough enough?" Grubert asked.

"I can't get enough," McClendon retorted, adding that Chesapeake's backlog of acreage surely would prove valuable.[2]

Grubert remained worried about the risks McClendon and Ward were taking. By the end of 2005, Chesapeake had nearly $5.5 billion of debt, up from less than a billion five years earlier.

Grubert didn't respond to McClendon's quip, however. There didn't seem any point in pushing back.

Chesapeake shares soared, and the company emerged as among the country's best hopes of staving off a seemingly imminent shortage of natural gas. Investors couldn't get enough. Chesapeake's stock doubled in just a year, moving past thirty dollars a share.

Few benefited from the move like McClendon and Ward. By late 2005, McClendon's nearly eighteen million shares were worth over $500 million. Ward had over fourteen million shares worth over $400 million. Each executive had also purchased between 1 percent and 2.5 percent of every well drilled by Chesapeake, adding to their wealth. The ability to buy a piece of the company's wells was granted to the executives when the company went public, a perk that executives at a few rival companies enjoyed.

Like adrenaline junkies, McClendon and Ward seemed desperate for new kicks—and ways to expand their already swelling fortunes. By then, they had already become investors in a hedge fund run by veteran oilman T. Boone Pickens, stakes that proved lucrative. They wanted more, though.

McClendon and Ward arranged lines of credit at major Wall Street firms, including Goldman Sachs and Morgan Stanley, which allowed the Chesapeake cofounders to borrow millions of dollars to make trades in personal accounts at the firms. The pair weren't required to seek approval from Chesapeake's board of directors for the trading, nor did they ask for it.

Every week or so, McClendon and Ward called trading desks of the Wall Street firms and placed bets on natural gas futures, among other investments. Traders at the firms say McClendon and Ward usually showed prescience in their trading moves, leading to hefty gains. Ward, for one, made tens of millions of dollars most years, and sometimes more than $100 million, he later acknowledged.

It's very unusual for top corporate executives to actively trade stocks and commodities. It's hard enough to run a big company, let alone also try to figure out where various markets are headed. Such activity can also raise questions about conflicts of interest and whether an executive has access to information that other investors aren't privy to. Indeed, Ward led a group that oversaw Chesapeake's trading in oil and gas that was done to hedge, or protect, the company from big price swings. As such, he and McClendon played an active role in daily trading by Chesapeake.

McClendon and Ward often bought natural gas futures contracts—financial contracts obligating them to buy gas at a future price and day—while Chesapeake tended to sell natural gas futures as part of its hedging program. As a result, there was no conflict of interest in the activity, at least in their minds. Also, McClendon and Ward didn't trade gas in large enough volumes to easily move markets, nor did Chesapeake itself.

While corporate executives operate under strict guidelines when it comes to buying and selling stocks, and insider trading is heavily scrutinized, commodities markets are different. Buyers and sellers are allowed to use their knowledge to hedge against price swings and trade commodities, as long as their trading doesn't manipulate prices.

McClendon and Ward never shared details of their trading with shareholders or members of Chesapeake's board of directors. Technically, they didn't need to. The public never found out about the lines of credit provided by the Wall Street firms or about the calls both executives were placing to do their trading. McClendon sometimes called hedge fund traders and others to discuss where natural gas prices were headed, but the traders usually didn't know he was placing trades in his own account.

The Chesapeake cofounders didn't seem to go out of their way to hide their trading, though, and some senior colleagues were aware of it. Whenever they were asked about the trading, McClendon and Ward downplayed it, a strategy that seemed to head off potential news stories about the activity.

"I just play the stock market, I play the commodity markets," McClendon told a reporter for the *Wall Street Journal* in 2006. "Sometimes I win, sometimes I lose."[3]

As for Ward, he viewed the trading as an extension of moves he had made in personal accounts for years. Heck, his personal e-mail address was made up of his initials and the word "trading," as if he were publicizing his hobby.

"I had trading accounts at several banks . . . I've always traded, I'm not afraid of markets," Ward says. "It started out of college."

The busy personal trading still wasn't enough for McClendon and Ward. In 2004, the executives started a hedge fund called Heritage Management Company to trade coffee, agriculture, livestock, grains, and oil. About 10 percent of the trading in the fund, which was based in a small two-room office in a 1930s skyscraper on Fifth Avenue in Manhattan, was in natural gas, Chesapeake's key product.[4]

McClendon and Ward sent daily e-mails to help set the fund's trading tactics and participated in weekly thirty-minute strategy calls that could be "exhaustive," according to the fund's head trader, Peter Cirino.

The fund eventually grew from about $40 million to over $200 million, half of which was the personal money of McClendon and Ward.

"I often marveled at their attention to detail," Cirino says. "They could multitask with the best of them."

To McClendon and Ward, the hedge fund's activities, and all their side trading, were no big deal. Chesapeake was on a roll, providing evidence that they weren't distracted. Besides, they were just advising the hedge fund, not running it.

It was hard to judge the impact of McClendon and Ward's trading, however. Chesapeake was responsible for about 10 percent of the activity in the natural gas market, traders estimate. There were times when Chesapeake's moves, including its hedging and its decisions on natural gas production, influenced markets, traders say. That raised the possibility that the hedge fund could have benefited from the activities of Chesapeake, though no such evidence emerged.

Either way, McClendon and Ward didn't spend too much time worrying about how the trading by the hedge fund, or their own speculation in the brokerage accounts, might look to outsiders. In their minds, commodity markets were so big that their knowledge of what Chesapeake was up to didn't give them any kind of trading edge or unfair advantage.

By then, there were few executives around to level much criticism at McClendon and Ward. They were among the most powerful men in American business, and it was hard for staff members to question their decisions, according to a former Chesapeake executive. Later, though, news of the hedge fund would emerge, leading to stinging criticism at an inopportune time for the Chesapeake cofounders.

Chesapeake was discovering new sources of natural gas and McClendon was making money hand over fist. He was determined to spend his newfound wealth like the true energy mogul he had become.

Making a lot of money and accumulating evidence of wealth always seemed to drive McClendon. He was a scion of Oklahoma's Kerr oil family, but his parents never enjoyed an outsized fortune like others in the clan. Some who knew McClendon said he had a chip on his shoulder and was determined to accumulate wealth that topped anything the Kerrs had. Other friends said he simply developed an appreciation for the finer things in life and tended to be desirous of what others had. Either way, McClendon was driven to have it all.

First there were the homes. McClendon built a 9,000-square-foot stone mansion in Nichols Hills, an exclusive enclave near Chesapeake's Oklahoma City campus. Then he paid $700,000 for the home behind it.

Aubrey and his wife also bought an $8.6 million home on Bermuda's so-called billionaire's row near homes owned by New York City mayor Michael Bloomberg, billionaire Ross Perot, and former Italian prime minister Silvio Berlusconi. The McClendons spent $12 million to give the property a makeover. Later, Katie McClendon paid $11 million for the next-door house overlooking a spectacular cliff on Windsor Beach. McClendon also paid $20.8 million for an eight-acre estate nearby that once was owned by descendants of industrialist Henry Clay Frick, but he later sold it for a small profit. They also owned properties in Minnesota, Maui, and near Vail, Colorado.[5]

McClendon was always on the lookout for new ways to spend his cash. Riding a Jet Ski on Lake Michigan on vacation, he spotted "the prettiest home along the lake," he later said.[6] He spent $40 million to buy the lakeside home and millions more to purchase 330 acres for a massive resort, including a hotel, nine-hole golf course, marina, and one hundred residences, a development that locals fought to stop.

Then there were his expensive hobbies and interests, including a $12 million collection of antique maps of Oklahoma and nearby states that "would be the envy of the Library of Congress," his adviser told Reuters. McClendon proudly hung the maps in Chesapeake's offices, to the cheers of employees and guests of the company. McClendon's wine collection would total more than 100,000 bottles and be worth millions. A highlight: a six-liter bottle of 1945 Mouton Rothschild valued at about $100,000.

"He went from not knowing anything about wines to having a basement of 1982 Bordeaux," says a longtime friend.

McClendon had so much going on that he set up a unit staffed with accountants, engineers, and supervisors, dedicated to managing his personal business. It was called AKM Operations and it was housed in an annex of Chesapeake's headquarters, a sign of how McClendon liked to mix personal with professional.[7]

While he enjoyed spending money on himself and his family, friends noticed that McClendon also seemed happy giving his money away, relishing the honor and prestige that came with the gifts. In 2001, he and his wife gave $2 million to the University of Oklahoma's athletic program. Three years later, he gave another half-million-dollar donation to the school, after earlier big gifts to Duke, McClendon's alma mater. He also became a big donor to his high school, Heritage Hall. His charitable donations eventually would approach $100 million.

McClendon began delving into politics, making a series of contributions to Republican candidates and conservative interest groups. He either viewed himself as above criticism or simply had no clue his actions might be viewed as controversial. During the 2004 presidential election, the *New York Times* reported that McClendon gave a quarter million dollars to Swift Boat Veterans for Truth, a controversial advocacy group running commercials questioning Senator John Kerry's military service in Vietnam, an effort that hurt his presidential bid against George Bush. The same year he and Ward gave over a million dollars to Americans United to Preserve Marriage, a group opposing same-sex marriages.

"I am for the concept that a marriage should be between a man and a woman; on the other hand, I am for civil unions for gay couples," McClendon said at the time. "In my opinion, that does not make me anti-gay at all. Instead, it makes me pro the traditional concept of marriage, and I do not believe the biblical sacrament should be between anyone other than a man and a woman."[8]

Aubrey McClendon's transformation into a true American oil tycoon was in the timeworn tradition of previous energy moguls who rose through the ranks and freely spent their winnings. Tom Ward emerged as a different kind of energy magnate, however, a baron for the Bible Belt.

Ward liked to spend money nearly as much as McClendon. He owned huge homes outside Oklahoma City and Scottsdale. He also had places in Bermuda and the Bahamas, and extensive land in western Oklahoma. At one point, he and McClendon were among the biggest owners of cattle and feedlot in the country.

But Ward didn't flaunt his extravagances. Few friends were aware of how extensive his holdings of real estate and other assets were. McClendon tweaked a reporter for not appreciating an expensive bottle of wine; Ward spoke of his love of steakhouses. "I eat steak about six nights a week," he told a reporter. "My favorite smell in the world is to walk through the back door and smell a steak grilling."

Ward spent and made money at prodigious rates, but he placed an equal emphasis on his faith, a seeming incongruity. On Sunday mornings, he and his family went to services at their evangelical church, the Crossings Community Church. Ward then headed to the office while his wife took the family home.

Ward donated millions to Anderson University, a Church of God–affiliated school in Anderson, Indiana. He also hosted a regular Bible study group in his Chesapeake office.

Despite his overloaded schedule, Ward and his wife helped found a home called White Fields to care for abused and neglected boys. They also began to take care of a young man, Frank Alberson, who had become friends with their son, Trent, in college.

"My parents refused me when I was born, and I was adopted," Alberson says. "My adoptive parents placed me into the custody of the state of Indiana, and after I graduated military school, I was on my own. I still had quite a few problems. Trent invited me home for the holidays in 2000, and I met Tom. They offered me a job working at their house. I was informally adopted as part of the family."[9]

If Aubrey McClendon was in the tradition of J. Paul Getty, who made a billion dollars, sought to influence politicians, and had an appreciation for fine art, Tom Ward was in the mold of more enigmatic energy tycoons like John Rockefeller, the sharp-elbowed monopolist who overcame humble beginnings and spent much of his life focused on philanthropy.

In late August 2005, Hurricane Katrina struck the American Gulf Coast and New Orleans, resulting in the costliest natural disaster in American history. Massive shut-ins of natural gas production sent prices climbing from nine dollars per thousand cubic feet in August to a record $15.40 in December 2005.

The price move didn't help Chesapeake shares very much. They flattened out at around thirty dollars as investors viewed the price surge as something that would ease when the nation and industry fully recovered from Katrina. Stress grew within Chesapeake's headquarters as competition for shale acreage grew, led by Chesapeake's aggressive rival, XTO, and up-and-comer EOG Resources. The companies were finding it easy to raise cash from investors, and they were plowing it into their own acreage.

When McClendon heard that a rival had acquired valuable land somewhere in the country he immediately picked up the phone to call Ward. "Why aren't we in that?" he asked his partner. McClendon was polite and respectful, but also persistent.

Ward often explained why the new field seemed too expensive. McClendon didn't push further, but it was clear he wanted Chesapeake to make the purchases, not rivals. "He has an issue of watching other people's plays and wanting to be in them," says a longtime colleague.

Pressure grew within the company. McClendon and Ward paid higher salaries than most competitors, but they expected employees to work as hard as they did, from early morning to late at night, with more work over weekends. When a visitor came to the office one day, he asked an executive about big bottles of aspirin he saw on various desks. "We call them 'Chesapeake vitamins,'" the executive said, explaining they were there to help employees deal with the stress.

Ward felt special pressure. He didn't want to let McClendon down and he felt a need to match McClendon's intensity. But he became uncomfortable spending big sums on acquisitions. The headlong charge was taking a toll on him. He kept his emotions in check. There weren't any blowups between the Chesapeake cofounders, who almost always treated each other with respect and warmth.

"There were virtually no arguments," Ward recalls. "I'm not an arguer by nature."

For those paying close attention, though, it wasn't hard to see a rift growing between them. Those close to McClendon sensed he was becoming frustrated with Ward's reluctance to pursue some new shale plays like the Bakken.

"Let me first see some established production" in the Bakken, Ward said in one meeting. "Then we can buy around it."

McClendon dropped the subject, knowing he wasn't going to get anywhere with Ward.

Ward often chose a passive-aggressive way to get his point across. In the middle of a discussion about new acquisitions, he sometimes closed his folder and bolted from the room, leaving few doubts in the minds of colleagues about his true feelings. Other times, he wouldn't show up for an important meeting with McClendon and others, colleagues recall.

"You could tell what was happening with his body language," says a former Chesapeake executive. "I never heard a cross word, but it wasn't hard to see."

For Ward, it wasn't so much the debt that was piling up as the prices the company was paying. The new shale plays didn't have much existing production, making Ward uncomfortable paying up for these new fields. "Don't we have enough" acreage and wells in an area? he asked McClendon one time in a group meeting. Another time he told McClendon, "I don't know how we're going to" afford a new acquisition.

Ward began to look tired and stressed. He and his team examined thousands of reports detailing the oil and gas reserves of wells up for sale all over the country, and it was beginning to wear him down. Friends and colleagues reached out to help.

"Hey, man, you should delegate that," Larry Coshow, the Chesapeake landman, told Ward one day, referring to a task that seemed to consume Ward. Another time, Coshow said to Ward, "Is there anything I can do?"

Ward turned red in the face, as if he was ashamed to ask for assistance.

For the first time, Ward began to show anger around the office. In late 2005, McClendon e-mailed Lindell Bridges, a senior geologist who had helped build Chesapeake's huge position in Arkansas's Fayetteville region, to ask a series of questions about wells in the area. Bridges made sure to include his bosses, including Ward, in his answers to McClendon. Bridges knew Ward wanted to be informed about all of his crew's communications with McClendon.

McClendon didn't seem to care as much about protocol and e-mail etiquette as Ward did. Or maybe he was hoping to stir things up with Ward. Either way, McClendon kept sending more questions to Bridges, almost always without including Ward or others in his e-mails. Bridges responded to them, adding his bosses to the replies. Finally, though, Bridges sent McClendon a note suggesting that they meet to discuss the company's activity in the Fayetteville region, to avoid all the back-and-forth. McClendon agreed and they set a time to meet.

Bridges realized he had made a big blunder by neglecting to include Ward or his other bosses in the last batch of e-mails to McClendon. He quickly forwarded the latest e-mail to Ward and the other senior executives, informing them of his plan to meet with McClendon.

A few minutes later, he got an e-mail message from Ward:

"We need to have a meeting about this in my conference room."

Shit, I'm in trouble, Bridges thought.

The next day, he walked into the main conference room of Ward's office building. There were a half dozen senior executives from Ward's team waiting for him. At the head of a long table was Ward, looking furious. He ripped into Bridges with unusual venom, admonishing him for his disregard of protocol. "Don't let that happen again."

Bridges was devastated. For weeks, he worried about getting fired. He beat himself up for forgetting to include Ward in the e-mails, knowing how sensitive he was about the issue. "I thought my career was over," recalls Bridges. "I was really low and depressed . . . you could tell he was unhappy."

Bridges kept his job, but the change in Ward's behavior was viewed by some as evidence he was under growing stress. By then, he was arriving at the office at 4 a.m. and not getting home before 10 p.m. at night. Then he'd go through a batch of e-mails that had piled up.

In 2005 alone, Chesapeake spent nearly $5 billion to buy energy properties, acquiring 1.4 million acres in the year, including 500,000 acres in the fourth quarter alone. The company had spent over $10 billion since 1998. The push for acreage, a land grab the likes of which the nation had rarely witnessed, seemed to be paying off. In 2005, Chesapeake produced over 422,000 million cubic feet of natural gas, up from 116,000 in 2000.

It was making Ward miserable, though. He ran the company's operations, including nearly a thousand landmen around the country. And he felt a need to stay involved in the details of all the drilling.

Prices for acreage and wells were climbing, McClendon was pushing Chesapeake to outspend rivals, and Ward couldn't take it. Every day, he had to make dozens of decisions about new plays, new acquisitions, and new spending proposals by McClendon.

"I couldn't keep up, we were spread out around the country, I just couldn't vet the opportunities," Ward recalls. "All day, then e-mails, nights and weekends, it didn't stop, it was getting harder to run the company."

By then, the economy and housing market were rolling, and even energy giants were beginning to throw money at acquisitions in the United States, making it hard to find bargains. In December 2005, Conoco-Phillips agreed to pay nearly $36 billion to buy smaller oil and natural gas producer Burlington Resources, which owned shale acreage in the Barnett. It was the biggest acquisition in the energy industry in years.

McClendon was sure the industry had changed and Chesapeake had to adjust with it. A backlog of acreage, no matter the cost, would reward Chesapeake down the road when it produced needed gas, McClendon argued. The key was to grab assets as soon as possible, he said.

Ward couldn't figure out how Chesapeake could compete without paying through the nose and jeopardizing its health, though. The shale formations Chesapeake now was examining had little existing production, making Ward even more uncomfortable. The land they were accumulating was an asset to McClendon, but it seemed a liability to Ward, because they'd have to pay to drill it and there was no guarantee it would work.

"Capital was free and he was overwhelming me with deals," Ward says, referring to how easy it was for Chesapeake to raise money from investors. "It was insane."

Ward was torn—he wanted his company to stop paying so much for deals, but he had no appetite for a showdown with his partner and friend.

No one at Chesapeake knew it at the time, but Ward was dealing with more than McClendon's spending spree. One day, he caught his youngest son, James, drunk, bringing back painful memories of the addictions of his father and grandfather.

One Thursday in February 2006, it all became too much for Ward. Grabbing a moment of quiet at home, he came to a decision, without consulting anyone else. The next morning, he walked straight into Mc-Clendon's office and gave him stunning news: He was quitting.

McClendon, his partner of twenty-four years, looked shocked. He said little to try to stop Ward, who soon was on his way out the door. That day, Ward sent a note to Chesapeake's employees, saying he had resigned and thanking employees for their work. Some were agape while reading the e-mail, the first Ward had ever sent to the entire company. Ward's staff was filled with sadness.

McClendon remained in a daze most of the day, colleagues recall. He knew he and Ward had drifted apart and weren't seeing eye to eye on many decisions, but he always figured they'd work out their issues, he told a senior executive. The disagreements didn't seem like a big deal and McClendon never anticipated Ward quitting so abruptly.

"It was kind of a shot to the gut," according to McClendon. "While I thought it might happen someday, I didn't think it would happen that day."[10]

A day or so later, however, McClendon snapped back, his natural optimism overwhelming any feelings of disappointment. He quickly began to focus on which executives to elevate to replace Ward and how he would run the company on his own.

Saturday morning, the day after he quit Chesapeake, Ward awoke with an unfamiliar feeling. He had nothing to do. He had generally sat with his drilling team and conducted a weekly meeting on Saturday mornings. This week he got up early, as usual, but had nowhere to go. He was devastated and confused.

What do I do now? he thought.

The melancholy feeling didn't lift for weeks. Ward no longer received many e-mails, and he didn't like it. It was as if he had been forgotten. He joked to Sch'ree that "the only thing worse than getting three hundred e-mails a day is getting three." The decision to quit had been an abrupt move, one that was very unusual for the deliberate executive. Now he didn't know how to deal with it.

"What am I going to do with the rest of my life?" he asked his wife.

Harold Hamm's mood was sky high as he set his sights on oil in the Bakken formation in North Dakota. Just a cursory understanding of the region's history would have brought him down to earth, however. Past oil booms had been fun but the busts quite painful, especially for locals, who had to stick around when roughnecks and others hightailed it from the region.

North Dakota's first producing well was discovered on April 4, 1951, when Andrew Davidson, the drilling superintendent on a well operated by Amerada east of the city of Williston, set a rag on fire and threw it in the air. Davidson watched as it met an invisible stream of natural gas flowing from the ground and shot thirty feet in the air. By nightfall, legend has it that the rag was still in the sky, a full ten miles away, proof to Davidson and others that energy was below the surface just trying to get out.[11]

News spread, prospectors crisscrossed the state, Davidson became a local hero, and *Time* magazine put the breakthrough on its cover.[12] New formations were uncovered and wealth created. Over the years, however, the state's energy production never quite lived up to the heady expectations, leading to dashed hopes and dreams.

Hamm's friend Mike Armstrong, for example, had 500,000 acres in North Dakota in 1989 but he let the leases run out. "I didn't have enough to buy a used Volkswagen," he recalls. "I was pretty jaded."

The region long had disappointed drillers. In the mid-1990s, majors including Royal Dutch Shell, Gulf Oil, and Texaco shut down operations in Montana's high plains, abandoned hundreds of nonproducing wells, and let leases to mineral rights expire.

The Bakken layers, in particular, were a perennial tease. The formation, created over 360 million years ago, was discovered in 1953 and named after a farmer, Henry Bakken of Tioga, North Dakota, whose property sat on an outcrop of the rock.

At one point in the 1960s, George Mitchell became a Bakken enthusiast and his company leased hundreds of thousands of acres for as little as four dollars an acre. Mitchell Energy saw promising early results but he too raised the white flag. "We couldn't get the damn thing to work, we failed, I don't know why," Mitchell recalls. "So we gave up on it, we just let the acreage go."

In 1999, a geochemist named Leigh Price in the Denver office of the U.S. Geological Survey took a good look at over one hundred old wells in North Dakota. He estimated that 413 billion barrels of high-quality crude were packed between the two layers of the Bakken shale, a huge amount of crude that should have excited the country.

Some in the industry considered Price, an avid weightlifter who wore his hair in a ponytail, off his rocker. Before his study was peer-reviewed or published, he died of a heart attack in 2000. That year, Montana's Board of Oil and Gas Conservation predicted the state's oil production would decline to zero.

"I thought my job was going to be turning out the lights," Jim Halvorson, a geologist for the state, told the *Wall Street Journal*.[13]

In the 1990s, companies like Burlington Resources did some drilling in the formation. But things got so bad late in that decade that Richard Findley, the son of an accountant for a chain of grocery stores, was making just $45,000 a year as a wildcatter. Unsure how long he could last, Findley and his wife sat down at their kitchen table to figure out what to do.

"We always came to the same conclusion," Findley later said. Geology "is what I know. This is what I love. So we just kept going."

Findley was able to convince some geologists that huge amounts of oil were stored in the middle layer of the Bakken, the dolomite section. But most drillers viewed the Bakken as a "bailout zone," or an area to extract a bit of oil if lower regions proved failures and a driller wanted to salvage a bit of oil from their time and expense. It just seemed too expensive to try to get more than that trickle of crude from the Bakken layers, at least with the vertical drilling everyone did at the time.

"The Bakken became a four-letter word," Findley says. "No one wanted to hear about it."

As horizontal drilling was perfected, Findley teamed up with Lyco to lease over 100,000 acres in Montana and target the Middle Bakken layer. They used computer-controlled directional drilling motors to bore laterally. By the early 2000s, they found surprising success in the Elm Coulee field stimulating the rock with hydraulic fracturing.

Findley and Lyco were meeting success in Montana, but Findley remained skeptical of the Bakken in North Dakota. "I could never see a trend as good as Elm Coulee in North Dakota," says Findley, who retired from working the region a few years later, though he later focused on the top of the formation, in Canada. "I thought [promising acreage] stopped at the border."

Hamm and Continental were out to prove North Dakota could be just as successful as Montana, if not more so.

O n a cold, clear March morning in 2004, as snow piled so high that Continental's crew couldn't make out nearby fenceposts, they began to horizontally drill and then fracture their first well in North Dakota.

To save the still tiny company money, Hoffman had located a well called the Robert Heuer 1-17 in Divide County, which had been drilled, plugged, and abandoned twenty-three years earlier. Using a preexisting well meant Continental wouldn't have to incur the expense of drilling vertically into the ground; they'd only have to extend the existing well horizontally and then pump it with fracking liquid a month or so later.

The Continental team did have to stop periodically to chase away spies from rival companies watching to see how they made out, but that didn't cause the Continental men too much of a delay.

They were drilling along the Nesson Anticline, a subterranean ridge stretching 150 miles. Hoffman and his colleagues figured oil might have accumulated along this bump. "Maybe we can get some help from Mother Nature," Hoffman told Hamm before they started drilling.

Hamm remained wary of spending a huge amount on speculative drilling in rock so dense that it looked like concrete, so he reached out to some friends to chip in and become partners in their effort.

He needn't have worried so much. The first well was another Bakken winner, and Hamm quickly ordered his landmen to lease even more acreage in North Dakota. It seemed the state was going to generate more oil than Montana, just as Hoffman and Stark had predicted.

But in 2005, after another well was drilled, the oil turned into a bare trickle and water flooded the wellbore. The same disappointing results were seen from additional wells. Their first North Dakota well had been a fluke.

"We knew there was something there but we couldn't get it to work," Hoffman says.

It turned out Bakken rock in North Dakota was very different from Bakken rock in Montana. When they tried putting a wellbore in the top layer of the shale, the rock caved in and collapsed. It was so brittle it crumbled in the hands of roughnecks in the field.

"We were shocked," Jack Stark recalls.

The way the men fractured the rock wasn't working. Their liquid concoction went anywhere and everywhere. Sometimes it even exited the Bakken formation, where it was useless or even opened up fractures in a water-producing zone. The results were frustrating because they wanted oil, not water, from their fracking efforts. Those manning the wells began calling them "Hail Mary fracks" because the odds of success were so low. It was like they were taking a sledgehammer to a surface and creating cracks all over the formation, including where they didn't want them.

The cost of all the leasing, drilling, and fracking began to add up to millions of dollars, and Continental made just $2 million in 2003. The company was producing around two hundred barrels a day in the area, and oil prices were zooming past sixty dollars a barrel, but each well cost about $6 million and the company had little to show for the huge expense.

In 2004, Continental ran out of cash for its Bakken project in North Dakota and Hamm decided to shut it down. He had to stop throwing good money after bad or he'd jeopardize his company. The decision caused grumbling from Hoffman and the rest of the staff in the area who believed in the rock's potential, but they knew Hamm had no choice.

"It was pretty frustrating," Hamm recalls.

At the same time, Hamm was dealing with growing troubles in his personal life. In 2000, a routine health screening showed that he had type 2 diabetes. He began to research the disease, eat more grains, vegetables, and salads, and donate money for research and prevention, aware the disease was a local scourge.

"It [is] a very pervasive disease, particularly in Oklahoma," according to Hamm. "All of us probably have Native American blood; there are thirty-nine tribes here. And it seemed to be more prevalent among [them]."[14]

Things got worse for Hamm in 2003 when his wife, Sue Ann, left their home in Enid and moved to Oklahoma City, a hundred miles away, with their two daughters. Friends said Sue Ann wanted to put her girls in private school in the bigger city. Harold would follow his wife to Oklahoma City, but their problems seemed to grow. Sue Ann filed for divorce in October 2005, though the case was dismissed two months later, according to court records.

Later, Hamm would say the couple separated in the fall of 2005, leading separate lives, according to court documents, though that would be disputed by Sue Ann. That year, Sue Ann began relying on electronic surveillance to monitor her husband's activities in their home, according to court documents, seemingly convinced he was having an affair, or a series of them.[15]

Hamm's personal life was in turmoil, he had run out of money for North Dakota's Bakken, and he risked losing his unhappy men to competitors as the energy industry sprung to life. He had to come up with a solution.

He swallowed hard and decided to sell half of his North Dakota position. He didn't like the idea but he figured he'd at least get enough money to enable his men to keep searching for a solution in the Bakken.

Continental hired an investment bank and reached out to twenty-five oil companies. Big companies, medium-sized companies, and smaller ones all were offered the Bakken holdings. Hess, Burlington, and others pored over the results Hamm's team had achieved. But no one cared and no one made them an offer.

Hamm lit into his bankers, convinced they were failing on the job.

Why else wouldn't rivals be interested in this choice land? The bankers redoubled their efforts but everyone kept passing on Continental's acreage.

"They didn't believe in it," Jeff Hume says.

There was one spot in North Dakota that the Continental team ignored. It was an 800,000-acre block in Mountrail County, near the town of Parshall and east of Continental's acreage. There were good reasons to dismiss the area.

In the late 1980s, Marathon Oil, a major company, had drilled at least five wells in the region but decided it was a loser. Barely anyone ventured there over the following decade. But in 2003, a seventy-seven-year-old geologist named Michael Johnson decided the time was ripe to give the spot a new look.

Johnson shared a similar family history with George Mitchell—they both were sons of poor Greek immigrants who had changed their names shortly after arriving in America. (Johnson's father, Efstathios Giannakopoulos, had adopted the name Sam Johnson.)

Michael Johnson was well aware that Mountrail County was considered the dregs of the Bakken. But the septuagenarian also knew horizontal drilling was helping all kinds of ugly fields become beauties. To Johnson, the area was a lookalike to the Elm Coulee field in northeastern Montana that was working so well for Continental and others.

In 2005, at the age of seventy-nine, Johnson, along with his partners, Henry Gordon and Bob Berry, leased forty thousand acres around the Parshall area. They paid as little as three dollars an acre, reflecting how worthless most judged this land to be. In one auction, they were the only ones even bidding.

The trio of energy-patch veterans didn't have enough money to drill all their land, which they called the Parshall Prospect, so they offered it to a dozen companies, including one called EOG Resources, run by Mark Papa, which had become active in the area. They hoped EOG would buy it from them and agree to give Johnson and his partners a small, retained interest.

Six years earlier, EOG had been a division of Houston-based power giant Enron that was called Enron Oil and Gas Company. As Enron grew, its executives became disdainful of Papa's unit, which was putting effort into searching for oil and gas rather than just trading it, which Enron viewed as a sexier and more lucrative business.

In the summer of 1999, Enron spun the unit off in what Papa calls "a very ugly divorce." After gaining its independence, EOG had a tough time earning respect from investors. By then, Enron was the hottest company in the country. If EOG was being discarded by Enron it must mean it wasn't worth owning, Wall Street figured.

"We were viewed as damaged goods," Papa recalls. "Something must be wrong with us, most people thought."

After Enron's collapse in 2001, investors warmed to EOG, but the company was no energy power. For that matter, Papa, a Pittsburgh native, was no natural-born wildcatter. In school, he only began to focus on petroleum engineering because he wanted to flee his cold city.

"It seemed most energy companies were located in warm-weather places, so that sounded good to me," he says.

Papa had a curious mind, though. By 2003, he was among those following George Mitchell's lead and getting natural gas from the Barnett Shale. He also grew excited about other spots that seemed full of gas. "You know, half of this country has shale under it," he told a conference in Boston in early 2004.

Papa began to wonder if this rock might yield oil as well. "Industry dogma was that pore spaces of shale rock were so small and oil molecules so much bigger than gas molecules" that it didn't seem possible that oil could flock from shale rock, he says. "We decided the prize of finding billions of barrels of oil" was worth examining whether that hypothesis was valid.

That year, while poring over pictures of a forty-foot-long shale-core sample that had been run through a CT scanner, Papa and his geologists detected a network of passageways. They determined that the pore spaces in shale were large enough to let oil flow, even though this rock looked as impermeable and thick as a slab of marble. Soon, EOG established a presence in the Elm Coulee field in Montana.

By 2005, EOG also had become a significant leaseholder in western

North Dakota. After hearing Johnson and his partners pitch their 40,000-acre block, EOG agreed to buy it and take a chance on the area.

EOG drilled its first well in Parshall in April 2006. At that point, Papa and his company were much more focused on searching for natural gas around the country than on finding oil. They were simply taking a flyer on Johnson's acreage.

The first well in the Parshall field was a huge disappointment, producing less than one hundred barrels of oil a day. The crew kept at it, though, and spectacular results burst forth almost immediately after the first clunker. Some of EOG's Parshall wells were true gushers, producing more than five thousand barrels a day, at least initially. Rising oil prices meant some of the best wells paid back their costs in less than a year, a remarkable turnaround. It turned out that crude trapped in the rock in the area was under especially high pressure, making it perfect for oil production.

Many of Johnson's leases were due to expire in less than four years, so the pace of EOG's drilling picked up. EOG was a public company, unlike Continental, so it had good access to money from equity and debt investors, aiding the effort. The company was especially tight-lipped about its results, partly because it was gun-shy after the embarrassing implosion of its former parent, Enron.

Still, some of EOG's results from the Parshall field leaked out, and a few industry members began to wonder if getting oil from the Bakken and other challenging shale formations in the country was within the realm of possibility.

Harold Hamm and Continental had struck out on their wells in North Dakota, just as the experts and rivals had predicted. Now they also had a serious competitor on their doorstep.

By 2003, Charif Souki was hustling to raise financing to build his LNG import facility in Sabine Pass, Louisiana, just over the Texas border. Cheniere shares traded a bit higher, moving up to three dollars a share. No one really thought Souki would be able to build such a huge and costly terminal, though.

Souki was about to get some good news from an unexpected source.

During the first half of 2003, members of Congress had become worried that rising natural gas prices would weigh on voters. The previous winter had been especially cold, and concern grew that another cold season would send energy prices higher. The politicians turned to Federal Reserve Board chairman Alan Greenspan, who had earned the nickname of "the Maestro" for conducting U.S. monetary policy, for advice.

Testifying on June 10, 2003, before the House Committee on Energy and Commerce, Greenspan said the country's situation was dire. The Fed chairman lamented the "seeming inability of increased gas well drilling to significantly augment net marketed production," and he bemoaned the spike in natural gas prices, which by then were at $5.28 per thousand cubic feet, up from $3.65 a year earlier and just $2.55 in July 2000.

"In the United States, rising demand for natural gas, especially as a clean-burning source of electric power, is pressing against a supply essentially restricted to North American production. . . . We are not apt to return to earlier periods of relative abundance and low prices anytime soon."

Greenspan's comments struck an immediate chord. His argument that it was getting harder to tap cheap gas supplies echoed growing concern that the United States, and the world itself, was running out of energy. Back in 1956, an irascible but brilliant Texas-born geologist working for Shell named Marion King Hubbert had created a model that seemed to prove the world was running out of oil and natural gas. Hubbert, who earlier in his career was active in a movement that pushed for democracy to be disbanded, allowing scientists and engineers to take the reins of government, had predicted that oil production would peak in the United States by the early 1970s and global oil production would level out around 2006.

"It's quite a simple theory and one that any beer drinker understands," explained prominent British petroleum engineer Colin Campbell, a Berlin native and a fan of Hubbert's theory. "The glass starts full and ends empty and the faster you drink it the quicker it's gone."[16]

For years, Hubbert's argument, a Malthusian notion that became known as the peak oil theory, was dismissed by the energy establishment. But when U.S. oil production *did* peak in 1970 and an energy crisis ensued, some began to suspect that Hubbert might have been on to something. By the time Greenspan issued his own warning, a consensus was emerging that the heyday of oil and gas production was over.

Greenspan had a solution, though. The United States needed to become a major importer of liquefied natural gas, he told Congress. The country then could get its hands on cheaper natural gas being produced around the world.

Then Greenspan uttered words that could have come out of Charif Souki's own mouth: "Access to world natural gas supplies will require a major expansion of LNG terminal import capacity," he said.

Alan Greenspan's speech changed everything for Souki. It was a seal of approval, out of the blue, as if Martin Scorsese had pointed to a struggling actor, ready to give up on Hollywood, as the next Marlon Brando. Within weeks, investors and potential customers gained confidence in Souki's strategy, helping Cheniere's stock jump to six dollars a share. If Greenspan backed the idea, then the regulatory process was unlikely to be very arduous, investors figured. And if Greenspan agreed the country was running out of natural gas and would have to import LNG, it had to be true.

"All of a sudden I didn't have to go through the first part of my spiel that gas prices were going past three or four dollars," Souki recalls. "I could skip the first forty-five minutes" of the pitch.

"People stopped thinking we were completely crazy," he says.

When Souki and his team approached the Federal Energy Regulatory Commission to seek permission to build their Louisiana plant, the agency said it hadn't seen such an application in years and needed time to study the issue. Souki was encouraged nonetheless. A few companies even expressed interest in acquiring Cheniere. Souki turned them down, more confident than ever that he could build his facility, maybe even a few of them.

In December 2003, when Michael Smith and his new company signed contracts with Dow Chemical and ConocoPhillips to turn their LNG into natural gas at Souki's original terminal, Cheniere stock climbed anew because it still owned 30 percent of Smith's facility.

Less than a year later, in the summer of 2004, Cheniere reached deals of its own with France's Total and the United States' Chevron to import one billion cubic feet of gas a day each for twenty years at Cheniere's Sabine Pass facility. According to the terms of the agreement, the two giant global energy companies agreed to ship liquefied natural gas to Cheniere's terminal. There it would be deposited into one of Cheniere's massive 170-foot-tall, or seventeen-story-high, storage tanks, each the size of a football field and almost big enough to contain Madison Square Garden.

Cheniere would return the supercold liquid to its gaseous form, or "regasify" it, using sixteen vaporizer modules that act like giant Jacuzzis to warm the gas. Then the natural gas would be injected into a pipeline connecting the terminal to nearby pipelines, allowing Total and Chevron to sell their gas throughout the United States. Cheniere agreed to charge each company $125 million a year to convert the one billion cubic feet of gas, the approximate amount needed to meet the heating and cooking needs of a city the size of Chicago for four days and about 3 percent of the nation's daily gas needs.

It was similar to lugging a pocketful of change to a Coinstar machine at a supermarket and paying a fee to have it converted to more useful dollar bills. The only difference was that the terms of the deals Souki had crafted were for the energy giants to pay Cheniere to provide the gas conversion service, even if it was never used. That was the only way Cheniere could afford to build the project.

Within months, Souki had gone from an obscure dreamer with an oddball idea to the head of a company that might save the country. By then, there were plans for forty new or expanded LNG terminals under consideration in North America. Cheniere set the price terms that others copied, though, and Souki was at the vanguard of an emerging LNG revolution.

Investors who once shunned Souki's company now threw money at it. By the end of 2004, Cheniere's stock had soared to more than sixty dollars a share. The company quickly sold $300 million of new shares and borrowed $800 million through the sale of debt, putting it on its way to raising the financing necessary to build its LNG receiving terminal. In March 2005, the company broke ground on its facility and split its stock in half, a sign of the continued exuberance of investors. Finally, Souki was on his way.

When two huge hurricanes, Katrina and Rita, hit the Gulf Coast in the summer of 2005, Cheniere's construction efforts were set back, as the design was altered to prepare for future big storms. But the hurricanes also sent U.S. natural gas prices soaring, encouraging additional energy producers to examine ways of shipping gas from foreign locales to the United States. Soon, Souki was meeting senior executives at Exxon, Conoco, Shell, and other major companies about using Cheniere's prospective site to regasify LNG the giants wanted to shift to sell in the United States.

Souki excited Wall Street investors and analysts by saying his company would build a string of LNG terminals around the country. The United States was producing nearly fifty billion cubic feet a day of natural gas, and expectations were that the level would drop to nearly forty billion in just five years. Cheniere's imported gas would be crucial to keeping the nation's lights on and its cities warm.

By the end of 2005, it didn't seem as if anything could stop Charif Souki. As the company's largest shareholder, he now owned shares worth over $100 million. He bought a 4,000-square-foot home in Houston's tony Memorial Park neighborhood, some real estate in Colorado, and his own eleven-seat Bombardier Challenger jet.

"It all felt great," Souki says.

CHAPTER TEN

Aubrey McClendon never wanted Tom Ward to quit Chesapeake Energy. McClendon always figured their disagreements could be overcome. He never had a true understanding of the stress Ward was feeling and was astounded when he turned in the company keys.

Now that Ward was gone, however, McClendon had a new freedom to run the company exactly as he wished. Ward had dragged his feet on many of McClendon's expensive moves. Now McClendon had the green light to buy even more acreage in shale formations.

It still wasn't clear whether drilling in this rock would pan out. Major oil and gas companies still ignored most shale formations, and for good reason. Some, such as the Barnett, were producing impressive amounts of natural gas, but these fields weren't exactly changing the world or even the country. Less than 5 percent of the nation's gas production came from shale fields in 2005.

That didn't stop McClendon from upping his bet. In 2006, Chesapeake spent nearly $4.3 billion on acquisitions, most of it on acreage that had no proven production, the kinds of deals Ward was most uncomfortable with.

Chesapeake drilled over a thousand wells that year, while adding nearly $2 billion more debt to push its debt load up to $7.4 billion. Not only that, but much of the acreage Chesapeake purchased was not yet "held by production." In other words, the company had to drill productive wells or the leases would revert to landowners. That put pressure on Chesapeake to speed up its drilling.

By early 2007, some Chesapeake investors had become uncomfortable with McClendon's land rush. He agreed to slow things down. But McClendon couldn't resist when he met a local petroleum engineer named Ronnie Irani at an IHOP restaurant in January 2007.

Irani placed a group of maps on the table in front of McClendon and explained why he was so excited about an area in Wyoming called the Powder River Basin, a field he was convinced could produce more than five billion barrels of oil from the Niobrara Shale formation.

It was a "Bakken look-alike," McClendon says.

McClendon was smitten. At that point, less than 10 percent of Chesapeake's production came from oil. Irani was presenting McClendon with a chance to become a true energy king by adding oil to his huge position in natural gas. He called his assistant to cancel his next meeting so he could keep talking to Irani.

Chesapeake and Irani reached a secret deal to buy one million acres in the formation, relying on third-party brokers to try to hide their interest. Word of Chesapeake's buying leaked out nonetheless, and prices soared. Land that sold for as little as eleven dollars an acre in early 2007 now rocketed past $900.

McClendon was swinging for the fences, like a true wildcatter.

"When you look at the sweep of history in this industry," McClendon later explained, "those who move first to lock in big new acreage positions when technology changes emerge as the winners."[1]

Nevertheless, members of Chesapeake's board of directors began to raise questions about the headlong push into shale formations. Conventional wisdom held that it was becoming harder, not easier, to produce oil and gas in the country and the world. While shale wells seemed to show impressive early results, the production seemed to decline quickly. Betting on America and this challenging rock appeared risky, and the debt was piling up on Chesapeake's balance sheet.

Charles Maxwell was among those with concerns. The seventy-year-old energy industry analyst, who had joined Chesapeake's board of directors in 2002, was a firm believer in the peak oil concept. Maxwell wasn't convinced Chesapeake's costly acquisition spree would produce a surge in gas.

"I spent eleven years at Mobil Oil and my big-oil crowd was unconvinced" about whether meaningful amounts of gas could come from shale at reasonable cost, Maxwell recalls.

McClendon brought in evidence from the field to try to show that new techniques were allowing Chesapeake's exploration and production team to extract natural gas like never before. McClendon agreed that others were finding it harder to produce energy. But he argued to Maxwell and others that Chesapeake was capable of extracting significant new supplies of gas.

"We can do it and it will be worth it for us," McClendon insisted to the board.

With prices sure to rise further, McClendon said the company had to seize a "once-in-a-lifetime" opportunity by making more deals. Not only that, but natural gas produced almost half the carbon dioxide tonnage of comparable amounts of coal and two-thirds as much carbon dioxide as oil, making it cleaner.

Maxwell and other board members eventually came around. "This felt like the beginning of a whole new industry," he later wrote.[2]

There was only one problem: McClendon was spending more money than Chesapeake had, and more than his chief financial officer, Marc Rowland, had allocated to new purchases.

Rowland delivered the bad news to McClendon in the spring of 2007. "Aub, we're about to run out of money," he told McClendon.

"Okay, go do a billion-dollar issuance," McClendon said.

With that, Rowland got in touch with global banking giants Credit Suisse and UBS. They quickly reached out to a group of mutual funds and other large investors. Almost immediately, the investors committed to forking over $1 billion to buy Chesapeake's senior debt. They had such faith in McClendon and the Chesapeake team that they didn't ask to meet the Chesapeake executives or even require them to hold a conference call to explain why Chesapeake was a safe credit.

Marc Rowland knew investors were running for their checkbooks. But the Chesapeake CFO also knew the company was taking in less than it was spending to lease land and drill on it. And he was well aware that McClendon wanted to spend more to expand in shale and other acreage. Chesapeake would have to find new ways to raise cash from investors, he concluded.

Rowland and McClendon arranged a phone call with McClendon's fraternity brother at Duke, Ralph Eads, who was a senior investment banker at Jefferies & Co., to try to come up with an answer to their problem.

"I can get the assets," McClendon said. "You have to get the money."

Working with a group of Wall Street bankers, they turned to a sophisticated financing scheme called a volumetric production payment, or VPP. In this kind of deal, Chesapeake would receive up-front cash in exchange for oil and gas production to be delivered over a number of years to a vehicle set up by Wall Street investment banks. The banks would sell interests in the vehicle to hedge funds and other banks.

Chesapeake's banks reached out to investors to test their interest in stakes in future production from some gas wells in Kentucky and West Virginia. McClendon and Rowland hoped to pull $550 million from the deal. There was so much demand, though, that Chesapeake pocketed a cool $1.1 billion.[3]

Chesapeake began turning to VPPs on a regular basis, raising over $6 billion from nearly a dozen of these kinds of deals. More than ever, McClendon and Chesapeake had become reliant on Wall Street bankers and investors since the company wasn't taking in enough cash to meet its spending needs.

McClendon assumed an even more public position, emerging as a spokesman and cheerleader for natural gas, fracking, and shale drilling, a shift that helped the company attract even more investors. While many energy executives try to keep a low profile, worried about being tarred as greedy polluters, McClendon embraced the spotlight, extolling the wonders of "clean-burning, domestically produced onshore natural gas."

Chesapeake adopted a new tagline, "America's Champion of Natural Gas," and erected billboards in the Fort Worth area featuring actor Tommy Lee Jones. The company even made plans for an online television station to promote their gas drilling efforts, called Shale.TV. The more McClendon could help grow demand for natural gas, the more it would help Chesapeake's bottom line while also sparking enthusiasm for the company on Wall Street.

McClendon saw himself as a booster of America and its natural resources. He confessed to never having visited Asia, Africa, or Australia.

"I'm as parochial of a guy as you'll find. . . . There's a lot to see in the U.S. and I've been satisfied with that," he said.[4]

McClendon also seemed to enjoy poking fun at giant oil companies, who still scoffed at shale formations in the United States, even as they cut deals with foreign governments to explore abroad. "When we go to bed every night, our assets aren't subject to a coup or a new tax regime or something like that," he told a reporter.[5]

McClendon was an all-American energy star, representing a new wave of wildcatters absolutely sure they would shake up the country, and the world.

As McClendon's confidence and ambition grew, so did his appetite for extravagance. Growing Chesapeake was his focus, but building Oklahoma City into a world-class city became a new passion. In the summer of 2006, McClendon and Ward joined forces again, this time to back a local friend and power broker, Clayton Bennett, in a $350 million deal to buy the Seattle SuperSonics pro basketball team.

At the time, Bennett claimed to have no interest in moving the team to Oklahoma City, assuaging residents of Seattle scared of losing the franchise. But McClendon, a 20 percent owner of the team, couldn't help himself when the *Oklahoma Journal Record* asked him about the purchase. "We didn't buy the team to keep it in Seattle; we hoped to come here," he said. "We know it's a little more difficult financially here in Oklahoma City, but we think it's great for the community and if we could break even we'd be thrilled."[6]

For speaking his mind and acknowledging that the owners held out hope of moving the team after all, National Basketball Association commissioner David Stern slapped McClendon with a record $250,000 fine. Two years later, the team did move to Oklahoma City.

McClendon wanted his company to be first-class and he wanted to help Oklahoma City gain its own recognition. The city had gone through its share of difficult times. In 1995, Timothy McVeigh, a man brimming with anger at the U.S. government, set off a bomb so powerful that it destroyed Oklahoma City's Alfred P. Murrah Federal Building, as well as destroying or damaging over three hundred buildings nearby, taking 168 lives, including nineteen children under the age of six.

McClendon cared about Oklahoma City and was determined to help it reinvent itself as a cultured and world-class city. And he also knew that as Oklahoma City's reputation improved, Chesapeake would be able to recruit better talent. It also didn't hurt that McClendon received kudos and thanks for his focus on the city.

"You can't attract first-rate employees, especially young ones," McClendon says, "if your city is a dud."

McClendon turned Chesapeake's campus into one that a Silicon Valley mogul might envy. There was a 72,000-square-foot fitness center, an Olympic-size swimming pool, and an elaborate health center. Young, clean-cut security men patrolled the grounds, questioning visitors and lending the campus a Big Brother air.

The company bought local stores and built an upscale shopping venue called Classen Curve, featuring a gastro sports pub and a gourmet restaurant serving raw vegan fare. McClendon even persuaded organic food chain Whole Foods to build a store in the area. Chesapeake would spend over $240 million to buy at least 109 local properties, according to data compiled by Reuters.

Local politicians couldn't have been more appreciative. "We don't have the money to do it and he does," said Sody Clements, the mayor of Nichols Hills, the suburb of Oklahoma City that was Chesapeake's home.[7] The *Daily Oklahoman* called McClendon a "one-man economic boom."

One day in 2007, Art Swanson, an acquaintance from high school, approached McClendon with a grand idea. "Every billionaire owns a golf course," Swanson told McClendon. "What you need is a private racetrack."

McClendon loved the idea and was agog when Swanson introduced him to Fritz Regier, a driver from Porsche's factory in Germany, who came to Oklahoma to discuss the project.

Soon McClendon was buying up tracts of land in Arcadia, Oklahoma, half an hour from Oklahoma City, to serve as his very own racetrack. He still wanted that golf course, though, so he hired famed golf-course ar-

George Mitchell, the father of shale fracking, celebrating his company's three thousandth well in 1976. Mitchell and his team would spend seventeen frustrating years trying to fracture challenging rock so that gas would flow.

Cliff Roe

Mitchell, who in 1998 finally figured out how to get significant amounts of natural gas from shale, after selling his company for $3.1 billion.

Baylor School of Medicine

Aubrey McClendon, cofounder of Chesapeake Energy, at a well in Texas's Eagle Ford Shale formation. McClendon stood out for his vision of a shale revolution as well as comfort with heavy debt and aggressive trading.

Tom Ward, cofounder of Chesapeake Energy and founder of SandRidge Energy.

Harold Hamm in front of his childhood home in Lexington, Oklahoma.

Harold Hamm and his wife, Sue Ann, at a 2012 event honoring *Time* magazine's one hundred most influential people.

Charif Souki at his company's Louisiana LNG terminal.

Hydraulic fracturing creates pathways for oil and gas to flow from shale and other types of compressed rock.

Liz Irish (far right) and an oil-drilling crew in North Dakota's Bakken region.

A drilling rig near the Badlands of North Dakota.

Left to right: Buck Butler, the author, and Butler's son Rodney.

Doug Duncan, USGS

A drilling site in Pennsylvania's Marcellus Shale formation.

Geologist Terry Engelder, who shocked the world by determining how much gas was in the Marcellus Shale, in front of the home of a fracking fan in the region.

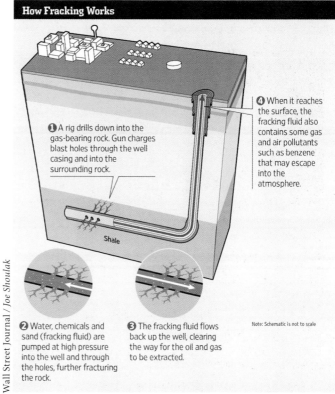

How Fracking Works

❶ A rig drills down into the gas-bearing rock. Gun charges blast holes through the well casing and into the surrounding rock.

❹ When it reaches the surface, the fracking fluid also contains some gas and air pollutants such as benzene that may escape into the atmosphere.

Shale

❷ Water, chemicals and sand (fracking fluid) are pumped at high pressure into the well and through the holes, further fracturing the rock.

❸ The fracking fluid flows back up the well, clearing the way for the oil and gas to be extracted.

Note: Schematic is not to scale

Wall Street Journal / Joe Shoulak

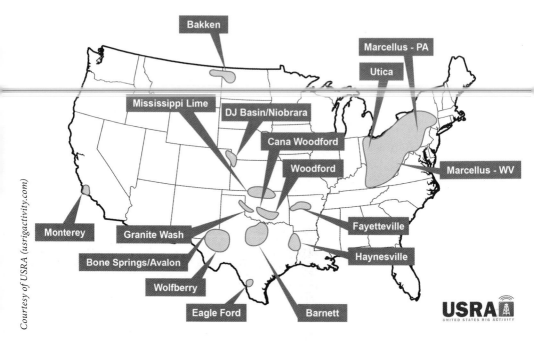

Significant oil-and-gas shale formations in the United States.

chitect Tom Fazio to design a course for the racetrack's infield. They began work on creating an American version of the Nürburgring, the motorsports complex around the German village of Nürburg.

Natural gas prices drifted a bit lower in the summer of 2007 amid signs of growing supplies. Chesapeake trimmed its gas production by 6 percent in September, joining some other gas producers in easing production. McClendon told investors that he would slow the company's spending once again, promising to reduce capital spending by 10 percent in 2008 and 2009.

"It's now prudent to pull in the reins and let the market rebalance," McClendon told analysts in a conference call.[8]

At the time, most experts were sure falling prices and rising supplies had resulted from the relatively mild weather that summer and the previous winter, which reduced demand for gas for air-conditioning and heating. Some thought the price weakness might also be attributable to consumers and businesses that were easing their reliance on gas after several years of high prices.

In September, the government reported that domestic gas inventories hit nearly three trillion cubic feet, a record high for that time of year. Few thought anything remarkable was going on, though. Supplies seemed high because demand was weak, not because anything special was happening in the nation's shale gas production. The previous year had also seen a temporary glut of gas, and natural gas prices soon rebounded. It seemed far-fetched to think any kind of revolution might be brewing that the experts were missing.

Sure enough, natural gas prices rebounded by late 2007 while oil prices surged, amid fears that demand was outstripping supply.

In late 2007, shale-acreage prices shot into new territory. Chesapeake began competing with an upstart company called Petrohawk run by Floyd Wilson, another energy veteran. McClendon and Wilson both relied on Wall Street for funding, and both were convinced that a field straddling northern Louisiana and eastern Texas called the Haynesville Shale would produce huge amounts of gas. They began a bidding war, fighting to lock up attractive Louisiana acreage, convinced a surge of gas production was around the bend. There was no turning back for McClendon and Chesapeake.

T om Ward suffered withdrawal pains after quitting Chesapeake Energy in early 2006. He desperately missed the adrenaline rush that came from searching for energy deposits, not to mention the potential financial windfalls. Almost immediately, he decided to restart his career. He was forty-six years old, certain that overlooked gas deposits remained in the United States, and eager to prove he could build his own energy empire, without any help from Aubrey McClendon.

Ward took an office in Oklahoma City's Valliance Bank Tower, one of the city's few skyscrapers, and began searching for ways to invest his cash hoard. Within weeks, he reached an agreement with Malone Mitchell, an Amarillo, Texas, oilman, to spend $500 million to buy effective control of Mitchell's company, which boasted a huge asset, the Piñon gas field in West Texas.

Ward was back in the game, but he was determined to chart a very different course from McClendon, who was chasing gas in shale. The Piñon field held a lot of gas, but it was in chert, a "conventional" kind of rock that wasn't as challenging and costly to drill as shale. Ward renamed his company SandRidge Energy to signal that he was going after sandstone and other types of rock that had been drilled for decades. Let McClendon and Chesapeake spend all that money on shale, I'm going to find enough energy in old-fashioned rock, Ward was saying.

Several months later, Ward approached Carl Icahn, the billionaire New York investor, who had cobbled a collection of energy companies into an entity called National Energy Group. Ward asked if Icahn was interested in merging his operation with Ward's new company.

Icahn was hesitant. He liked Ward on a personal level and thought he was bright, but Icahn was wary of all the money Ward and McClendon had spent at Chesapeake. "Their spending is just *way* too much," Icahn had told someone in the energy business at one point.

After the lukewarm response, Ward came back to Icahn with a different idea. "Let me buy you out," he said.

Icahn hadn't been looking to sell his company, but Ward seemed so eager to buy his energy assets that Icahn decided to throw a huge number out and see if Ward went for it. Icahn didn't think his assets were worth more than $1 billion, so he came up with a much larger number, to test Ward's level of interest.

"A billion and a half," Icahn told Ward, according to someone in the industry.

Ward didn't flinch. "I'll get it for you," he told Icahn.

Eventually they agreed that Icahn would get $1.2 billion in cash and a $300 million stake in SandRidge. The price was so high, and Icahn was getting so much in cash, that he didn't mind retaining a stake in Ward's company, at least for a little while.

Ward had been the quiet one in Chesapeake, silently plotting the company's strategy at various executive meetings. Now he proved outgoing, convincing, and surprisingly smooth as he sold his new company's story to Wall Street. Investors lined up to provide the necessary financing for his deal with Icahn.

"He's not as quiet as people think," McClendon told a reporter for Oklahoma City's *Journal Record*.

Soon, investment banks were knocking on the company's door to help underwrite an initial public offering of Tom Ward's new company. Some saw it as a way to bet on a second coming of Chesapeake. Icahn remained wary, though. He decided to sell his 20 percent stake as part of SandRidge's IPO.

In the early fall of 2007, as Ward and his new team counted down to SandRidge's IPO, Ward called Icahn with some news. "I want to be straight with you," he told Icahn. "The stock is valued at eighteen dollars a share, but it's a hot issue, I think it will open at thirty dollars."

"That's amazing," Icahn responded.

Ward was going out of his way to caution Icahn that he was leaving money on the table because the billionaire planned to sell his shares at eighteen dollars in conjunction with the IPO. Icahn said he still wanted to sell, though. If he didn't sell in the IPO, he knew securities laws barred him from getting rid of any stock for the first six months after the offering. He had no interest in taking the chance on holding SandRidge for that long. He decided to sell the stock.

Sure enough, Ward's company debuted above thirty dollars a share, giving it a market value of about $3.5 billion. Tom Ward's shares were worth $1.2 billion. It was clear that Ward didn't need Chesapeake Energy or Aubrey McClendon.

Harold Hamm had run out of money to drill in the Bakken. He gave up on the idea of selling portions of his acreage to raise crucial financing, and he faced growing competition from EOG, a publicly traded company.

Rather than abandon North Dakota altogether, Hamm decided to go on drilling and fracking but to do it at a much slower pace. Continental would conserve cash and only spend from extra money generated by the rest of its business.

There was little reason to expect any kind of breakthrough. By the spring of 2006, U.S. oil production had dropped to about 5.4 million barrels a day from 9.6 million barrels in 1970 and the country imported a record 60 percent of its oil. President George W. Bush set a goal of replacing 75 percent of the nation's Mideast oil imports by 2025 with ethanol and other energy sources. Even Bush—a Texas oilman—wasn't willing to waste his breath urging more U.S. oil production.

"America is addicted to oil," the president warned in a major speech in 2006.

By that point, the Montana Bakken was producing forty-eight thousand barrels of crude a day from more than three hundred wells, making it the highest-producing onshore field discovered in the lower forty-eight states in more than five decades. But most major oil exploration companies saw the area as an aberration and had little interest in what was going on there or in North Dakota.

All the negativity should have discouraged Hamm. But he had studied what George Mitchell went through before his own breakthrough in the Barnett Shale, and the hurdles Mitchell had overcome inspired him. American technology had transformed dozens of industries; Hamm was convinced a breakthrough could transform the energy business, too.

"We have to go back to experimenting," he told his men, hoping they somehow would manage to create fractures, enabling oil to flow in North Dakota.

When Mike Armstrong, Hamm's partner on hunting and fishing expeditions as well as on some North Dakota wells, cast doubt on their drilling, Hamm told him not to give up hope. "Mike, it's a technology play . . . the technology will catch up with this," Hamm told him.

Armstrong didn't buy it. He had spent more time in North Dakota than Hamm, and the disappointments had embittered him. He kept his leases but didn't invest more money. It didn't seem worth the expense. The failures "shattered my confidence in the area, I was chicken shit," Armstrong says. "I was at a real low point."

In early 2006, the *Wall Street Journal*'s John Fialka began working on a story about the Bakken production. Colleagues snickered, viewing drillers in the region as wide-eyed optimists. "The whole energy group at the *Journal* told me I was nuts," Fialka recalls.

Fialka convinced editors to run the piece on the paper's front page, where it appeared in April of that year. In the article, Fialka predicted that "the Bakken could eclipse Alaska's Prudhoe Bay as the largest recent U.S. oil find."

But the article focused on eastern Montana and the work of Richard Findley and a few smaller companies. The 2,100-word story didn't even have a stray mention of Hamm or Continental, though it mentioned its rivals.

"What's unusual about this boom . . . is that small companies like Headington and Lyco have most of the key acreage tied up in leases," Fialka wrote.[9]

Hamm's team still couldn't figure out how to effectively fracture the rock in the North Dakota Bakken. Each time they shot their mixture of water, sand, and chemicals into the formation they managed to cause fractures in the rock closest to the vertical part of the wellbore, just after it curved and went horizontally into the ground. But the fluid wasn't reaching rock farther out. It was hitting the "heel" of the well, in the lingo of the drilling team, but not the "toe." The fracking fluid shot anywhere and everywhere in the rock and didn't make it to the end of the wellbore, resulting in too little oil flowing to the surface.

Hamm liked to mix it up with his geologists and engineers, and he loved keeping in touch with others in the business to track the latest technology being developed. One day, he told Hoffman and his colleagues about a fracking method he heard rivals like EOG and others were beginning to use in the fields. It was one that Nick Steinsberger and others in the Barnett had experimented with a few years earlier, but now a greatly improved version was catching on in the Bakken area. Instead of just drilling a horizontal well and sending fluid into the rock, hoping against hope that it created proper fractures, some drillers were using an innovation Mark Papa's engineers at EOG had invented to seal off certain zones in the wellbore so fluid could be focused on specific areas in the rock.

Once a section was sealed, or "packed" off, with a special gasket or other material that swelled to close it off, fracking fluid was pumped in to create a concentrated liquid shot to fracture rock in a select, pressurized area. Once that was done, drillers would focus on the next section of the well, which they'd seal off with another "swell packer" that would inflate to block the fracking fluid from reaching other sections of the well, to give the new area its own pinpointed dose of fracking fluid. Then they'd move on to still another segment in the well, a few hundred feet away, and do it all over again.

The idea was to isolate a section of the horizontal wellbore, frack it, and get oil flowing. After a dozen or more "stages," the fracking would be complete. By segmenting the well and fracking it in stages with concentrated fluid, drillers were getting impressive amounts of oil from dense rock that had proven especially stubborn, like the Bakken shale.

The new method was expensive. Drilling and fracking a North Dakota well in this new way cost about $8 million, compared with less than $4 million for a well that wasn't fracked in stages. But Hoffman and the Continental group instantly understood how much more crude they might be able to get.

Around the same time, specialists were developing ways to drill horizontally for as much as two miles, or about twice as long as they had been doing, without losing track of the shale layer. This "long lateral" was a perfect complement for the new, multistaged fracking.

Nicholas Steinsberger and his colleagues at Mitchell Energy had developed the ideal fluid to fracture shale and other tough rock a few years earlier in the Barnett. Others, including specialists at Oryx Energy, had perfected methods to drill horizontally to get more oil and gas from shallow, wide rock formations. By combining fracking and horizontal drilling techniques, early adopters like Devon and Continental saw oil and gas production soar.

Now Continental's pros realized that a new leg in the industry's technological revolution had arrived: a way to get oil and gas to flow by fracturing even cementlike rock with short, concentrated blasts, thanks to EOG's multistaged fracking innovation.

In late 2006, Continental's engineers began to employ the new methods to stimulate the Bakken shale and limestone. From the outset, they got much more oil from the wells, encouraging the executives. They also seized on an intriguing formation directly below the Bakken called the Three Forks. It was too early to be sure, but this formation also seemed full of oil. During a meeting in a Continental conference room, a geologist wrote "New Upside!" in bright red pen on a map of the new layer, a sign of their growing excitement.

In 2006, Continental made $253 million from its oil and gas production. Less than 10 percent came from the Bakken, but it was enough for Hamm to pursue an initial public offering. Hamm knew his company would need cash down the road to drill the Bakken. Being a public company would help it get better borrowing rates, while also enabling the company to compete with EOG and others for employees. An IPO also would enable Hamm to cash in on some of his holdings in the company, a reward for four decades of hard work.

As Hamm, Hume, and Stark met with mutual funds and other investors, the executives began to field tough questions. Continental was still a small company with fewer than three hundred employees, and it needed oil prices of about fifty dollars a barrel to make its expensive Bakken drilling worthwhile. By late 2006, crude prices were above sixty dollars a barrel, but there was no telling if they'd slip again.

There was another problem with the Bakken's oil production. There weren't any wells flowing with 10,000 barrels of crude a day, let alone 100,000 or more, like some famous foreign and off-shore fields. A few started at around four thousand barrels a day, but that was rare. After a burst of about eight hundred barrels of light sweet crude a day during their first year of production, most Bakken wells settled at between two hundred and eight hundred barrels of crude a day. The falloff raised questions about what the wells would look like down the road.

A few pros realized the Bakken wells were steady and dependable producers, rather than gushers, much like those producing oil and gas from other tight and dense rock, like shale. Continental's engineers expected the Bakken wells to produce more than one hundred barrels a day for many, many years. They also knew the vastness of the Bakken basin meant wells could be dotted throughout the territory, boosting the accumulation of crude.

That's all well and good, some investors said. But Continental was only producing seven thousand barrels of crude a day from the Bakken. Almost all of its 342,000 acres remained undeveloped. "What evidence do you have" that production won't slow to a trickle in the Bakken? a mutual fund executive asked Hamm at one meeting.

Stark and Hume received encouraging results from staff in the field as they joined Hamm in meetings with investors during the ten-day road show in early May 2007. But it wasn't the kind of formal data the Continental executives could share with investors. Passing a BlackBerry device with raw corporate data to a hedge fund executive during a meeting isn't considered proper, let alone legal.

"We knew production was growing," Hume recalls. "But they couldn't see it, and they didn't know if it would grow."

Other investors doubted the new drilling and fracking technique could be applied successfully throughout Continental's North Dakota acreage. "Is it repeatable?" an investor asked Hamm and his team during another meeting.

Halfway through the road show, Hamm seemed to have won over most investors. The two investment bankers from Merrill Lynch leading the sales effort had introduced Hamm to investors as one of the nation's last wildcatters. Hamm's plainspoken, straightforward speeches earned him high marks with prospective investors, and the bankers said an IPO was expected at between seventeen and eighteen dollars a share.

Back at Continental's headquarters in Enid, key staff members like Brian Hoffman counted down to the IPO. They were set to receive shares as part of the offering. Word within the company was that there would be so much demand for the stock that it was sure to begin trading above twenty dollars a share. Maybe it would go even higher, Hoffman thought, like other hot IPOs.

Things were looking up when the Continental executives and their bankers reached Denver for a dinner at the elegant Brown Palace Hotel and Spa after another investor meeting. That's when a phone call to the Merrill Lynch banker interrupted their meal. They were told that public companies Bill Barrett Corp. and St. Mary Land & Exploration had run into trouble drilling in the Bakken. The two bankers, Christopher Mize and Aaron Hoover, leveled with Hamm—it was really bad news for Continental's IPO. Now it was going to be hard to sell the shares.

All eyes at the table shifted to Hamm.

"That's some luck," he responded glumly. He looked discouraged.

"Do we keep going?" one of the Continental executives asked him.

The bankers said it was up to Hamm. A poorly received IPO would be an embarrassment that could weigh on the company and make it harder to raise financing at attractive rates.

Hamm didn't answer. He couldn't believe what was happening. He knew their drilling was going to work. Continental was targeting the Nesson Anticline, a subterranean ridge that seemed full of oil. The two rivals with the bad luck were west of the anticline, Hamm knew.

Hamm returned to his room for the evening, weighing whether to go forward with the offering. The next day, he decided to proceed with the IPO. They had come too far to turn back now. "Guys, let's go ahead," he told his colleagues. "We'll have to show" Wall Street.

Continental succeeded in selling its shares, but only at a reduced price of fifteen dollars a share. "We felt like we were getting there," but investors didn't seem to agree, Hamm recalls. "It was a bitter pill."

The lowered price disappointed Hamm and his staff. But it was a shock to Hoffman. When the stock fell to $14.10 on the very first day of trading, Hoffman became really concerned. He knew all about geology, but not nearly as much about the stock market, so he walked over to Continental's legal counsel to ask why investors were turning their backs on the company's new shares.

The lawyer, who didn't seem any more of a stock market expert, told Hoffman the subdued reaction was due to lower commodity prices, even though oil prices were rising at the time, not falling.

Hoffman, who had introduced the North Dakota Bakken to Harold Hamm, immediately sold all his shares. Just like investors, he held serious doubts about Continental and its future, even though they seemed to be making headway. After speaking with the lawyer, Hoffman decided the "shares wouldn't go any higher because the hype was over," he recalls.

Shortly after the IPO, Hoffman quit Continental, though it mostly was to get away from living in Enid, the company's headquarters. "Enid was just too small a town," he says. "Everyone knew each other's business."

Hoffman never received any kind of special bonus for encouraging Hamm to push into the North Dakota Bakken, nor did he get any kind of thank-you from the Continental chief executive, he says. Soon, however, he was on a flight to the Rockies to work for another energy company that let him live in Denver, fulfilling his dream of moving to that city. He made enough money from the proceeds of his sale of Continental shares to afford a down payment on a home.

Hoffman had finally reached Denver. But Harold Hamm's own goal of finding enough oil to change the country and creating "ancient wealth" for himself seemed as fanciful as ever.

Charif Souki and Cheniere seemed set to help satiate America's growing hunger for natural gas. The company had inked deals with Total and

Chevron to convert two billion cubic feet of LNG per day to natural gas. Souki owned Cheniere shares that were worth over $120 million, and he was fast becoming the toast of Houston's business establishment.

By 2006, the idea that the world was in danger of running out of energy had grabbed hold of the nation's psyche, despite the enthusiasm of McClendon, Hamm, Papa, and others for shale drilling. By then, the United States was importing about 16 percent of its natural gas, close to an all-time high.

Those subscribing to the peak oil theory included most mainstream politicians, businessmen, and investors. A flood of new books appeared, some predicting deep economic troubles, maybe even popular unrest, anarchy, starvation, and disease, all triggered by expected oil and gas shortages. Some of the semi-apocalyptic titles included: *The Long Emergency, High Noon for Natural Gas,* and *The Empty Tank.*

"We face an era of scarcity that involves higher prices for oil and fiercer competition for what's left," argued Peter Maass in *Crude World: The Violent Twilight of Oil.* "We are a foggy-headed boxer on his knees, unaware of the blow that awaits us."

Another book, the best seller *Twilight in the Desert,* by Matthew Simmons, a wealthy investment banker and adviser to George W. Bush during his first presidential campaign, argued that Saudi Arabia "clearly seems to be nearing or at its peak output." Some experts, such as Daniel Yergin, the Pulitzer Prize–winning author of *The Prize,* disputed the peak oil theory, saying it was based on incomplete data. He was in the minority, however.[10]

With Russia assuming the mantle of the world's biggest natural gas producer, President Vladimir Putin spoke of using his nation's energy resources as a political weapon. Around the same time, Chevron made plans to sell oil and gas properties in the United States, including those in Texas's Permian Basin, to focus on what it believed was more attractive acreage outside the country, another sign of how energy power seemed to be permanently shifting away from North America, no matter what the shale enthusiasts were saying.

With all the doom and gloom, Charif Souki realized he had an extraordinary opportunity to bring relatively inexpensive gas to his energy-hungry country. The plant Cheniere was building in Louisiana was slated to convert 2.6 billion cubic feet of LNG each day into natural gas, or over 4 percent of the nation's daily gas demand. Cheniere already had sold two billion cubic feet of that capacity to Total and Chevron, but the country's hunger for gas was such that Souki was sure he could find new customers if he built more capacity at the import facility.

In 2006, Souki and his team decided to expand the Sabine Pass terminal to allow four billion cubic feet of LNG to be converted to gas each day, or nearly 7 percent of the nation's daily needs. At that size, the facility would be able to receive about five hundred LNG ships each year, though the expansion also meant pushing the cost of the plant up to $1.5 billion, with another $500 million needed for the pipeline connecting to the facility.

Cheniere borrowed money to pay for the expansion, pushing its debt to a whopping $2 billion. It seemed worth it, though. The $250 million Cheniere was going to get each year from its first two customers was nearly enough to meet its annual debt payments. If they could sell the full four billion cubic feet of capacity, Souki knew the company would be solidly profitable.

Souki didn't lose much sleep over the decision. U.S. demand for natural gas was growing and foreign companies were furiously building natural gas production. Their supplies could easily be sold to the American market. Even after the expense of shipping the liquefied natural gas to Souki's terminal in Louisiana and converting it back to gas, the foreign companies seemed sure to score profits because U.S. natural gas prices were so high. If Souki built his bigger plant, he was sure the foreign gas would come.

Souki anticipated limp domestic natural gas production and strong U.S. prices. He had good company. In February of 2007, private equity powers KKR and Texas Pacific Group, along with investment bank Goldman Sachs, joined forces to pay $45 billion to acquire a huge Texas utility called TXU Corp. The deal, the largest leveraged buyout in history, was a mammoth bet that natural gas prices, already past seven dollars per thousand cubic feet, were headed higher, likely pushing up prices for wholesale electricity. Famed investor Warren Buffett spent about $2 billion on TXU bonds—even he seemed to believe energy prices would remain strong.

In some ways, Souki had no choice but to expand the terminal. Investors had bid up shares of Cheniere in the expectation that more contracts would be sold and more plants would be built. If Souki veered from the vision he had sold Wall Street, his stock would tumble. He couldn't shift gears now.

An odd thing happened in the second half of 2007, however. As the Cheniere staff tried to find new customers to ship LNG to America, they began getting the cold shoulder. Meg Gentle, the head of strategic planning, found it almost impossible to sign up global natural gas producers, as if foreigners were becoming wary of sending gas to the U.S. market.

"The mood had changed," Gentle remembers. "Something seemed to be happening."

Gentle got in touch with an industry consultant who said U.S. natural gas production was slowly growing, after bottoming out in the summer of 2005, perhaps helped by production from new shale formations. Not only that, but the cost of producing gas from shale formations was dropping sharply, and in some places was lower than the cost from more conventional rock formations.

"That's nuts," Gentle responded. She told the consultant that it was "geology 101" that the cost of producing gas from compact, challenging shale couldn't be lower than producing it from more conventional reservoirs.

The consultant also was skeptical that anything big was happening that would affect the U.S. supply-demand equation. Yes, seven years of high natural gas prices meant drillers actually could make money focusing on high-cost shale formations. And sure, many were managing to reduce production costs. But it still seemed too costly to get gas from shale wells. A wave of new U.S. supply seemed unlikely. Imported natural gas seemed as necessary as ever.

Souki, Gentle, and others at Cheniere eventually concluded that foreign producers were wary of shipping gas to America simply because few of them had extra gas to sell. A number of projects were set to be completed, however, so sales to the United States likely would pick up. Souki's decision to borrow more money to build additional import capacity still seemed like a no-brainer.

Late in 2007, Cheniere's stock weakened a bit, falling to about thirty-three dollars at year's end, as some investors fretted about the company's debt. Souki, usually ebullient around the office, turned more pensive. The housing market had begun to weaken and subprime mortgage lenders were cratering. The overall stock market held up, but Souki had an inkling that trouble might be around the bend.

Cheniere already had raised over a billion dollars from investors and lenders. Souki didn't feel they'd have the appetite to buy additional shares or debt from his company, even though Cheniere needed to drum up an additional $300 million to complete construction of his Sabine Pass terminal.

He decided they'd wait until early 2008 to get the financing. He crossed his fingers, nervously hoping markets would hold up.

Mark Papa also was becoming scared, though for a different reason than Charif Souki.

Papa ran EOG Resources, the energy company spit out of Enron years earlier. It already had accumulated 320,000 acres in the Bakken formation, buying some of the most promising chunks from Michael Johnson, the septuagenarian oil explorer. By 2007, EOG was making a bit of progress extracting oil from challenging rock in the region.

But Papa and his fellow EOG executives weren't very focused on the Bakken or even on oil. EOG was a natural gas company, after all. Seventy-six percent of the company's revenues came from gas. Thanks to an early embrace of horizontal drilling and advanced fracking techniques, EOG was seeing surging production from natural gas wells in the Barnett and in other areas around the country and improving on George Mitchell's techniques. In 2007, EOG's gas production and reserves grew by about 15 percent, management boosted the company's dividend by 50 percent, and there didn't seem much to worry about.

"We were busy patting ourselves on the back," Papa recalls. "We were geniuses, finding more gas than we ever dreamed of."

In October 2007, nearly twenty of the company's top executives gathered at the Hyatt Regency Resort and Spa in Scottsdale, Arizona, prepared for a few days of meetings, sun, and hopefully some fun.

Mark Papa's mood was anything but cheery. Every month, a different rival was announcing a new shale play or an extension of an existing big field, and touting how much gas was on the way. To most in the industry, the progress tapping shale and other tough rock was heartening. To Papa, the news couldn't have been more troubling.

Papa kept especially close tabs on what his biggest competitor, Aubrey McClendon, was up to. Papa kept reading about how Chesapeake was finding more gas in shale deposits in the Barnett, the Haynesville, the Marcellus, and elsewhere, and how the company was paying higher prices for additional acreage. Papa went on Chesapeake's Web site to listen to Chesapeake's quarterly conference calls, trying to learn more about the company's strategy.

Even if they're exaggerating, that's a ton of gas, he thought.

One day it hit Papa hard—a glut of natural gas was imminent, no matter what the peak-oil believers thought. Shale drilling was for real. He was gripped by fear—he ran a gas company and prices were sure to tumble. He had to find a solution. As the EOG managers settled into their seats in the resort's meeting room, Papa got right to the point.

"Guys, we need to talk," he began. "We're all euphoric right now and everything looks rosy, but we need to make some radical changes."

Papa saw confused looks on the faces of his staff, but he kept going. "I'm here to tell you, natural gas prices are going to be weak for twenty or thirty years," he told the group. "We're gonna have to convert this company to an oil company or we're dead ducks. Tell your people to stop looking for gas, right away."

Complete silence.

No one in the room really knew how to react. Gas wasn't a cinch to extract from shale and other dense rock. But they all knew oil was much, much harder. Everyone else was buying shale gas acreage and their boss was saying they needed to look for oil? It was as if legendary New England Patriots coach Bill Belichick, in the locker room after winning the Super Bowl, told his team they were going to ditch football and learn to play another sport.

"People thought I had lost my mind," Papa says. "Their whole lives it was about finding more gas, gas is good, the nation needs more of it."

Papa said EOG would begin searching for oil in earnest, both in the Bakken and elsewhere in the country. His right-hand man, Bill Thomas, had been pushing them to focus on oil. Now Papa was on board.

He just hoped he wasn't too late.

CHAPTER ELEVEN

Aubrey McClendon welcomed his guests with a warm smile.

It was a brisk March evening in 2008 and some of the most powerful men in the world were walking down the staircase at the '21' Club, one of New York's oldest and classiest restaurants, to join McClendon for dinner.

Legendary investors George Soros and Stanley Druckenmiller were there. So was Sadad Husseini, a former senior executive of Saudi Aramco, the largest oil company in the world, and the evening's invited speaker. Matthew Simmons, an influential banker, author, and adviser to President George W. Bush, also made his way downstairs, rounding out a who's who of energy and finance.

McClendon and his cohost that evening, stock research specialist Kiril Sokoloff, greeted their guests in a private dining room surrounded by racks of top-notch wine and near a private cellar that once held the private collections of John F. Kennedy, Ivan Boesky, and Frank Sinatra. McClendon exchanged a few words with each man, rattling off a personal detail that made each feel he was the most important guest that evening. Everyone took an appointed seat, with McClendon at the head of a long rectangular table, his confidence and swagger unmistakable.

The stock market had turned a bit rocky, housing was weakening, and some wondered if the economy could hold up. A month earlier, the market for risky "subprime" mortgages had collapsed. But investors, convinced that energy supplies were running dry, had sent natural gas prices over ten dollars per thousand cubic feet and crude above $100 a barrel, putting the energy men in the room in good spirits, especially McClendon.

Once, McClendon had been the industry's bad boy, a self-promoting salesman who seemed to spend and borrow too much to grow Chesapeake. He had nearly imploded his company not once but twice. Now he was virtually a spokesman for the oil and gas business, promoting the wonders of homegrown natural gas and leading a group of upstart companies with dominant positions in shale formations that experts finally were beginning to appreciate.

Chesapeake shares were up 50 percent since Tom Ward quit Chesapeake a year or so earlier, making McClendon a billionaire, just like some of his dinner mates. More accurately, McClendon was a multibillionaire, including shares of his company and his other investments. George Soros wanted to have dinner with him. It didn't get much better than that.

As waiters converged on the table, McClendon chose an assortment of top-of-the-line wines to accompany a specially ordered three-course meal featuring sautéed lobster tail and risotto with asparagus.

"This is the best wine you'll ever drink," Simmons whispered to an investor seated next to him.

Husseini rose to address the group, quickly painting a picture of gloom. Saudi Arabia's oil production was slumping and the country had invited ExxonMobil, Total SA, Royal Dutch Shell, and others to search for new pockets of natural gas. Their efforts weren't worth the cost, however, another sign of how much harder it had become to discover energy.

Oil and gas demand from China, Brazil, and other emerging-market nations was growing, Husseini noted, even as global supply came under pressure, sending prices higher. The Western world would just have to adjust to a lower standard of living, he argued.

From all the head-nodding around the table it didn't seem as though anyone disagreed. Two years earlier, Simmons had made a similar argument in his best-selling book *Twilight in the Desert*. Oil and gas production in Mexico, the North Sea, and elsewhere was in decline, and U.S. oil fields had been written off as a hopeless cause. Shale gas seemed promising for only a few companies, like Chesapeake; shale oil from places like the Bakken wasn't on anyone's radar.

As Husseini finished, one of the dinner guests, Jason Selch, raised a hand to make a comment. Selch, a veteran investor who worked for real estate mogul Sam Zell, was known for being a bit cheeky. Two years earlier, he had been fired by a previous employer after dropping his pants and mooning an executive who had fired one of Selch's colleagues.

Selch knew McClendon's company was making rapid progress extracting gas from U.S. shale formations. It gave him an idea. "You should ask Aubrey to help," he told Husseini. "He's been finding lots of gas in places no one thought" it could be found.

All eyes turned to McClendon. He gave a shake of the head and a modest smile. "Thanks, but we're too busy in the U.S.," he told the group.

McClendon and his guests thought they had it all figured out: The world was running out of oil and gas. A few players able to find some would survive. Even fewer managing to tap exciting shale formations, like Chesapeake, would get super-rich.

Just a few weeks earlier, McClendon had told Chesapeake shareholders that his company was in such a good position that he was done buying acreage. They controlled more than thirteen million net acres and had plans to drill thirty-six thousand wells over the next few years.

The "modern-day equivalent of the great Oklahoma land runs of the 1800s" was "largely over for Chesapeake," McClendon wrote in the company's annual report. "We believe we have emerged with a superior position."

McClendon's shares were growing more valuable with each trading day. He also was profiting from natural gas futures positions in his brokerage account, according to his friends. It seemed that he and Chesapeake couldn't be stopped.

On a Sunday morning in May 2008, McClendon drove to an Oklahoma City restaurant to meet Art Swanson, the small-time energy operator who was a high school acquaintance of McClendon's and was working on their ambitious racetrack/golf course development. By then, the price tag for the project had topped $200 million, a sum that seemed outrageous, even for McClendon. Over pancakes and omelets, he and Swanson decided to scrap the idea.

They began to discuss business. Two months earlier, venerable investment bank Bear Stearns had collapsed, amid growing jitters about a potential bursting of the nation's housing bubble and its impact on the financial system. The mortgage mess didn't seem to matter to the energy world, though, as crude prices approached $120 a barrel and natural gas was close to twelve dollars per thousand cubic feet.

McClendon didn't seem quite as upbeat and cocksure as usual, though. He leaned over to Swanson, as if to share a secret. "You know what my big fear is?" he asked.

Swanson couldn't venture a guess. McClendon never seemed to fear much of anything.

"We could break the gas market," McClendon said.

He confided that Chesapeake was discovering so much natural gas that their new supply, along with the gas others were extracting from shale, might overwhelm demand, sending prices tumbling and jeopardizing McClendon's empire.

Swanson had an instant solution. "Aub, why not just pay off everything and go to cash?" he asked. "Why not hedge things?"

If McClendon sold some of his shares and gas futures positions, he could pay off his debt and walk away a rich man, like a gambler cashing in a portion of a huge stack of chips, Swanson was saying. Chesapeake also could slow or hedge some of its gas production. The company also might look for some oil, if only to spread its bets out.

Do something to reduce your risk, Swanson urged his friend.

McClendon gave Swanson a dismissive look, as if the small-fry, local producer didn't understand the predicament he was in. "He looked at me like I was a dumbass," Swanson recalls. "Like I just don't get it."

Swanson didn't realize it, but McClendon was boxed in. He had spent years as Chesapeake's biggest booster, and as the most vocal advocate of drilling for natural gas in the United States. Bailing out now, with Chesapeake shares soaring, risked sending a message to the market that extraordinary amounts of gas couldn't be extracted in the country after all. After selling investors a story about Chesapeake's growth, even someone as smooth as McClendon might have found it challenging explaining why he was taking money off the table.

McClendon also was in a bind because Chesapeake had negative cash flow—it spent more than it took in—and owed $11 billion. To make ends meet, the company regularly sold shares and debt to Wall Street investors. The only way McClendon could continue to do that was by assuring the investors that the future was bright and Chesapeake's growth would allow it to pay off all that debt. If he turned the spigot off and ordered Chesapeake to slow its production, the company's stock and bonds likely would crumble and McClendon wouldn't be able to raise new cash to keep the company going. It was a game of musical chairs. If McClendon turned off the music, he'd suffer a painful fall.

Even if McClendon had wanted to pare some of his holdings, it wasn't clear he could. Like other companies, when Chesapeake sold stock it came with conditions barring senior executives from selling shares, at least for a limited time. Besides, McClendon still believed in his company and in his country. More than 75 percent of his $18.7 million in compensation the year earlier had been in stock awards.

When he spoke with colleagues and friends, McClendon was as confident as ever in the shale revolution and in his own ability to navigate any challenges. He was the nation's gas king, after all. "I thought that our employees and most importantly, our investors, needed to see me as a strong leader believing one hundred percent in what our company stood for," he said.[1]

Indeed, the caution he expressed to Swanson proved fleeting. Both McClendon and Tom Ward were so bullish that by June each controlled nearly identical wagers on natural gas derivatives worth around $2.3 billion, according to trading data later disclosed by U.S. senator Bernie Sanders of Vermont and later reported by Reuters. Among three hundred banks, hedge funds, energy companies, and other traders identified by the Commodity Futures Trading Commission, only four held bigger gas bets than McClendon and Ward. McClendon held additional contracts on oil worth another $240 million, according to the data.[2]

McClendon had vowed to rein in the company's spending, but a buzz was growing about two new plays. One was the Marcellus Shale, a broad region stretching through Pennsylvania and up into Ohio and New York. A few years earlier, a geologist named Bill Zagorski at a rival company, Range Resources, had convinced his employer to take a shot at leasing and drilling this formation. After a fitful start, they were beginning to make progress.

Earlier in 2008, Terry Engelder, a noted geologist at Penn State University, shocked the industry by determining that the Marcellus held fifty trillion cubic feet of natural gas, an estimate that suggested it was one of the world's largest fields.

Range was leasing wide swaths of Pennsylvania, but Chesapeake had caught up and passed competitors in earlier shale plays. McClendon was determined to do it again in the Marcellus. The company quickly leased Marcellus acreage, but drilling and extracting gas required additional cash Chesapeake just didn't have, causing McClendon new headaches.

The Haynesville Shale, straddling northern Louisiana and East Texas, seemed at least as huge as the Marcellus. Chesapeake believed it was the largest natural gas field in the Unites States. Lease rates jumped as drillers such as Floyd Wilson's Petrohawk, funded by eager Wall Street investors, descended on the area. Each well cost as much as $10 million, but some spewed enough gas in a single day to power eighty-four thousand U.S. homes. "We had never seen something like that," Wilson says.[3]

Chesapeake didn't have enough money to get gas from both the Marcellus and Haynesville, but McClendon couldn't resist competing for the new areas. "In these plays, if you snooze, you lose," he told investors. "And with over four thousand landmen in the field every day buying new leases, I can assure you that Chesapeake is not snoozing."[4]

Chesapeake managed to get its hands on promising acreage in the Marcellus by focusing on parts of the state Range hadn't yet locked up. It also was leasing land in the Haynesville. Coming up with money to drill new gas wells in the area before the leases expired and reverted to landowners was an entirely different story, however.

McClendon sat down with his old friend Ralph Eads, the senior investment banker at Jefferies & Co., to figure out how to find a cash infusion. They took out a giant piece of paper and began to list all the known and emerging shale plays and how much money it would take to become a dominant player in each. There was the Marcellus, and the Haynesville. The Bakken looked interesting, as did the Fayetteville and a few others.

Eyeing the list, Eads estimated that it would take more than $500 billion for companies like Chesapeake to become major players in the various regions. There was just no way McClendon's company, or anyone else, could raise that kind of cash from investors, however. Selling stocks and bonds no longer would cut it.

"You've got to come up with another plan," Eads leveled with McClendon, who agreed. They quickly developed the idea of selling portions of Chesapeake's existing acreage to companies that had missed out on becoming players in the shale revolution but were eager to get involved.

Eads reached out to Jim Flores, the chief executive officer of Plains Exploration & Production, to discuss the Fayetteville Shale. "Aubrey and I have calculated it, and it might be the largest gas field in the world," Eads told Flores, citing early results from a test well that showed unprecedented gas flows.[5]

On July 1, 2008, Chesapeake and Plains announced a joint venture called a "cash and carry" deal. Plains agreed to buy a 20 percent stake in the 550,000 net acres Chesapeake had accumulated in the Haynesville Shale for $1.65 billion in cash. Plains also agreed to spend another $1.65 billion over seven years to cover half of Chesapeake's drilling costs in the area, in exchange for just 20 percent of the future profits.

The deal was a coup for Chesapeake because it instantly raised billions in cash to pay for new wells. Just as important, the transaction proved what McClendon had been telling investors—shale acreage was really valuable. Chesapeake had spent an average of $7,100 an acre on its drilling sites in the Haynesville, but Plains paid Chesapeake the equivalent of $30,000 an acre. McClendon had instantly locked up nearly $23,000 of paper profit for each of its 550,000 acres, or almost $13 billion.

Even skeptical members of Chesapeake's board of directors were wowed by the deal, deciding that McClendon's push into shale formations was as shrewd and lucrative as he had promised it was going to be.

In July 2008, the Energy Department reported that U.S. gas production had jumped 8.5 percent to 1.86 trillion cubic feet, the highest monthly output since May 1974.[6] It was becoming clear that shale formations might change not just companies, but the country itself.

Despite growing production, natural gas prices leaped to $13.58 per thousand cubic feet in the first week of July, while crude hit a record $145 a barrel amid frenzied speculative trading by investors sure that supplies wouldn't be able to satisfy demand from emerging markets.

McClendon shared the belief. On a conference call to discuss his company's second-quarter results, he predicted that natural gas prices would stay in a "$9 to $11 range."

"We're pretty confident that much below $9 you'd see a drop-off in drilling activity," he said, something that would keep prices elevated.

Rising energy prices and enthusiasm over the Plains deal helped Chesapeake hit an intraday record of seventy-three dollars a share on July 2, up from forty at the beginning of 2008. Chesapeake took advantage by selling over $1.5 billion of new shares to investors, part of the $2 billion it sold that year to bolster the company's balance sheet. McClendon was the largest buyer of these new shares.

One day that summer, Marc Rowland, Chesapeake's chief financial officer, left his office in search of his boss. He walked past the cubicles of three executive assistants they shared and found McClendon's door closed, as usual.

Rowland knocked and entered. After discussing some corporate business, he delivered some advice. "Aubrey, you should sell some stock," he told McClendon.

Rowland had been paring his own cache of Chesapeake shares, selling over a hundred thousand shares in the second quarter of 2008 alone, as if he had a premonition some bad news was coming. He also had been pushing the company to issue more shares, to raise cash while the stock was still hot. He urged McClendon to do the same, or at least enter into a transaction with a bank to hedge his position using financial derivatives. "You never know" when something bad could happen, Rowland told McClendon.

Rowland had been gently repeating this advice to McClendon for several months. Once more, McClendon waved him off with a smile. You're being "too conservative," he teased.

By then, McClendon was telling Wall Street analysts that the entire energy industry had been transformed. Instead of drilling down and just hoping to find gas, shale production had become as reliable as an assembly line. "We consider ourselves to be in the gas manufacturing business," he told analysts. "We think that's pretty impressive and hope you do as well."

Mark Papa and EOG were looking for crude. Two months earlier, Bob Simpson, the hard-charging executive running XTO, Chesapeake's biggest natural gas competitor, had spent almost $2 billion to buy over 350,000 acres full of oil in the Bakken from Tim Headington's Headington Oil.

But McClendon believed in natural gas. When Eads offered Chesapeake an opportunity to buy some oil wells around that time, McClendon turned it down. Over the spring and summer of 2008, McClendon doubled down, spending about $200 million for another four million shares of his company, like a salesman backing up the truck for his own product. McClendon predicted to Swanson that the stock would hit $100 a share.

In early July, McClendon's thirty-three million shares were worth a staggering $2.3 billion—all from an initial investment of just $50,000 made with Tom Ward nineteen years earlier. Including his other investments and holdings, McClendon was worth more than $3 billion. He had become one of the wealthiest men in the nation. He was all in.

Tom Ward, Chesapeake's departed cofounder, was just as confident about the outlook for energy. Shares of Ward's new company, SandRidge Energy, were soaring as investors bet it was the new Chesapeake.

Ward began 2008 with more than thirty-three million shares, worth a heady $1.2 billion. It wasn't enough, though. He spent about $200 million to buy nearly five million additional shares of his company in the first six months of 2008, showing as much self-assurance as McClendon.

Even Ward didn't have that kind of cash lying around, so he borrowed much of the money for the shares. He had grown up without much money, had come to rely on debt, and didn't view it as any big deal. "If I wouldn't have taken on debt I would still be on a tractor," he explains.

Ward also continued to actively trade in his brokerage accounts, even as he ran his new company. One day, when a SandRidge employee phoned an investment bank to discuss a potential hedge of the company's gas production, he got an odd response.

"That's funny," the banker told the SandRidge staffer. "I just got off the phone with Tom. He was trading his own book," or making moves in his own trading account.

"Please don't tell me that," the employee told the banker. "Ahhhhh." He indicated that he was putting his fingers in his ears to prevent learning any more about his boss's wheeling and dealing.

The staffer knew Ward had done nothing illegal by playing commodities with his own money. Nonetheless, it made him uncomfortable that his boss was trading in the same markets as his company, the employee says. It didn't feel right.

By then, SandRidge had become the country's fifth most active driller of new gas wells. Production was up nearly 100 percent in the first six months of the year and shares were soaring. Ward's own stock was worth more than $2.5 billion, boosting his net worth to more than $3 billion. All that borrowed money seemed well spent.

Indications were growing that significant amounts of natural gas might come from U.S. shale formations. Oil from shale still seemed like a waste of time, though, no matter what Hamm and Continental were doing up in North Dakota.

After all, the U.S. Geological Survey, a government arm, had estimated that just 151 million barrels of oil were buried in the Bakken, enough to satisfy less than eight days of the nation's consumption of oil and petroleum products. That figure hadn't been updated since 1995, so it was quite stale. But everyone knew North Dakota only was producing about 140,000 barrels of crude per day. That was up from about eighty thousand five years earlier, making the state the eighth largest producer in the nation. But it wasn't like the state had tapped a gusher.

Even some of those drilling in the Bakken, such as John Hess, didn't seem especially enthusiastic. He ran Hess Corp., a company that had been early to discover oil in North Dakota. It began building its holdings in the Bakken in 2005, controlling about 570,000 acres by 2008. But in March, when Hess, the son of company founder Leon Hess, spoke with *Newsweek* columnist Fareed Zakaria about "new frontiers" in oil, he didn't utter a word about the Bakken.

Instead, Hess, who had studied Arabic and Farsi, pointed to the Gulf of Mexico, Brazil, and West Africa. Hess had become an advocate for energy conservation and alternative fuels, and he seemed concerned about supplies, no matter what was going on in the Bakken.

"It's hard to see how we can meet" growing demand, Hess told Zakaria. "It's hard to see any relief in sight."[7]

But news emerged on April 10, 2008, that shocked those skeptical about U.S. oil. That afternoon, the USGS released a fresh survey estimating that North Dakota and Montana had 3 billion to 4.3 billion barrels of "undiscovered, technically recoverable oil" in the Bakken formation, a twenty-five-fold increase from the government's previous guess. The Bakken had more crude than any formation in the lower forty-eight states, the USGS was saying. Harold Hamm, Leigh Price, and others who believed in the region were right after all.

Jaws dropped, even in North Dakota. Kathleen Neset had graduated from Brown University with a geology degree and moved to the Williston Basin in 1979 to become one of the few women in senior positions in local fields. She fretted that her town of Tioga, with about one thousand residents, was withering. Schools had closed and young people were moving away.

New drilling by Hamm and others in Williston and other cities in the area had raised hopes of a revival, but locals had seen so many boom-and-bust cycles that many remained dubious. The government's updated numbers, which earned front-page coverage and a banner headline in the local *Tioga Tribune,* seemed proof that a rebirth for the region was imminent.

"It was wow!" Neset recalls. "We knew it was big but this confirmed it. . . . It was extraordinary."

The figures caught the attention of large domestic companies and international giants; stock investors began searching for companies with big positions in North Dakota. In the middle of July, Continental's stock soared past eighty dollars from thirty-five. A year earlier, investors had balked at the company's IPO and Brian Hoffman had sold his shares, fearing what was ahead. Now Wall Street was throwing money at the company.

For Jeff Hume and Jack Stark, the Continental Resources senior executives who had spent more than eight years trying to unlock the mysteries of the Bakken, the updated numbers from the government felt like validation. "We were pumped," Stark says.

But many of those working in the Bakken fields were stupefied by the government's estimate. They weren't seeing "recovery rates," or levels of oil flowing from wells, indicating that such a high estimate was possible. Continental was getting well under a thousand barrels a day from each of its Bakken wells, about half as much as some thought they should be seeing from the most promising ones. The government's figures seemed too high to some of these field hands.

Back in Oklahoma, the USGS's figures irritated Harold Hamm. He was sure investors and industry experts were still underestimating the Williston Basin's potential. He insisted to a friend that there was more crude in the Bakken than the government's geologists were saying.

Hamm was an upbeat guy, but his confidence stemmed from something else. Continental was getting more excited about the Three Forks formation, a zone just below the Bakken. Hamm and his staff suspected these new layers might be full of oil, just like the Bakken.

He ordered his landmen to lease even more acreage in the formation. It still was expensive to drill wells there, and they still weren't pumping a tremendous amount of crude, but oil prices were zooming so it seemed worth it.

"We felt very confident," Hamm recalls.

Mark Papa was beginning to focus on oil, trying to catch up to Harold Hamm.

Throughout the first half of 2008 Papa, the chief executive of EOG, sat in his Houston office poring over data showing climbing natural gas production from various U.S. shale formations. But when he read reports from Wall Street analysts, or spoke with others in the business, almost no one was worried that growing supply might outstrip demand. Most were reassured by rising natural gas prices.

Papa thought they were all wrong. He couldn't figure out why Aubrey McClendon and others weren't shifting away from natural gas like he was. EOG seemed to be making progress in the Bakken, but an industry member cautioned Papa that the formation was a one-of-a-kind "freak discovery," and that other shale oil regions in the country were unlikely to be very productive.

Papa bit his lip, unwilling to share that his staff was scouring the country for shale oil formations that seemed similar to the Bakken. He hated divulging what he and his company were doing. He knew McClendon was bragging about how Chesapeake was gobbling up shale acreage, but that wasn't Papa's style. He remembered how Enron, EOG's former parent, was full of show-offs who boasted that they were going to change the world. The Enron executives made sure everyone knew they were the smartest men in the room, but the company had suffered an embarrassing demise.

Papa's employees guarded corporate information like it was a state secret. Local officials in North Dakota griped that EOG wasn't quick to share how the company was faring with local wells, unlike the more outgoing Harold Hamm. That made it more difficult for the officials to trumpet the region's progress. Papa didn't really care. Only recently had EOG begun to tell its own shareholders about the activities in the Bakken. Even then, the details were few and far between.

EOG needed to be more clandestine than ever, Papa told his troops. His geologists had identified an area in South Texas called the Eagle Ford Shale that seemed worthy of drilling. EOG aimed to lock up 200,000 acres. After accumulating that kind of holding, Papa and Thomas were sure rivals would make their own push into the area, sending prices higher.

"The rest of the industry doesn't need to know," Papa told a member of his exploration team.

But as EOG's landmen accumulated leases in the Eagle Ford, paying an average of just $400 an acre, few competitors emerged. Papa couldn't believe it—everyone was still chasing gas, not oil. He raised his goal, telling the landmen to lease 300,000 acres and then 400,000.

EOG hadn't drilled a single well in the Eagle Ford, though. The executives thought they could get a lot of oil from the region's rock, but they hadn't tested their theory. Rival company Petrohawk had drilled a well in the Eagle Ford, but Papa and Thomas didn't want to stop and test their own wells. It would be too expensive to lease it later on, they calculated. They had to grab acreage now.

EOG spent over $100 million on land, hoping it would yield meaningful amounts of oil. It was a dangerous move in any market. It would turn out to be a huge risk in one about to crumble.

In early 2008, leasing and drilling activity grew heated in the Marcellus Shale, a region stretching across New York's picturesque Finger Lakes region to northern and western Pennsylvania, eastern Ohio, and through parts of Maryland, West Virginia, and Virginia. The area seemed to have more gas than any other that drillers had come across, making it an obvious target for Aubrey McClendon and others in the lead in the shale race.

They'd find something else in the Marcellus: scrutiny and criticism as they never before had experienced.

Excitement over the Marcellus Shale can largely be attributed to two people, Bill Zagorski and Terry Engelder, who focused on the formation when no one else cared about it.

Zagorski was a middle-aged geologist at a subsidiary of a small gas company called Range Resources. In 2003, he was charged with finding "the next Barnett," but Zagorski and his team met disappointment drilling a deep layer of rock in the Appalachian Basin. Frustrated and eager for a diversion, he took a trip to meet a friend who was meeting some success with a shale formation in Alabama and wanted Zagorski's advice. Zagorski quickly realized that his crew had been drilling through a similar layer of Appalachian shale called the Marcellus.

"Holy cow, we have same thing!" Zagorski said.

Back in Pennsylvania, Range began drilling test wells in the Marcellus. *Holy shit, this is big stuff,* Zagorski thought.

In 2004, he convinced his bosses to spend several million dollars to frack the rock in the Marcellus, using the same techniques Mitchell Energy had used in the Barnett. The early results were even better than Zagorski and Range could have hoped for, so the company began leasing additional land in Pennsylvania. It took them another three years to crack the code and coax huge amounts of gas from the Marcellus layer by combining fracking with horizontal drilling.

Some others followed Range into the Marcellus, but activity was still limited. Land in the town of Dimock, in the lush green mountains of Susquehanna County, in the northeastern corner of the state, didn't go for much more than twenty-five dollars an acre in late 2005 and 2006. It wasn't much money, but the area was struggling, many farms were on their last legs, and the cash came in handy.

The wild roar of the diesel generators was deafening as the first wells were drilled; most used fracking techniques that had not yet been refined. The process involved pumping over a million gallons of water into the shale a mile below, along with some chemicals and sand. Between 10 percent and 30 percent of it usually came back up, carrying a range of chemicals used to create the fractures in the rock, as well as natural components including iron, radium, and oil-laced drilling mud. The flowback in Dimock and elsewhere usually collected in large, lined ponds.

Just before Christmas 2007, Engelder, a sixty-one-year-old professor of geosciences at Penn State University and one of the few focusing on this rock layer—together with a colleague named Gary Lash from the State University of New York—calculated how much gas was buried in the Marcellus by examining the production of Range's wells. Their figure was staggering: fifty trillion cubic feet of recoverable gas, enough to put the Marcellus among the largest gas fields in the world and roughly twenty-five times the government's previous estimate. It turned out that Zagorski had discovered one of the largest energy fields in modern history, one that was ten times the size of the Barnett.

In 2008, Chesapeake and others raced to lease land—paying as much as $5,000 an acre—and Engelder was an instant celebrity. Later, he'd boost his estimate to 489 trillion cubic feet, or the equivalent of twenty years of gas consumption for the entire nation

Alta Resources—the company launched by Joe Greenberg and Todd

Mitchell with backing from eighty-eight-year-old George Mitchell—was among the most convinced the Marcellus area was packed with enough natural gas to change the country.

Alta sold its holdings in Arkansas's Fayetteville formation for $580 million, giving George Mitchell a new, $100 million windfall. The company immediately began searching for acreage in the Marcellus region. They called Nick Steinsberger to help develop a strategy to fracture the rock and let the gas flow. Soon, Mitchell's new company had leased thousands of acres in Northeast Pennsylvania, not far from the town of Dimock.

Until 2008, drilling and fracking around the country hadn't attracted much controversy. There were few reported examples of harm and much of the production was in less populated areas in the country, or in states like Texas with a long history of supporting the energy business.

Pennsylvania wasn't nearly as welcoming to the high-volume hydraulic fracturing cropping up around the state. Although Pennsylvania has a rich history of oil and gas drilling and was the site of the nation's first oil well, in the city of Titusville, the Dimock area and many other parts of the state atop the Marcellus Shale hadn't seen a great deal of recent drilling. In early 2008, some locals became wary of the activity, especially those who had moved to the area to get away from noise and pollution. Drillers rushing to the Marcellus to tap the gas, such as a company called Cabot Oil and Gas, were honing their techniques and working to make them safer. But many were still figuring out how to drill the challenging rock in the region.

Soon, Dimock's residents—and their growing concerns—would be heard around the world.

Charif Souki was working hard to join McClendon and Ward atop the energy world. Souki had bet it all on a Louisiana terminal that would transform imported liquefied natural gas and pipe it around the country.

He was beginning to have a harder time of it, though. Shares of his company, Cheniere Energy, began to weaken in the latter part of 2007 amid concerns about all the debt it had piled up to build its plant.

Things got worse as 2008 began. The stock dropped from nearly thirty-three dollars a share to just over eleven dollars in mid-April as doubts grew about whether the expensive facility would ever be built. Souki began to hear chatter that short sellers, or investors who profit by wagering against a stock's price, were zeroing in on his company, eager to push it lower.

The loss of confidence came just as Souki was trying to raise the last few hundreds of million dollars of financing to build the terminal. With the stock weak, Souki knew it now would be almost impossible to raise the money, especially with financial markets turning wobbly. He'd have to figure out another way to come up with the cash.

"I started to get really nervous," he says.

Cheniere executives at least had one event that might bring them good fortune. On Monday, April 21, 2008, senior government officials, local politicians, and energy regulators were scheduled to fly to Cameron Parish, Louisiana, for a grand opening of the terminal. The ceremony had the potential to remind investors that Cheniere was on its way and would actually be importing LNG soon enough. It also figured to improve lagging morale among employees, many of whom were coming from the company's headquarters in Houston.

The morning of the ceremony, Meg Gentle, Cheniere's head of strategic planning, got into her car with a colleague to make the nearly three-hour drive to the Gulf Coast. Along the way, she got a call on her cell phone from an unfamiliar New York number. It was an analyst from Moody's Investors Service, the debt rating company, with some bad news.

"We're putting you guys under review" for a possible ratings downgrade of debt sold by the regasification terminal, the analyst told Gentle.

The news startled Gentle and her colleague. Cheniere had contracts with Total and Chevron that seemed sufficient to meet the interest payments on the debt.

Gentle was told she and her colleagues had an hour to dispute anything factual in a memo that Moody's was about to distribute to investors.

"We're in a car," she told the analyst. "Can you just do this tomorrow when we're in the office to deal with this?"

Sorry, that won't be possible, the analyst said. The news went out, signaling to investors that the debt was riskier than it had appeared, the latest blow to the company.

When visitors arrived at the site, just over the border from Texas and adjacent to the Gulf of Mexico, they witnessed an industrial facility that had been transformed by Cheniere workers. A couple of years earlier, the spot had been little more than marshland, so thick and deep it seemed like quicksand. Once, a civil engineer working on the grounds had even lost a tractor in dense marsh.

Then there were the animals that usually made the site their home, the variety of which might have made a zoo jealous. There were bobcats, coyotes, hogs, raccoons, rattlesnakes, huge rats, and free-range cattle, all walking around doing their thing, unconcerned with Charif Souki's big plans. Migratory ducks, geese, and almost every other kind of bird made regular appearances. Most of the animals had been shooed away by Cheniere employees, except for a newborn alligator in a ditch of standing water just outside the terminal's entrance.

Souki's team seemed to think of everything to try to make the event a success. Each guest was handed insect repellent to deal with the mosquitoes on the warm, humid day. Carpeted trailers filled with a Louisiana-style lunch were brought in, as if it was the Super Bowl. There was a Cajun band and luxurious Porta Pottys that gave the Louisiana locals who worked on the facility a good laugh.

Samuel Bodman, the U.S. energy secretary, got the festivities off to an upbeat start, saying the day was "an important moment for southwestern Louisiana and for the entire country." He noted that the Cheniere facility was "the first domestic onshore regasification terminal built in more than twenty-five years," predicting that it would "help us diversify our energy supplies, suppliers, and supply routes."

As Bodman spoke, Cheniere employees on the side of the room began whispering about the debt downgrade and how the stock had begun the day's trading with a troublesome dive.

Souki didn't seem aware of the chatter as he walked to the podium to begin his own address before the array of senators, congressmen, and businessmen. He began his speech, stressing how America needed to import natural gas. But he soon became distracted. Souki noticed a few audience members focused on their BlackBerry phones and other devices. Some glanced down, shaking their heads, as if something bad was happening.

A buzz grew, even as Souki continued. He tried to stick with his upbeat remarks, but he could sense what was happening. Shares of his company were plunging, on the very day of his long-awaited triumph.

Later, Souki checked the stock for himself. Cheniere had opened for trading at just over eleven dollars a share. By the end of the celebratory event, the stock was well under ten. By the close of trading, Cheniere shares were barely above seven dollars, a remarkable collapse of 36 percent in a single day.

Just months earlier, Souki had been worth $150 million. His plan to import gas seemed foolproof. Even Goldman Sachs and Warren Buffett seemed to agree with his approach. Now Souki's company was on life support and his wealth was evaporating. A few weeks later, Standard & Poor's cut its own rating on Cheniere to junk levels that suggested the company might not make it.

Every day, Souki heard a new rumor in the market that he had to refute. One day it was that Total and Chevron would stop making payments, something Souki was sure wouldn't happen. Another day it was that Cheniere was defaulting on its debt, though by Souki's calculation he had a cushion of more than two years.

"It was a vicious rumor campaign, but it panicked investors," he says. "Things looked completely desperate."

Friends called to see how Souki was holding up amid all the bad news. "How does it feel?" one well-wisher asked.

Souki tried to make light of the dramatic turn of events. "Well, I've never lost $150 million before."

Inside, though, Souki was confused and angry. He was disgusted with the fickle stock market and furious at short sellers who he believed were ganging up on Cheniere even though its terminal remained on track and nothing much had changed for its business. He called a lawyer to see if there was anything he could do about the short sellers, but realized he didn't have many options. He also blamed himself for not anticipating the troubles and securing the crucial financing.

Suddenly, he was in a personal bind. Souki had been unwilling to sell any of his nearly three million Cheniere shares, worried about the message it might send investors and confident that bigger gains were ahead. So he had borrowed money to pay for his expensive lifestyle, pay his taxes, and pay other expenses, using the shares as collateral.

As the stock plummeted, Souki received a margin call from his lenders, forcing him to sell over two million of his shares to pay back the debt. By the end of June 2008, he was left with just 600,000 shares of his company, worth nearly $3 million. A few weeks later, the stock was below three dollars a share. Souki was worth a few million dollars, but it mostly was in real estate, which wasn't liquid and was tumbling in value. He sold his private jet and a boat to raise some cash.

Adding to Souki's frustration was that friends and employees were investors in his company. The stock tumble meant that they too were suffering.

Almost everything Souki had touched in his career had turned golden. He didn't know how to react to this setback, and it sapped his confidence. "I felt shitty and seriously doubted myself," he says. "You question your intelligence . . . it was shocking."

An old friend called to comfort Souki. "Charif, you did something spectacular," he said. "You did your best, you never lied to anyone."

Souki felt horrible, though, and his self-doubt grew. Shares of other energy companies were soaring, but Cheniere dropped below three dollars in July 2008.

Souki had to take drastic measures to keep the company afloat. He laid off more than half of his 360 employees, in an attempt to preserve cash.

"I felt terrible," he says. "I still thought it would work, but I wasn't sure anymore."

Souki had been pummeled by the market and his energy dreams were in tatters.

It was a prelude of what was to come.

CHAPTER TWELVE

Oil and gas prices slipped in July 2008, after soaring for much of the year. By month's end, Chesapeake Energy was trading for fifty dollars a share, down from seventy earlier in July. It was becoming clear that the crumbling of the housing market was going to impact the overall economy, a worrisome sign for the energy market. The broader stock market had also begun to weaken, dropping over 10 percent in just six weeks.

Chesapeake's stock remained 25 percent higher on the year, but reasons for concern were mounting. In late August, natural gas prices, which are more directly tied to the health of the U.S. economy than oil, fell below eight dollars per thousand cubic feet, and oil dropped to about $115 a barrel.

Aubrey McClendon had resisted taking steps to prepare for any bad times, seemingly oblivious to the tempest building around him. Now he felt compelled to do something, just in case things got worse.

McClendon got in touch with a group of banks that had underwritten the sale of new Chesapeake shares a few weeks earlier. He asked the banks for permission to hedge his huge holdings of Chesapeake stock. McClendon hadn't soured on his company and he didn't want to sell his shares, he told the banks. He just wanted to buy some protection, an umbrella in case of rain.

"At that point, he recognized the potential for failure," says a colleague of McClendon.

The banks, which had the right to approve such a move according to the terms of the offering, refused. McClendon wasn't allowed to reduce his exposure to Chesapeake. If things turned stormy, he would have to deal with the consequences.

By then, McClendon had personally borrowed a total of more than

$500 million from banks, including Goldman Sachs, J.P. Morgan, and Wells Fargo, using Chesapeake stock as collateral. He had used the borrowed money to buy additional Chesapeake shares, finance gas wells he jointly owned with the company, and pay for all his real estate purchases and commodity trading. He had borrowed millions more from other parties, including Centaurus Advisors, the personal investment fund of John Arnold, a billionaire Houston hedge fund trader.

McClendon had become accustomed to borrowing one dollar for every three dollars of stock he held in his company, a ratio that he thought was quite conservative. But as Chesapeake fell to forty-five dollars a share in early September, the value of McClendon's collateral—which amounted to about half of his borrowing at each of the three banks—shrank, raising some concerns among McClendon's lenders.

Financial markets quickly spun out of control. Investors fled from almost every financial- and housing-related company, and the U.S. government was forced to bail out mortgage giants Fannie Mae and Freddie Mac.

On Sunday, September 14, the American financial system was shaken to its core when venerable investment bank Lehman Brothers Holdings filed for bankruptcy protection and brokerage giant Merrill Lynch & Co., in its own weakened position, agreed to be sold to rival Bank of America. In a tailspin, the Dow Jones Industrial Average dropped to around 11,000 from just over 13,000 in early May. Investors, suddenly scared about the nation's economic outlook, pushed natural gas prices to about seven dollars per thousand cubic feet, down by almost half in just over two months.

It wasn't just the financial meltdown that was weighing on gas prices. Some investors also were starting to suspect that the nation might have an oversupply of natural gas, thanks to the accelerated drilling by Chesapeake and others in shale formations around the country.

Marc Rowland, Chesapeake's chief financial officer, wanted to do something to put the company on a firmer footing. He called Chesapeake's lenders and converted the company's $4 billion credit line into cash. He figured the money might serve as a cushion if the economy took a turn for the worse. The move caused some nervousness among Chesapeake shareholders, but so many companies were considering similar moves at the time that investors didn't dwell on what Chesapeake was doing. The company also announced plans to reduce spending and sell assets, trying to assuage shareholder worries.

Despite the carnage on Wall Street and the hand-wringing in corporate suites around the country, McClendon seemed remarkably buoyant around the office. He worked on cutting new energy transactions, as if he were sure the downturn in prices would prove fleeting.

In early October, Chesapeake finalized a deal to buy gas fields and an eighty-mile pipeline in the Barnett Shale owned by ExxonMobil. Chesapeake also closed in on an agreement to raise $3.4 billion by selling a 32.5 percent interest in its Marcellus Shale wells to Norway's Statoil ASA, a transaction similar to Chesapeake's earlier, successful joint venture with Plains Exploration.

Wall Street was in its death throes at the time, so few paid much attention to the Statoil deal, but it was one more sign that global energy companies were becoming convinced the shale revolution was for real. For McClendon, the Statoil agreement was reassurance that all was well with his company. "Everything except macro events" was "really going our way," according to McClendon.[1]

But Chesapeake shares kept falling, setting off alarm bells at the banks that had lent McClendon hundreds of millions of dollars. During the first week of October, as an intensifying housing crisis seemed likely to lead to a U.S. recession and pressure energy prices, Chesapeake tumbled to just twenty-two dollars a share.

The lenders decided they needed to act. Late in the afternoon of October 8, a Goldman Sachs executive in the firm's private wealth group called McClendon, reaching him in his Oklahoma City office. The Goldman manager relayed disturbing news: The Chesapeake shares that McClendon had used as collateral had dropped so dramatically that they no longer held enough value to back the approximately $300 million that McClendon had borrowed from the investment bank.

Come up with more collateral fast or we're going to have to sell your shares to help satisfy your debts, the Goldman executive told McClendon. The Chesapeake cofounder was getting a "margin call" from Goldman, as if he were a gambler under pressure to repay debts to his bookie after a bad losing streak. Only in this case, the bookie was facing his own strains caused by the plummeting market.

McClendon had to do something to stop Goldman. If they sold his shares, Chesapeake likely would sink even further and he'd suffer the kind of embarrassment few chief executives had ever endured. But McClendon was in a corner. He didn't have enough cash or investments to boost his collateral. He had another idea, though. He picked up the phone to call a Goldman executive to ask for help.

Bill Montgomery, a senior investment banker in Goldman Sachs's Houston office, was the usual intermediary between the bank and Chesapeake. Over time, McClendon and Montgomery had become close, so he might have been expected to help McClendon out of his jam. But Montgomery was on a retreat with energy executives in a remote part of Wyoming where there was no cell phone service.

McClendon didn't have time to wait for Montgomery's return. Instead, he called someone even more senior at the bank, Jon Winkelried, Goldman Sachs's copresident. A few months earlier, McClendon had hosted Winkelried at the Deep Fork Grill, a restaurant in Oklahoma City co-owned by McClendon. They had forged a relationship and there was good reason to think Winkelried might be sympathetic to McClendon's plight.

Only a few weeks earlier, Winkelried had been in a state of panic about his own finances. Born in suburban New Jersey, Winkelried had become a gentleman rancher, spending millions to build a facility to raise and train horses and to buy ranches around the country. He paid nearly half a million dollars to purchase a mare aptly named "I Sho Spensive." Like McClendon, Winkelried had most of his net worth tied up in his company's shares, so he had borrowed money to live out his cowboy fantasies.

As Goldman Sachs stock plunged in September, pressure grew on Winkelried. He became so desperate for cash that he turned to his bosses for help. Winkelried had been a loyal and accomplished executive, so Goldman agreed to buy his stake in investment funds run by the firm, handing Winkelried a check for nearly $20 million.[2]

Goldman had come to Winkelried's rescue in a situation that shared similarities with McClendon's. It made sense to think Winkelried now might save McClendon, a longtime customer of the investment bank.

Winkelried took McClendon's call. The Chesapeake executive, aware he didn't have much time left, made an urgent case as to why Goldman shouldn't close his trades out. He told Winkelried that hedge funds were teaming up to short Chesapeake shares, hoping to profit as the stock tumbled.

"Hedge funds are killing the stock," McClendon insisted.

Chesapeake was still doing great and its outlook was bright, McClendon argued. The stock was bound to rebound. Just allow a few more days before closing out the account, he asked Winkelried. It wasn't fair that Goldman was cutting him off.

"Y'all are just killing me," McClendon pleaded.

Winkelried listened patiently. He even sounded sympathetic. He told McClendon that he'd speak to others at Goldman about granting him a reprieve of a few days. Maybe the stock would rebound enough to prevent the need for them to act.

"We have to do what's right for the firm but if I can get you some more time I will," Winkelried said. It wasn't clear if he was going to advocate for McClendon or if he was just being polite. But now McClendon had a chance at staving off disaster.

An hour or so later, Winkelried called McClendon back with a decision—the investment bank couldn't do anything to help. Winkelried was polite but firm. In fact, Goldman already had begun selling McClendon's shares that it held as collateral, as had many of his other lenders. It was too late for the Chesapeake executive.

"Those are the rules," Winkelried told McClendon. "I'm sorry, Aubrey."

McClendon hung up the phone, rose from his desk, and opened the door to his office. He slowly made his way down the hall, past the three female executive assistants, and into Marc Rowland's office.

When Rowland looked up he was surprised by what he saw. McClendon looked pale and shaken. The two executives had spent years working together, but Rowland had never seen his boss like this.

Rowland knew McClendon had borrowed a lot of money to buy Chesapeake shares. But he and others at Chesapeake had no clue how much he had borrowed or what kind of pressure McClendon was under. He was about to find out.

"They just sold my stock," McClendon confessed to Rowland.

"What the fuck do you mean?"

"The banks," McClendon responded. "I'm in trouble."

It dawned on Rowland what was happening. "Do you have anything left?" he asked.

"Yeah, but I still owe" money to lenders, McClendon replied, still looking dazed. He told Rowland that other banks would also have to sell his Chesapeake shares to satisfy his remaining debts.

It was a week of intensifying pain for McClendon and Chesapeake. On Wednesday, October 8, 4.6 million of his shares were sold at $22.68 a share, according to securities filings. The next day, another 11.4 million shares of his stock were sold at between $17.56 and $24. On Friday, October 10, 15.48 million shares were sold at prices ranging between $12.65 and $16.16 a share.

In one horrible week, the forty-nine-year-old had been forced to sell more than thirty-one million shares, or 94 percent of his stake in Chesapeake, the company he helped start and had led to the upper echelon of the energy business. He was left with fewer than two million shares. All the selling put new pressure on Chesapeake. In a single week, the stock fell 43 percent to $16.52, down from nearly $70 in early July.

"What I never dreamed could happen, did happen," McClendon later said.[3] "I honestly did not feel it was risky to have one dollar of margin debt for every three dollars of stock value."

On Friday, October 10, McClendon tried to explain what had happened to shareholders. The company issued a statement saying that McClendon had "involuntarily" sold "substantially all" of his Chesapeake stock over the previous three days to satisfy margin calls, blaming the "extraordinary circumstances of the worldwide financial crisis."

"In no way do these sales reflect my view of the company's financial position or my view of Chesapeake's future performance potential," McClendon said in the statement, adding that he was "very disappointed" by the turn of events.

For a few days, McClendon appeared down and discouraged. His face looked so haggard that colleagues wondered if he'd managed any sleep. They were shocked to see the dramatic change in a man so ebullient for so long.

"It was a tsunami," according to McClendon. "It was a prairie fire."[4]

It probably didn't make McClendon feel any better, but he wasn't the only energy entrepreneur forced to sell huge stakes in their companies to cover debt collection calls from brokers during that brutal period. On the same Friday that McClendon acknowledged his margin call, XTO Energy, the company vying with Chesapeake for key shale acreage, announced that its chairman and chief executive officer, Bob Simpson, had to sell nearly three million shares, worth more than $100 million. Other energy bigwigs, such as the chief of Tesoro Corp., also had to sell shares to satisfy stock loans that suddenly came due.[5]

Eighteen minutes after issuing the press release revealing McClendon's stock sales, and as employees struggled to digest the shocking news, McClendon sent a company-wide e-mail urging his staff "just to ignore" the huge decline in the stock and focus on their work. He attributed the precipitous decline in Chesapeake shares to fallout from the global financial collapse and said the price was "ridiculous."[6]

McClendon told employees they had entered a new world. "What was a fair price 90 days ago for a lease is now overpriced by a factor of at least 2x given the dramatic worsening of the natural gas and financial markets," he wrote in his e-mail.

Days later, Chesapeake tried to back out of various deals to buy oil and gas rights, including an agreement with a company called Peak Energy over land in the Haynesville Shale formation. Chesapeake's effort would lead to a lawsuit and eventual damage awards to Peak of almost $20 million.[7]

"Some were in areas where the acreage values dropped between agreeing to a deal and closing, some were condemned due to poor drilling results, and some we just ran out of money and needed to delay," Rowland recalls.

Charles Maxwell, the board member who had been convinced that McClendon's spending was prudent, seemed to change his mind now that the stock was tumbling. "I didn't think it would come to this, Maxwell told the *Wall Street Journal.* Though he still supported McClendon, Maxwell said Chesapeake had suffered from "an excess of enthusiasm."

Chesapeake wasn't in danger of going under, but it seemed clear the company no longer would be able to borrow and grow like it had in the past. Analysts began to question how the company could find the money to drill all its land before the leases expired. Chesapeake was forced to halt almost all of its activity in the Haynesville Shale in Louisiana, as well as in the Marcellus region in Pennsylvania.[8]

Aubrey McClendon appeared out of sorts. To some, it looked as though he'd suffered a knockout blow.

The funk lasted only a few days, though. In late October 2008, McClendon snapped out of it, as if billions of dollars of wealth hadn't really evaporated from his accounts, no matter what his brokerage statements told him. When he dealt with bankers, including those who knew details of his painful margin call, they were shocked by how calm and upbeat he was. Around the office, he regained his composure and positive outlook.

It was as if he had a secret plan for a comeback, for both himself and his company.

Every time Tom Ward checked on the market, gas prices were falling, almost without a letup. By the end of August, gas had fallen 40 percent in just two months.

Ward and his wife, Sch'ree, got on a plane to fly to Santa Fe, New Mexico, to attend the wedding of a child of a good friend, Rob Braver. Braver was a grade school classmate of Aubrey McClendon's, but over the years he had become closer with Ward. Their friendship was cemented nearly a decade earlier on a golf trip to Cabo San Lucas, Mexico, when Braver, who was Jewish, asked Ward if he believed that Jesus was the messiah. Ward said he did and that he was willing to take the time to explain his view to Braver. They ended up spending several months discussing theology. Eventually, Ward invited Braver's friends to join their weekly conversations at Ward's office at Chesapeake. Over time, Braver became inspired and converted to Christianity.

The Braver wedding was joyous, but some guests noticed that Ward didn't look himself. There was music and dancing around him, but he didn't seem to be having a good time. He was strained and didn't seem well, some recall.

Back in Oklahoma City, Ward shared a confession with a friend: He had lost a billion dollars in his trading account over the previous month, due to the tumble in natural gas futures prices. It was a staggering sum, even for an executive who had become one of the nation's wealthiest men. Ward's losses soon became a topic of conversation in local energy circles.

In his wildest dreams, Tom Ward never expected natural gas to fall so dramatically. For a decade, demand had been growing, just as Ward had anticipated. The economy was using more energy and the population was still expanding. It all suggested to Ward that consumption and prices would keep rising.

Now gas prices were plunging, as investors buckled up for a bumpy ride for the economy, and it was costing Ward. "I missed the enormity of the change," he says.

SandRidge shares, which hit nearly sixty-eight dollars at the beginning of July, finished August at thirty-five dollars, a drop of nearly 50 percent. At the company's Oklahoma City headquarters, employees were shell-shocked. Many owned SandRidge stock and literally couldn't take their eyes off the falling shares—the price flashed on each employee's computer screen when it was turned on each morning, a cruel daily update on how much of their wealth was evaporating.

As the staff watched SandRidge shares fall below eighteen dollars by the end of September, a gloom settled in around the office. Ward seemed tense and anxious, reflecting the mood of his staff.

By October, Ward realized he had bills piling up but not enough ready cash, even though he still had a great deal of paper wealth. He moved to sell working interests he owned in SandRidge's gas wells to the company for $67 million, a deal that later would raise the ire of investors.

The price was less than what the well interests were worth, in the view of a fairness opinion offered by an outside party, so Ward viewed the deal as fair for SandRidge shareholders. But $67 million was a huge chunk of change for a company that would end the year with less than $1 million on hand, another reason the deal later would irk shareholders.

That month, the financial system tottered. For several days, Sand-Ridge couldn't get the short-term funding it relied on to keep its doors open, causing fear within the company's boardroom.

The panic lifted a bit by the end of 2008, as the government took steps to support financial firms and markets. But energy prices remained weak and SandRidge fell to $6.15 a share, down a sickening 91 percent from July of that year. Tom Ward's stock was worth over $200 million and he had at least that much in assets outside the company. That still was a lot of money, of course. But his shares had been worth over two and a half billion dollars just six months earlier.

More troublesome for Ward was that he had borrowed hundreds of millions from various lenders, just like McClendon. Ward's collateral held up better than his former partner's, so he didn't face margin calls from lenders. He came under pressure, nonetheless, because he owed millions of dollars in taxes and other obligations but didn't have the cash to pay for it all, even after selling his well interests to his company.

"Banks and liquidity were drying up and I had a short-term liquidity crunch," Ward says.

Ward didn't want to sell his company's shares on the open market, figuring such a sale would crush SandRidge's shares, just as McClendon's forced selling had pressured Chesapeake. Ward needed a different answer.

Looking for help, he reached out to Tulsa billionaire George Kaiser, the chairman of Bank of Oklahoma. On Christmas Eve, they worked out a complex deal for Ward to sell Kaiser 23 percent of his SandRidge stock, or nearly nine million shares, for $50 million. It was enough money to help Ward to pay his debts and other obligations. Kaiser also received a warrant to buy more shares in the future, though this deal also would turn into a point of controversy, partly because Bank of Oklahoma did some lending to SandRidge.

Ward told SandRidge investors that he was disappointed to have to sell his shares, but relieved he didn't have to dump them on the open market, something investors had fretted about. He tried to keep employees focused on their work and not on SandRidge's tumbling shares.

Ward had a vague understanding of trouble brewing in his company's boardroom. He had much less awareness of what was going on in his own home, however.

Every day at 3:30 p.m., Harold Hamm received a note from his marketing department updating him on what oil prices had done that day. Throughout the fall and early winter of 2008, the news got worse each day.

By the beginning of December, crude prices had tumbled to fifty-four dollars a barrel, down over 60 percent in five months. A week later, Continental shares fell all the way back to thirteen and a half dollars a share, *below* the level where they had debuted a year earlier, an embarrassing round trip.

Unlike Aubrey McClendon and Tom Ward, Harold Hamm's personal finances were fine. He owned so many shares of his company, and had made so much money over the years from trucking and other businesses, that he remained a billionaire. It also helped that he hadn't borrowed millions to trade in markets and buy stock, like McClendon and Ward.

The price collapse still had a huge impact on Hamm—it dashed his hopes of discovering huge amounts of oil in the Bakken. Between the staged fracking and horizontal drilling, costs in the region were so high that it didn't pay to do any drilling unless crude prices were at fifty or sixty dollars a barrel.

Some competitors in the Bakken pulled out, figuring it no longer was worth focusing on the region. In North Dakota, locals figured the inevitable bust they had long feared had begun.

"It was frustrating," Hamm says.

Stress grew within Continental's offices as it became clear that the company's executives had blown it. They had become so overconfident that they hadn't hedged any of their oil production. In other words, they hadn't sold oil in the futures markets, as many companies do, to lock in gains. Instead, Continental was "running naked," figuring oil prices would keep rising.

"The market had been moving up and we had been enjoying the rise," says Jeff Hume, Hamm's key lieutenant.

Now that prices had crumbled, Continental's revenues fell sharply. Hamm convened a meeting of his exploration group and made a tough decision: Continental would shut down almost all its drilling in the Bakken.

It was a huge risk—it wasn't clear the dismantled crews would be around if Continental ever wanted to restart work in the area. Not only that, but by not drilling wells, the company risked losing all the leases it had purchased. Hume calculated that if they remained inactive for about eighteen months they would lose their acreage in the Bakken and be back to zero.

There wasn't much they could do, though. Continental wasn't making a penny from the Bakken. It was forced to slash its projected spending for 2009 for the entire company by 31 percent.

Hamm insisted that some leasing continue, however. Continental had managed to horizontally drill a successful well in the Three Forks formation, the zone just below the Bakken. It encouraged him, despite the gloom in the markets.

Mike Armstrong, Hamm's old friend in Dickinson, North Dakota, thought Hamm was crazy to buy leases with oil crashing and the financial world in flames. Hamm also spent more than $600,000 of his personal cash to buy more shares of his company.

"Are you fucking nuts?" Armstrong asked Hamm.

"Oil's going to come back, Mike," Hamm responded.

In late December, though, oil prices fell below thirty-four dollars a barrel, making Bakken drilling appear even more foolish. Hamm's dream of finding huge amounts of oil seemed over.

Mark Papa and his fellow executives at EOG Resources were becoming more enthused about their holdings in the Eagle Ford formation in South Texas. They focused on a section that was about sixty-five hundred square miles that they thought might have the most oil. The shale seemed to have a thickness almost triple that of the Bakken's best sections, another reason they thought they might have something special.

As an added bonus, there were pipelines in the Eagle Ford area, which figured to make it easier for EOG to transport any oil it found. There also seemed to be a lot of natural gas stored in the rock.

In many ways it was déjà vu for that part of the state. A decade or so earlier, Chesapeake and others had drilled in nearby acreage when they targeted the Austin Chalk rock layers. But by 2008 the area had become better known for cattle than crude. Now EOG and a few competitors were back to focus on a new rock layer and they were just as enthused as the original wildcatters who made quick money drilling in the Austin Chalk.

The collapse of oil and gas prices in the fall of 2008 threw a wrench into EOG's plans, however. Papa was relieved that he and his right-hand man, Bill Thomas, had pushed the company to shift away from natural gas, which was tumbling in price. But Papa and his staff hadn't expected oil prices to fall at least as much as gas. Now EOG's revenues were weakening and the company couldn't justify leasing any more land in the Eagle Ford area.

"We would have liked to have captured more, but we simply didn't have the cash flows," Papa says.

Papa and his team were determined to grab more land, but they'd have to wait for the world to calm down. And EOG still hadn't drilled any wells in its newly leased acreage, so Papa and his staff didn't even know if their big bet would pan out.

Charif Souki had experienced a pounding at the hands of financial markets in early 2008, well before the drubbing that McClendon, Ward, Hamm, and Papa now were dealing with. By the summer of that year, he was just trying to keep his company from collapsing.

In August, Souki managed to raise the $250 million of financing that Cheniere Energy needed to build its LNG terminal in Louisiana. He got the money from GSO Capital Partners, a hedge fund that's part of the huge investment firm Blackstone Group, a sign that at least someone believed in what he was trying to do.

Not only that, but hedge fund manager John Paulson's New York firm, Paulson & Co., bought Cheniere's debt. Paulson, who had anticipated the housing meltdown and pulled off the greatest trade in financial history in 2007 and 2008, a haul of $20 billion for himself and his clients, took time to personally visit the Sabine Pass facility in the fall of 2008. Paulson and his lieutenant, Sheru Chowdhry, agreed with Souki's strategy. The hedge fund bought more Cheniere shares late in the year, establishing an approximately 15 percent stake in the company's stock.

That was the good news for Souki.

The bad news: Terms of the financing that Cheniere had received from GSO were hard to swallow. Souki's company was in such a bad position that it couldn't sell shares, as he originally had wanted. Cheniere also couldn't borrow money at interest rates of about 7 percent, as it had in the past.

Instead, the company had turned to GSO for a "rescue loan" secured by the cash still coming to Cheniere from Total and Chevron. To gain the financing, Cheniere agreed to pay GSO an annual interest rate of 12 percent and to give the hedge fund a security that would convert into Cheniere shares if the company somehow rebounded.

Two GSO executives, Dwight Scott and Jason New, were placed on Cheniere's board of directors, as if they were Souki's new chaperones, keeping an eye on him.

"It was painful," Souki says. "I wanted to reduce debt but this added to it."

The August financing deal gave Cheniere enough cash to keep its operations going for about three more years, providing the company with time as it searched for someone to buy the extra capacity in its gas import terminal. But Cheniere was burning through about $50 million a year and had to repay its lenders about $1 billion in 2012. It was difficult to imagine how the company was going to pay that money back.

Souki had kept his engineers and development people around, despite the deep layoffs at the company, so its operations continued. His hope was that natural gas supplies might shrink enough by 2012 to let his idea of importing gas to the United States become popular once again.

By the end of 2008, though, the whole idea of importing natural gas seemed preposterous, much like it did at the beginning of Souki's long journey. So much gas was pouring out of shale wells that the notion of getting even more gas from foreign countries was absurd. In December, the U.S. government's Energy Information Administration estimated that LNG imports eventually would reach 1.2 trillion cubic feet a year, down from a forecast of 6.4 trillion cubic feet a year just two years earlier. Imported gas was expected to drop to 3 percent of U.S. natural gas consumption by 2030 from 16 percent in 2007.[9]

By the end of 2008, Souki had his own doubts about the country's need for foreign gas. "I started getting suspicious" that producers would be able to get so much gas from U.S. shale deposits that his company's services wouldn't be needed after all, he says.

Natural gas prices finished 2008 at about $5.60 per thousand cubic feet, close to their lows for the year, while oil closed at $44.60 a barrel. At those levels, gas was cheaper in the United States than it was abroad. It made absolutely no sense for anyone to even consider paying the cost of freezing gas, shipping it in huge ships similar to giant thermoses, and then having it converted at Cheniere's plant for sale in the United States. Global gas producers were gearing up to send their product anywhere but America. Souki wasn't the only one with a project to import LNG who now looked foolish, but that didn't make him feel any better.

Shares of Cheniere Energy finished 2008 at less than three dollars. It was such a dramatic comedown for Souki and his company that his friends became concerned about him.

"A lot of friends took me out and made sure I wasn't left alone," Souki remembers.

Charif Souki seemed down for the count, much like his fellow dreamers and wildcatters with grand visions of a new era for energy in America.

CHAPTER THIRTEEN

Technology is the real enemy.

—Saudi sheik Ahmed Saki Yamani

Aubrey McClendon and the Chesapeake Energy brain trust tried to reassure nervous investors that the company would be fine, despite lower energy prices and inhospitable capital markets. After they announced plans to sell as much as $3 billion of assets and reduce spending, the stock stabilized around sixteen dollars a share at the beginning of 2009, as the overall market bottomed out.

McClendon seemed in good spirits around Chesapeake's campus, perhaps because he had a new plan to score a huge windfall for himself. A bit later, an attorney working for McClendon approached the three members of the compensation committee of Chesapeake's board of directors. McClendon's lawyer had an unusual suggestion for board members.

McClendon had been stunned by 2008's margin calls and how he had been forced to sell almost all of his Chesapeake shares. Now he couldn't afford the cost of the 2.5 percent stake he was getting in his company's wells. Chesapeake gave McClendon ownership in the wells under what it called its Founder Well Participation Program, but only if he paid the wells' proportional drilling costs.

McClendon's lawyer told the three board members that the Chesapeake chairman and chief executive needed a cool $75 million so he could afford his bill for the wells. Since McClendon no longer had that kind of money, the lawyer asked that the board grant him that sum as a special, one-time bonus.

McClendon didn't have an obvious reason to claim a bonus, let alone one so outsized. In the corporate world, those kinds of payouts are usually for doing something special, or at least a job well done. But Chesapeake shares had dropped nearly 60 percent in 2008. Headlines about his margin calls, and how flat-footed McClendon had been before the market's meltdown, didn't suggest someone deserving of a huge reward, or even a pat on the back. The guy should have felt lucky just to keep his job, some said.

The three board members on the committee—longtime analyst Charles Maxwell, investment banker Frederick Whittemore, and former Oklahoma governor Frank Keating—were well aware of the criticisms leveled at McClendon. Some of them, such as Maxwell, had even tried to warn McClendon that his spending was over the top, though it wasn't like they challenged him very much.

The committee members were also quite cognizant of how much the stock had tumbled. The value of Maxwell's own Chesapeake shares had dropped from $2.5 million to $600,000 since the 2008 crisis.

Nonetheless, the three board members agreed to meet to consider the request from McClendon. As they began discussing the matter, the board members quickly adopted a very different perspective on the bonus idea than the outside critics.

In their view, McClendon had directed $18 billion of spending over the previous several years on shale acreage around the country. The joint ventures completed in 2008, including those with Plains and Statoil, placed a value of about $28 billion on that acreage, while generating over $5 billion of up-front cash for the company. So McClendon had created $10 billion in value, in the view of these board members.

Some of the committee members were especially impressed with the Statoil deal. They were amazed McClendon managed to pull it off, since the agreement was inked in the teeth of the financial meltdown and Statoil could have found a way to back out of it. The transaction clearly was a herculean feat by McClendon, in the eyes of these board members.

The more they debated it, the more board members were convinced that a $75 million bonus meant getting McClendon for a bargain. They began to feel outrage that anyone would question such a sizable payout.

"God damn it, if we can't give him one percent of what he's made the company, what balance is there?" Maxwell asked at one meeting. "This man has performed brilliantly."

Sure, the deals McClendon made were part of his role as the company's chief executive officer, rather than any kind of special additional chore. And the $28 billion value for the properties was just on paper. If Chesapeake actually tried to sell the acreage, there was no way it was going to get any sum close to that valuation in early 2009, given that financial markets had cratered and the economy was still crumbling.

That didn't seem to matter to the committee members. They agreed to cut the $75 million check. Aubrey McClendon's personal comeback was well under way.

Not only that, but McClendon, who agreed to keep his cash compensation flat for five years, also received nearly $33 million in Chesapeake shares as part of his 2008 compensation. Including some extras, it added up to a $112 million package that made him one of the nation's highest-paid executive that year.

It probably didn't hurt that members of the compensation committee had long-standing personal ties to McClendon. Several years earlier, for example, McClendon had donated thousands of dollars to the congressional campaign of Catherine Keating, Frank Keating's wife. Whittemore in the late 1990s had lent money to McClendon.

Each member of the compensation committee had made over $600,000 annually for their services on Chesapeake's board, with Keating netting over $760,000. By contrast, most directors at other companies made about $200,000 annually for their service on corporate boards.

Like all of the company's board members, the three also had access to Chesapeake jets. In addition, Keating's son and daughter-in-law both worked at Chesapeake, making over $130,000 a year each, according to securities filings.

The $75 million bonus wasn't enough for McClendon, though. He also worked out a deal to sell his personal collection of five hundred antique maps of the American Southwest to Chesapeake for just over $12 million, the amount he had paid for them.

It made a bit of sense for the company to purchase the maps, since they were being displayed in conference rooms and elsewhere on Chesapeake's campus, and the company was already insuring them. Chesapeake told shareholders that the dealer who had worked with McClendon to amass the collection judged their value to be $8 million more than the company was paying.

"The maps complement the interior design of our campus buildings and contribute to our workplace culture," Henry Hood, Chesapeake's general counsel wrote to investors in late April 2009, most likely with a straight face. "The company was interested in continuing to have use of the map collection permanently and believed it was not appropriate to continue to rely on cost-free loans of artwork from Aubrey."

That argument was something of a stretch. Natural gas prices were under pressure and Chesapeake was cutting spending and selling assets to raise cash. Yet it somehow had enough spare coin to buy McClendon's nineteenth-century maps.

Even McClendon's fans on the board blanched at the map purchase, though they didn't put up much of a fight to stop it. "Those things would have dropped like lead in an auction" at the time, Maxwell says, because the economic downturn was impacting art and collectibles. "Paintings at Sotheby's already were dropping. . . . It was too much to pay.

"I didn't like it, but what was I going to do about it?" Maxwell says.

In the past, McClendon had received some criticism centering on Chesapeake's outsized spending and use of debt, among other things. But the compensation package and map deal set off a wave of invective the likes of which McClendon hadn't experienced. Wall Street analysts, investors, and members of the media, many of whom had been deferential to the executive, began digging up a range of potential conflicts of interest involving McClendon and Chesapeake, forcing the company to respond, even when the perceived infractions were small potatoes.

In late April, for example, the company revealed that it had paid approximately $177,000 to an affiliate of the Deep Fork Grill, the restaurant in which McClendon was a 49.7 percent owner, for food and beverage catering services during 2008. The company said McClendon hadn't been involved in the decision to hire Deep Fork and that he had requested the company limit its use of the restaurant in the future.[1]

It wasn't the money that got people so upset, really. The purchase of the maps, which led to lawsuits from shareholders, was a drop in the bucket compared to Chesapeake's operating budget; the catering expense was even more insignificant. But to many investors, the deals were a sign that McClendon had too much leeway at the company, even when it wasn't yet clear that his costly shale push would pan out.

The payouts to McClendon also came as the country was affixing blame for the deepest economic downturn since the Great Depression. Many viewed the huge bonuses that had been handed out to already wealthy executives as among the reasons the markets and economy had hit the skids. Around that time, a separate furor had erupted over bonuses given to executives of insurance giant American International Group, which had been bailed out by the government.

Others resented the tumble in Chesapeake shares and the selling McClendon had done that placed additional pressure on the stock. Shareholders weren't getting special bonuses that year; it wasn't clear to many of them why McClendon was.

"I have never seen a more shameful document than the Chesapeake proxy statement," investor Jeffrey Bronchick wrote to the company's board of directors on April 23, 2009, adding that he was disgusted with the board's leadership. "If I could reduce it to one page, I would frame and hang it on my office wall as a near-perfect illustration of the complete collapse of appropriate corporate governance."[2]

At the company's annual meeting in Oklahoma City, some investors cheered McClendon. But others berated him and the board. At one point, McClendon appeared to choke up as he responded to criticism.

"You became so enamored with your own success that your greed and your ego took over and you bet the farm," said Jan Fersing, a Fort Worth, Texas, investor who said he planned to sell his Chesapeake shares.[3]

A series of lawsuits were filed by Chesapeake shareholders, including one from the New Orleans Employees' Retirement System that called McClendon's payouts "a personal bailout" that was "destructive of company value and opportunity."

McClendon seemed to shrug off the condemnation. Friends cited a strong marriage that gave him a source of comfort and support. An overriding confidence and positive outlook also likely helped, as did that $75 million bonus.

But those who worked with McClendon said he cared deeply about the insults, even if it sometimes was hard to see their impact. "He doesn't show it but he's very sensitive," says Maxwell, the Chesapeake board member. "He mulls it over in his mind and doesn't want to give you an advantage, but he talks about it later."

Amid the criticism of McClendon in 2009, something else was happening: The nation was finding a bottom. Economic growth remained weak, unemployment high, and natural gas prices dipped below three dollars per thousand cubic feet in August. But the stock market was rallying as investors breathed a sigh of relief that a collapse of the financial system was becoming more unlikely.

The tone of Chesapeake strategy meetings became more hopeful and enthusiastic. Soon, McClendon had announced his company's third "business model." He might just as well have called it his "Back to the Future" plan, since it entailed accumulating acreage once again.

It was as if his bad dream was over and McClendon was ready to get back to work. He made plans to expand in the Marcellus Shale, to gobble up land in new and exciting formations like the Utica Shale that was under parts of Ohio, New York, and Pennsylvania, and to catch up to EOG in the Eagle Ford. There also was the Mississippian Lime, a limestone play covering seven thousand acres between south-central Kansas and north-central Oklahoma, not to mention the Niobrara Shale beneath eastern Wyoming.

McClendon had genuine reason for optimism. Geologists were zeroing in on new and exciting shale deposits, while huge amounts of gas appeared to be coming from existing shale formations, including those that Chesapeake had placed expensive bets on. These new wells were producing more gas even while becoming less expensive to drill.

Most of the country was trying to recover from a debilitating housing collapse, but excitement was building within the energy patch. Just as huge amounts of oil were uncovered in East Texas during the Great Depression, setting off "a boom unlike any American had ever seen," as Bryan Burrough writes in *The Big Rich*, American frackers in 2009 had a chance to help the nation rebound from its economic meltdown.

In August 2009, natural gas prices fell to three dollars per thousand cubic feet, the lowest level in seven years, amid a record amount of gas in storage. Part of the excess supply was due to lower demand amid a still anemic U.S. economy. But there also were fewer than seven hundred drilling rigs operating in the entire country, down from nearly sixteen hundred a year earlier. And yet natural gas output was rising, suggesting that companies were doing a better job getting gas from shale and other rock.

Some of the most promising spots to drill seemed full of oil, not just natural gas. Bedrock with oil was especially attractive because crude prices were on their way to nearly eighty dollars a barrel in early 2010 from just above forty dollars a barrel in early 2009, as oil demand grew in China and other emerging-market nations.

McClendon wasn't the only one getting reenergized about shale. So were environmentalists like Robert F. Kennedy Jr., who had founded the Waterkeeper Alliance, which was fighting hard against coal mining. "A revolution in natural gas production over the past two years has left America awash with natural gas and has made it possible to eliminate most of our dependence on deadly, destructive coal practically over-night," Kennedy wrote in an op-ed in the *Financial Times* in July 2009. "America's cornucopia of renewables and the recent maturation of solar, geothermal, and wind technologies will allow us to meet most of our energy needs with clean, cheap, green power. In the short term, natural gas is an obvious bridge fuel to the 'new' energy economy." (Later, Kennedy became an opponent of fracking to find natural gas.)

A year or two earlier, when Chesapeake bought in a frenzied way, its competition was limited to upstarts like Bob Simpson's XTO. Now everyone was bidding up acreage, forcing Chesapeake to pay top dollar for land. Before the 2008 financial crisis, acreage in the Haynesville Shale went for $1,000 an acre. Now the Mississippian Lime was going for $4,000 an acre. Chesapeake borrowed more money to pay for the purchases, pushing its debt load to $12.3 billion.

Gas managed to push back above $5.50 per thousand cubic feet by the end of 2009, suggesting that Chesapeake could score big profits by pumping its existing fields and also adding new oil and gas fields. Some shale formations were profitable with gas as low as four dollars per thousand cubic feet. Could anyone really expect McClendon to ignore all these juicy targets?

"Gas prices came back up and we thought we were going to be okay," says a Chesapeake board member. "The temptation was too much."

Shares of Tom Ward's young company, SandRidge Energy, had suffered an embarrassing tumble from sixty-six dollars in July 2008 to six in January 2009.

Members of the company's board of directors wanted to know how Ward and his management team were going to set the company on a safer course and revive investor interest. Some on the board were especially nervous. They didn't want to have to deal with another touch-and-go situation like they had in October, when the company was hours away from running out of financing.

In board meetings, Dan Jordan, a veteran Oklahoma energy man, put up a spirited defense of Ward. Give him time, he'll figure out a plan, Jordan insisted. "You don't understand, Tom does his best work under pressure," Jordan told fellow board members. "This guy will do something. . . . You don't know him."

Jordan, who had worked with Ward and McClendon for decades, tried to explain why they could be trusted. "These guys are adrenaline junkies, it's calculated, they love risk. . . . They get right up to the edge of the cliff but never go off."

In February 2009, Ward came up with a new plan, just as Jordan had predicted. McClendon and others in the business still believed in natural gas, but Ward thought they had it all wrong. There was going to be too great a supply of natural gas for months, if not years, thanks to all the money being thrown at new shale fields, just as Mark Papa and EOG were saying.

SandRidge needed to shift into oil, Ward argued. Because oil is more easily shipped around the world than natural gas, crude prices tend to be based on global supply-and-demand dynamics rather than what's going on in the United States. American crude production seemed unlikely to have much impact on global oil supplies, especially since there wasn't clear evidence of an imminent rise in U.S. shale oil production. It all made oil a safer bet, Ward argued.

Drilling the Bakken seemed too expensive to Ward, who had turned down the chance to enter some of the plays at the end of his tenure at Chesapeake. Instead, SandRidge would focus on more "conventional" oil plays, or those with limestone and other rock that was easier to tap than shale. Let McClendon, Hamm, and Papa go after the sexy shale drilling, we'll make money on the boring stuff, Ward was saying.

Soon, SandRidge was selling natural gas acreage, much of which eventually would become worthless, and buying land in oil-rich formations, including in Texas's Permian Basin and the Mississippian Lime, where land could be purchased for just $200 an acre. Later, SandRidge would spend over a billion dollars to buy a Texas oil company called Arena Resources.

SandRidge's stock didn't show much of a response to the strategy. A year earlier, investors were sure SandRidge was set to become the new Chesapeake. Now they weren't convinced of Ward's plan to turn Sand-Ridge into an oil power.

Ward was sure he had saved his company, after being caught by surprise by the 2008 energy collapse, and he was confident that Wall Street eventually would appreciate his moves. He was about to be shocked in a new way, however.

Tom Ward drove home one evening in mid-January 2009, exhausted from another trying day. Shares of his company were still under pressure and Ward was working hard on the company's revamped strategy, but he instantly relaxed as he opened his front door in suburban Edmond, Oklahoma. Like most days, Ward was in time for dinner. His wife, Sch'ree, was both a good cook and the most upbeat person in his life. With markets under assault, Ward turned to her for comfort.

Before Ward could sit down, though, his youngest son, James, called out to him. The eighteen-year-old needed to speak to him about something. When Ward walked into James's room, his son looked troubled and unsteady. James told his father that he wanted his help—he needed a doctor to get him Suboxone tablets.

Ward didn't know what his son meant. He had never heard of Suboxone. James took a deep breath and explained that Suboxone was a drug used to treat addiction to Oxycontin, the pain medication.

"Do you think you have a problem?" Ward asked his son.

"I don't know," James responded.

A stunned Ward hugged his son, trying to digest what he was telling him.

Later that evening, the doorbell rang. It was Mike Harrison, who had once played lunchtime basketball games with Ward but now did odd jobs around SandRidge's offices. As he often did, Harrison had come by to deliver SandRidge paperwork for Ward to sign.

Most evenings, Ward was businesslike when Harrison dropped by. This time, though, when Harrison found his boss in his study, he couldn't believe his eyes. Tears were streaming down Ward's face and he had difficulty speaking.

"My son . . . it's bad," Ward struggled to tell Harrison. "He's got a bad, bad thing going on. It hurts."

The news was less shocking to Harrison, who had seen James and some of his friends come late to Sunday church services at a SandRidge auditorium. Harrison had also glimpsed James, usually with those same friends, in grittier parts of town, smoking and hanging out.

Ward hadn't suspected anything, though, and he was shaken. "Our family had always been like a Norman Rockwell painting," he says.

A month later, James entered a drug treatment facility to try to get well.

The revelation brought back dark memories for Ward of the alcohol addictions his father and grandfather battled. Ward couldn't stop thinking about what he could have done to help his son avoid his own addiction. He felt he should have been more focused on his son's needs and that he had failed to provide James with ample warnings. "It was my fault I didn't tell him how easy it was to become addicted in our family," he says.

In a quiet moment, Ward told a friend, Greg Dewey, that he felt "humbled" by the shift in his fortunes. He told Dewey that he felt he had become focused on "power and money."

"Pride crept in and he became consumed with some things that we're all tempted with," says Dewey, a SandRidge communications and community relations executive who doubled as the company's in-house pastor.

Ward admired his son's honesty and eagerness to begin the long, difficult road toward beating his horrible addiction. It gave Ward, a staunch Republican, new empathy for addicts. He learned that many former addicts can't obtain driver's licenses, making it hard to get a job. Ward's company would begin a program to hire more than two dozen former drug addicts, to try to give them a second chance.

Suspicions were growing about the drilling in Dimock, Pennsylvania, which sat atop the Marcellus Shale formation. It was hard to tell how many locals were worried about the activity. Many were grateful for the checks they were receiving to lease their land, money that allowed some struggling landowners to stay on their farms and properties. Others, however, grew sick of the noise and dirt kicked up by the drilling, or became wary that chemicals in the fracking fluid might invade their water.

An antidrilling movement congealed among residents of both Pennsylvania and New York. Members of the media, such as Tom Wilber of the *Binghamton Press & Sun-Bulletin*, began asking tough questions of Terry Engelder and others about the radioactive components of the flowback coming up to the surface from local wells. The growing questions resulted in New York's placing a moratorium on high-volume hydraulic fracturing in the state.

Meanwhile, a telegenic, articulate resident of Dimock named Victoria Switzer, who was frustrated by how noisy the area around her pretty wood-frame home had become, was making progress reaching out to the media and demonstrating how disruptive the drilling and fracking was.

On New Year's Day 2009, problems erupted. That's when an explosion in Norma Fiorentino's backyard in Dimock shattered the pump housing her drinking well. Pennsylvania's Department of Environmental Protection put the blame on nearby natural gas drilling by Cabot, the gas producer active in the area. After complaints, the department also determined that Cabot had polluted the drinking water of at least nine homes in the township by allowing methane to migrate.[4]

The explosion, and other water complaints in the area, came at a time when the nation was coming to grips with, and apportioning blame for, the housing collapse and resulting economic downturn, notes author Seamus McGraw. Some put the responsibility on reckless borrowers. But more fingered what they saw as indifferent lenders, greedy megacompanies, and complicit government officials. That helps explain why some were open to the argument that these same types of players again were conspiring to hurt innocent citizens, this time in the nation's gas fields.

The Fiorentino well explosion brought a young film director named Josh Fox to the Dimock area to investigate what was going on. Soon, Vera Scroggins also arrived on the scene.

About eighteen years earlier, Scroggins had moved to an 1880s farmhouse nearby after working as a teacher in Long Island. A matronly woman with time on her hands, Scroggins already had experience picking fights with the local stonecutting industry.

"They put trucks on my road where I liked to walk and bicycle and that bugged me," she says, referring to the stone cutters. "There was no zoning so they could blast away. This is America, you have to claim your rights."

When workers told her they couldn't help making the noise, which came with the job, Scroggins was dismissive, telling them they should find a different job that didn't cause so much of a racket.

Scroggins made some headway getting the stone cutters to lower their volume. When a friend in Dimock told her that she had to rely on bottled water, due to the drilling mishap, she knew she had a new cause. Scroggins loved to photograph and videotape her kids and grandchildren. She decided to chronicle local gas drilling with her video camera. Soon she began reaching out to the media, attracting attention to the drilling issues in the area.

The frackers weren't prepared for the scrutiny that would soon result.

On March 3, 2009, shares of Harold Hamm's Continental Resources traded for about fourteen dollars a share—even lower than when it went public nearly two years earlier. Oil was less than forty-two dollars a barrel, down 70 percent in eight months, and crews were pulling out of the Bakken.

Continental controlled over 600,000 acres in the Bakken, making it the largest leaseholder in the formation. The company's production had increased 130 percent in four years, to thirty-three thousand barrels a day. But Continental's Bakken wells still weren't wowing anyone. They were producing less than six hundred barrels a day during their first seven days online, well below the thousand or more a day that impressive wells pumped during their "initial production" phase, before production begins to level off.

Continental had been forced to pare its expensive drilling in North Dakota because it couldn't make money with oil prices so weak. Unless crude rose to at least fifty or sixty dollars a barrel, the company couldn't afford to ramp up again in the area. There remained a chance that Hamm's company would lose leases if they didn't go back to drilling soon, because the leases required drilling within a few years or they'd revert to landowners.

"Rates of return get pretty minimal below fifty dollars," Hamm told a *Forbes* reporter. "You wonder if we ought to be doing it or not."[5]

Wall Street remained dubious that Hamm and his team could squeeze much out of the Bakken rock. "We just don't know how much of their acreage is going to work," Leo Mariani, an analyst at RBC Capital Markets, said. "It's pretty early on in the Bakken and a lot of the acreage is not tested."[6]

Hamm had gotten used to being seen as the hopeless dreamer in little Enid, Oklahoma. The sixty-three-year-old enjoyed steak dinners at a local Applebee's, spending an extra buck on a double scotch to wash it down, much like his fellow patrons. But he still was worth a few billion dollars, so it was easy for him to ignore the naysayers.

The country seemed to have lost its will to drill, Hamm told the *Forbes* reporter. The Dow Jones Industrial Average had fallen over 45 percent in a year and the energy business was one of many on its back, but Hamm kept saying things would get better in the Bakken. He believed that his crew, along with others in the business, would further refine their drilling technologies and approaches.

Hamm was dealing with more pressing issues than Wall Street critics. A pipeline called the Keystone XL promised to add over 800,000 barrels of Canadian tar-sands crude each day to U.S. energy supplies, potentially lowering prices still further. "We're going to be drowned with Canadian crude," Mickey Thompson, the Oklahoma political veteran, told Hamm as they sat in his kitchen trying to map out a response to the pipeline.

"We've got to do something to stop it," said Hamm. They decided to enlist politicians and environmentalists to try to kill the pipeline project.

Hamm also began to worry that some of those pushing natural gas as a clean, homegrown fuel, such as Aubrey McClendon and T. Boone Pickens, were becoming a bit too effective in their plans and pitches. Pickens, a longtime oilman turned hedge fund investor, was spending big money to market a proposal to have the nation make electricity from wind instead of natural gas. The freed-up gas would be used as a transportation fuel, replacing diesel for commercial trucks, siphoning demand for crude. Pickens announced that his company would spend $2 billion on wind turbines for the Texas Panhandle, to create the world's largest wind farm.

Hamm said Pickens's plan wouldn't work. The government would have to spend too much to upgrade the nation's electrical transmission grid. The proposal also figured to reduce crude demand, hurting the price Continental could get for its oil.

Hamm called Pickens to try to get him to drop his idea. "Boone, your numbers are just wrong," he said.

But Pickens kept pushing his plan, and McClendon made inroads in his own public push for natural gas. To defend crude, Hamm decided to take a more public stance in support of domestic oil. Staying out of the spotlight made it easier for Hamm to lease land, and he still didn't relish public speaking, after years of working on his elocution, friends said.

But Hamm figured that if he could just get Americans to realize how close the nation was to energy independence, maybe they'd be more supportive of what he was doing in North Dakota, and less likely to change the tax code or enact laws to make it harder for those drilling for crude. And if he could spark some enthusiasm for domestic oil, companies and public bodies might be more willing to invest in housing, roads, pipelines, and other infrastructure in North Dakota, investments that would help Continental's drilling in the Bakken.

At a charity event, Hamm approached Rex Tillerson, the chief executive officer of Exxon, to try to get him to join his public effort to promote U.S. oil drilling. Tillerson was lukewarm to the idea, frustrating Hamm.

Soon, Hamm's concerns seemed less pressing. The Keystone pipeline's operator, TransCanada Corp., agreed to carry U.S. oil along with Canadian crude, a move that gave Continental a new way to ship its oil out of the Bakken. More important, the U.S. economy showed signs of rebounding and oil prices rose. Hamm quickly moved drilling rigs back into North Dakota as his team continued to try to make its wells pump meaningful amounts of oil.

Finally, Continental's crew figured out the last piece of the Bakken drilling puzzle. Through trial and error, they realized that fracking wells in about thirty stages helped extract huge amounts of crude. Fewer than thirty stages led to too little oil; more than thirty was too costly and not worth it. Soon, Continental's wells were giving up over one thousand barrels a day.

"Areas that weren't that good all of a sudden became commercial," Jack Stark recalls. "Now we knew we had a significant play on our hands."

Almost overnight, the Bakken went from good to great.

Shares of Cheniere Energy traded for just three dollars at the start of 2009, but Charif Souki was worried things could get even worse for his company.

At an energy industry conference, Souki saw Mark Papa, the chief executive officer of EOG, and Larry Nichols, the head of Devon Energy, and thought he noticed something wrong. Each executive had optimistic things to say about how natural gas prices would bounce back.

To Souki, though, the executives seemed almost in a daze. Their upbeat words didn't match the concern on their faces, as if Papa and Nichols knew some bad news that hadn't come out yet. "There was something about their body language that made me feel like I was missing something," Souki says. "They had a look in their eyes that wasn't good."

Back at the office, Souki told his staff to pull the results of every gas field in the country. He also discovered a Web site run by drilling company Baker Hughes that tracked the country's rig counts. Souki realized that natural gas production was soaring, even though there were fewer rigs in operation. Drillers were becoming more accurate, and better able to reach targets horizontally, than ever before. The shale revolution was for real.

"It was astounding," Souki says.

Now Souki understood why gas producers looked so worried—their fracking and horizontal drilling techniques in shale plays were working so well that the companies were beginning to worry that a glut of gas was going to send prices skidding.

It also made sense that investors were turning on Souki and his stock. Who needed a costly facility to import foreign gas when there was so much cheap natural gas in the United States? Expectations of an incoming wave of imported LNG seemed like folly. There were birds chirping and cows mooing near his Louisiana facility, but fewer than five LNG tankers had even arrived at the newly built terminal, and Cheniere still had unused space amounting to half of its processing capacity.

This is not working, Souki thought to himself.

Souki strived to maintain his usual upbeat, confident mien around the office. He didn't want employees to lose faith. He told reporters and investors that natural gas prices were bound to rebound, creating a need for imported gas, and he shared few doubts with his board of directors. If Souki told them that his company was in trouble they'd want to know what he was going to do about it. At that moment, though, he didn't have a clue.

A bit later, Souki got a call from Aubrey McClendon, whom he knew from industry events. After some chitchat, McClendon threw out an unexpected question.

"Hey, can you guys do liquefaction at Sabine Pass?" McClendon asked. He was suggesting that Cheniere's facility *could* turn natural gas into liquefied natural gas, so that it might be exported from the country.

"We'll take a look," Souki replied, unconvinced that McClendon was giving him any kind of hot tip.

Souki went home and ran the numbers. They didn't work. Natural gas was above four dollars per thousand cubic feet, and oil traded at around forty dollars a barrel. With crude that cheap, it didn't make sense for foreign companies to pay the cost of buying U.S. gas.

McClendon's question stuck in Souki's mind, though. A bit later, when Davis Thames, Cheniere's head of marketing, walked down the hall to make a cup of coffee, Souki yelled out to him from his office. "Guess who just called us? Aubrey McClendon," he said. "You won't believe what he just asked me—he said we should export."

"No way!" Thames responded.

Thames and his boss had a good chuckle, and Thames was on his way back to his office. They'd just built a company to import natural gas to America. They weren't about to throw it all away and start shipping gas out of the country.

"I thought it was ridiculous," Thames says.

A bit later, it dawned on Thames that Souki had mentioned the conversation with McClendon to get Thames to start thinking about the notion. The thought of exporting gas from the United States had crossed Souki's mind before McClendon's call, but it was hard to tell where oil and gas prices were heading, so it seemed too early to think about scrapping the idea of importing energy.

A month or so later, McClendon and some fellow executives invited Souki and Meg Gentle, Cheniere's head of strategic planning, to visit Chesapeake Energy. On the plane, Souki and Gentle laughed that they had no clue why McClendon had summoned them to Oklahoma City. A sit-down with the charismatic executive was hard to turn down, though.

As the meeting got under way, Souki updated McClendon on Cheniere's plans to court foreign producers hoping to ship gas to America. They discussed how the LNG process works and McClendon picked Souki's brain for details of his business.

That's when McClendon blurted out something radical: "Why don't you build me an export terminal?" He wanted a way to sell more of the natural gas his company was suddenly finding easier to produce, and he was eager to somehow get prices higher. He asked what it would entail to build a terminal to export gas to foreign markets.

Souki said they hadn't run the numbers on what it would cost to reconfigure the Sabine Pass facility so it could turn natural gas into LNG for export. Until then, exporting hadn't been in the realm of possibility.

On the flight back, Souki and Gentle couldn't stop thinking about their meeting. Here was a huge natural gas producer saying there was so much spare gas in the country that it should be exported. For Souki and Gentle, it was confirmation that shale drilling would overwhelm the nation with cheap natural gas for the foreseeable future. That was pretty much the worst news a company trying to import gas could get.

When they got back to the office, Souki and his team decided to start looking into the idea of exporting gas—even though they already had contracts with Total and Chevron to convert their imported LNG into gas to be sold in America. If Cheniere was going to export, the company had to be the first out of the gate, the executives decided. Souki's old partner, Michael Smith, and others working on importing gas also likely would see the logic of exporting, as natural gas supplies piled up.

At first blush, retrofitting the terminal seemed much too expensive. Yes, they had already built a plant, so they wouldn't have to build another one to house the turbines and equipment necessary to liquefy and purify gas so it could be turned into LNG for export. And Cheniere had tanks and shipping berths available thanks to all that extra, unnecessary capacity it had built for importing. Those, too, could be used for gas exports.

But there seemed to be way too many hurdles for the idea. Could regulators be persuaded to let little Cheniere export energy from the country even though no one had been doing it? Might a company that seemed on its last breath actually raise billions of dollars for a single "train," or a unit with the dozen gas turbines and other equipment required to create LNG? And could Cheniere stay alive during the four or so years it might take to build such an export facility?

Souki didn't share details of the McClendon meeting with his board of directors. Nor did he tell them that his people had begun to test the export idea on officials in Washington. The thought of shipping gas from the United States seemed too preposterous. Just a year or so ago the nation was scrambling for ways to get its hands on gas, and now there was so much it was going to export it?

"I didn't want to go to them until I knew what the costs were and that it would work," Souki says.

By then, Cheniere's board members weren't thrilled with Souki and his brilliant ideas. GSO, the Blackstone unit, just wanted to get its money back, somehow, as did John Paulson's hedge fund. They likely wouldn't have much patience for another harebrained scheme from Souki, especially one with so few details fleshed out.

"They knew I was looking at different ideas, but my credibility with the board was suspect," Souki says. "It wasn't a warm and fuzzy relationship."

In April, Souki was still reeling when he returned to Aspen and saw his old friend Geoff Tasker.

"This thing has wiped me out," Souki said. "What really bothers me is the fact I misread it and everything turned on me."

"How could you possibly have known" that U.S. gas production would surge? Tasker asked Souki, trying to boost his confidence.

"Yeah, but I missed it," Souki responded glumly.

By late 2009, however, Souki's mood had improved. Oil prices were soaring again, thanks to growing emerging-market demand, sending crude toward eighty dollars a barrel. But natural gas prices only rose a bit, to about $5.50 per thousand cubic feet.

Souki asked Bechtel Corporation, the big international construction company, to come up with a quote of how much it would cost to retool the Sabine Pass terminal for LNG exports. Souki didn't think he could ask permission of his board for such an expensive project until he had a quote on what it would cost.

As he waited for the estimate, Souki's natural optimism took hold and he became more confident the idea could work. Gentle, Thames, and others at the company were much more skeptical. The executives didn't want to pour cold water on his new idea, but their body language was unmistakable.

"I knew they weren't completely convinced," Souki recalls. "They were loyal and supported me, but you could tell they thought I was crazy."

In early December 2009, Cheniere's stock fell to a mere $1.90, less than a large cup of coffee. Souki felt more heat from his board and top investors. He told Sheru Chowdhry, a senior executive at hedge fund shareholder Paulson & Co., that the whole importing-gas thing wasn't working out. He said he thought he could find a new strategy, though, so he shouldn't lose faith.

"You are a delusional optimist," Chowdhry told Souki.

CHAPTER FOURTEEN

Everyone underestimates perseverance.

—Charif Souki

It was time to let the secret out.

Mark Papa, the Enron refugee who ran EOG Resources, had spent two and a half years directing a clandestine effort to lease land in South Texas. EOG's geologists and engineers were convinced that a huge amount of oil was buried in a 400-mile-long shale formation called the Eagle Ford stretching from the Mexican border south of San Antonio to north of the city of Austin.

Until January 2009, when EOG drilled its first Eagle Ford well, Papa and his team weren't entirely sure how fruitful their acreage would be. But that first well had initial production of almost seven hundred barrels a day, enough to cause a buzz—and more than a little relief—around the company's Houston headquarters. In late September, another well came in. This time, almost seventeen hundred barrels of crude poured out each day, a sign they were on to something really big.

Papa and EOG's president, Bill Thomas, put the pedal to the metal, leasing even more land. By April 2010, the company was sitting on over 500,000 acres after shelling out more than $200 million, or about $425 an acre.

Papa hardly visited EOG's crew in the Eagle Ford. He also didn't spend much time supervising drilling in the Bakken formation. Harold Hamm liked to fly to North Dakota to visit Continental's sites, but Papa was content to let his Bakken and Eagle Ford groups do their thing while lending support from afar. Most days, he walked the halls of EOG's offices in deep thought, arms folded, strategizing the company's next move. Papa was deliberate and even-keeled, like a kind uncle. The pens in the front pocket of his dress shirt made him look like any petroleum engineer in the industry, rather than a high-powered CEO making the bet of his lifetime.

If Papa didn't have much in common with Harold Hamm, he and his company were entirely different from Aubrey McClendon and Chesapeake Energy. EOG's executive floor was as quiet as a library, unlike Chesapeake's youthful and lively campus. EOG issued a single press release in 2009 unrelated to an earnings release, dividend payment, or public presentation. That year, Chesapeake's public relations machine churned out over twenty releases touting awards it had received, new strategies it had unveiled, and even an appearance by McClendon on a business television show hosted by CNBC's Jim Cramer.

Most industry members didn't have a full idea why EOG was spending so much money, nor did they believe the Eagle Ford would amount to very much. Papa didn't seem to care. On his office wall, he had framed a line from Theodore Roosevelt's famed "Man in the Arena" speech, delivered at Paris's Sorbonne in 1910: "It's not the critic who counts. . . . The credit belongs to the man who is actually in the arena, whose face is marred by dust and sweat and blood."

By the spring of 2010, EOG had drilled ten wells in the Eagle Ford and executives knew they had to stop playing coy. It was time to level with investors and the public. "We felt the data was on the verge of becoming public," Papa says.

On April 7, EOG hosted an "analyst day" at Houston's Four Seasons Hotel to update Wall Street analysts and investors tuned in to a public webcast. Addressing the crowd from the podium, Papa began detailing EOG's acreage and how pleased he was with the company's progress. Then he dropped a bombshell on the audience. He said EOG was sitting on nine hundred million barrels of oil in the Eagle Ford formation. Papa argued that this rock was going to rival the Bakken's.

There was tumult in the crowd. Some in the audience grabbed BlackBerry devices and other smartphones and began typing furiously. Others pulled out laptops or raced out of the room.

"You could see people making trades," Papa says. "It was pretty dramatic."

Papa also told the crowd that the company was getting reports that EOG's wells in the Bakken were brimming with new oil and its Parshall field was turning into the most prolific in the lower forty-eight states. A monumental change was afoot, he said.

"Horizontal oil from unconventional rock will be a North American industry game changer," Papa told the Four Seasons audience, or at least those who hadn't bolted from their seats to buy the stock.

Within hours, EOG's stock was shooting past the $100 mark. By the next day, it was at almost $107, up about 10 percent in a couple of days, putting the company's market value at $27 billion, up 80 percent in one year.

Papa's announcement did more than confirm EOG's ascendance in the energy world. It was proof that the Bakken wasn't a freak, one-off formation and fresh evidence that the country was beginning to pump enough oil and natural gas from shale to shake up the world's energy order.

By then, the big boys of the oil and gas world had taken belated notice of the American energy revolution, one that carried the possibility of American independence, this time from foreign oil. Now the giants had to get in, before it was too late. In 2011 and 2012, London's BP, Norway's Statoil ASA, and France's Total SA each spent billions of dollars for acquisitions, interests, and joint ventures in shale formations in Pennsylvania, Oklahoma, Texas, Arkansas, and elsewhere. So did the China National Offshore Oil Corporation, Italy's Eni, and Australian energy conglomerate BHP Billiton.

American energy goliaths also headed home searching for deals, grasping that they had overlooked something special in their own backyards. It was like a college student racing home to woo an old girlfriend after hearing that foreign guys had begun to court her.

ExxonMobil—which continued to maintain its headquarters on a spot above the Barnett Shale formation—finally made a huge bet on U.S. shale formations, playing catch-up in a very big way. Exxon shelled out $31 billion to buy natural gas driller XTO Energy, the company started by Bob Simpson that had been competing with Chesapeake for shale acreage. The deal, Exxon's biggest in a decade, made it the nation's largest natural gas producer.

Huge energy companies often lie back and wait for new fields to be developed before swooping in. They like to be sure new fields are big enough to justify their time and money. This time, the giants were refocusing on America because they now appreciated why George Mitchell had gotten so excited two decades earlier when he zeroed in on shale.

At the same time, it was becoming more expensive to find oil and gas reservoirs abroad. The rekindled love for American rock also resulted from the slick sales efforts of bankers, including Ralph Eads, Aubrey McClendon's old college buddy.

"This is like owning the Empire State Building," Eads told mega oil companies to try to get them to open their checkbooks for shale deals, he later recalled. "It's not going to be repeated. You miss the boat, you miss the boat."

There was some bluster in the salesmanship, Eads acknowledged. "Typically, we represent sellers, so I want to persuade buyers that gas prices are going to be as high as possible," he said. "The buyers are big boys—they are giant companies with thousands of gas economists who know way more than I know. Caveat emptor."[1]

The winners in this buying frenzy included those running exploration companies that had been early drillers in shale formations, such as Bob Simpson of XTO and Tim Headington of Headington Oil, both of whom sold out to mega oil companies.

Sometimes the windfalls were astonishing. Until 2004 or so, Terrence Pegula, who was born to a coal-mining family in Carbondale, Pennsylvania, and had borrowed $7,500 from friends and family members to start an exploration company called East Resources, was grinding out a living drilling vertically in conventional Pennsylvania rock layers. Pegula's acreage was atop the Marcellus layer, but no one thought there was much gas in that rock, so Pegula didn't do anything with it. Locals thought Pegula was a solid producer, but nothing special.

When it became clear that the Marcellus layer was filled with more gas than anyone imagined, Pegula quickly expanded his acreage. As the smart money zeroed in on the area in 2009, Pegula sold about one-third of his company to leveraged-buyout giant KKR, pocketing a hefty $350 million. A year later, Pegula and KKR flipped the entire company to Royal Dutch Shell for an even heftier $4.7 billion, turning heads in Pennsylvania.[2]

George Mitchell also decided to sell in 2010. By then, he had helped finance the purchase by Alta Resources—the company run by Joe Greenberg—of forty-two thousand acres in the Marcellus. Their wells were flowing with gas, but the Williams Cos., a large natural gas company, bid over $500 million for the acreage, an offer the group couldn't refuse. It was another remarkable score for George Mitchell, who turned ninety years old that year.

Many of those who sold out to major oil companies and became instant billionaires seemed eager to spend their money on pleasurable diversions after years of hard work in the oil patch. Pegula, for example, paid about $200 million to buy the Buffalo Sabres hockey team. Bob Simpson, the XTO chief who liked to hire employees who had once received corporal punishment from their parents, joined with a partner to pay nearly $600 million for the Texas Rangers baseball team.[3]

Tim Headington, who once had competed with Harold Hamm in the Bakken in 2008, became a Hollywood player, like oil barons before him, including Marvin Davis and Howard Hughes. Headington produced movies and television shows, including Martin Scorsese's *The Departed* and *The Aviator*, along with Academy Award winners *Hugo* and *Argo*, while continuing to search for new pockets of oil.

These executives garnered headlines, but an equally remarkable story was in the larger sums flowing to landowners in the hottest shale plays. So many farmers, ranchers, and homeowners became wealthy leasing drilling rights to their properties that a nickname was coined for them: "shale-ionaires." The industry said it paid out $6 billion from 2008 to 2010 just in Pennsylvania, the heart of the Marcellus Shale formation.[4] EOG calculated that it would pay over $30 billion over the life of its Eagle Ford wells to leaseholders in the region.

The Eagle Ford became the site of particularly rabid interest after EOG's news. By 2010, about thirty companies were chasing EOG, more than a thousand drilling permits had been issued—up from twenty-six in late 2008—and crude production had grown to seventy-one thousand barrels a day, up from virtually nothing in 2008.[5]

For some of those benefiting from the land grab, relief came in the nick of time.

William Butler was born in 1926 in the tiny South Texas city of Cabeza, the son of a local farmer. While Butler was serving in the navy, his commander decided to give him a new name, one that stuck the rest of his life. "We already have a Bill Butler—you'll be Buck," the commander, an obvious fan of alliteration, told the young man.

After completing his service, Butler told his father that he wouldn't be joining him in the fields. That life seemed too difficult to the young man. "I know one thing I won't do," Butler told his father, "that's farm."

Over the years, Butler worked on an oil pipeline, broke and sold horses, and served as a state cattle inspector. Eventually he purchased a "sale barn," or a barn where local cattle were auctioned, in Nixon, Texas. The business suited the gregarious Butler, who enjoyed conversing with anyone and everyone.

"I don't care if they're black or brown," he says. "I don't give a damn."

Some years were lucrative, others saw losses. Butler supplemented his income by tending to over a thousand cows and calves of his own, his true passion. Whenever he had spare cash, he used it to buy acreage. He didn't invest in stocks, his home needed work, and he hardly ever took a vacation. But he kept on buying nearby land, accumulating over five thousand acres around Nixon.

Some of the acreage was for his cows to graze, but other purchases seemed to feed another need. When he drove visitors around, Butler proudly pointed to his land, noting how far it stretched. "Over there's my country," he told a visitor, a reflection of the enduring streak of Texan independence in the area.

"I always believed in land, I always thought land was the best place to put your money," Butler says. "They ain't making more land but they're still making more babies."

Butler usually didn't tell his wife, Vera, about his purchases until after the deals were made, nor did he inform his kids, which sometimes was a point of frustration for them. Butler never expected rock below his land to be worth much. "My father used to say, "Oil's in there but it's too deep, no way they'll ever get to it," he recalls.

By 2009, Nixon had just about two thousand residents, an estimated household income of $28,000 dollars, and was thirty minutes from the closest movie theater. Butler's outgoing and generous personality helped him grow his auction business. He also received support from his wife, his two sons, and a daughter-in-law, all of whom helped with the auctions or worked with Butler's own cattle.

Butler placed his faith in too many people, however, leading to serious problems. One time, a bookkeeper who had become close to the family stole over a million dollars from the operation. Butler found $150,000 of it in a cigar box, but much of it was sent to the bookkeeper's boyfriend, a cow buyer in Mexico, Butler says. Some of the money was repaid, but not all of it.

Another time, while Butler was recuperating in a hospital in San Antonio after doctors diagnosed him with Guillain-Barré syndrome—an autoimmune disease that left him weak—two local young men paid a sick visit. They eventually smooth-talked Butler into signing over some of his mineral rights, he recalls.

In 2009, as oil leasing began in the area, Butler was having breakfast with a local bachelor and began to feel sorry for him. He agreed to lease some of his land to him, asking only that the man make sure an oil well and a water well were drilled on the property. The local man flipped the mineral rights to ConocoPhillips, which told Butler they were never told about any agreement to drill wells on his land.

"I didn't know anything about the oil business," Butler acknowledges.

After each setback, Butler was forced to sell some land or cattle to raise cash to keep the business going. But he also resorted to borrowing millions of dollars to maintain his operation, he says, turning to banks and sometimes a local Chevrolet dealer. It got harder as he aged. At one point, he was down to a thousand dollars in the bank.

Butler worked seven days a week into his eighties. He desperately needed a new truck, but he couldn't afford it. Butler and his wife, who years earlier had traveled the nation, hadn't taken a vacation in thirty-seven years.

"We scraped by," Butler says, especially when drought conditions sent the price of hay and feed climbing. "We were barely making our payments" on the debt.

By 2010, Butler was eighty-four years old and worried about the future. His middle-aged sons were hard workers, but Butler didn't know if they could handle running a business that owed several million dollars.

"I lay awake thinking about what the boys would do," he says. "I didn't think they could handle the pressure I was dealing with."

Vera Butler had a nervous breakdown at one point and several times had to call an ambulance to come to her aid. She began taking a low-grade Xanax pill every night to help with the strain.

"I'm a worrier," she says. "We owed on everything."

One day in 2010, Butler was on the phone in his tiny, ragged office, a deer with huge antlers mounted on the wall above him, when a visitor knocked on the door. He said he represented the giant oil company ConocoPhillips and he wanted to lease some of Butler's land to drill oil wells on his property.

"What would you give me?" Butler asked, as the man sat on a torn couch that once resembled the color yellow.

"I'll give you fifty dollars an acre for a two-year lease," the ConocoPhillips landman replied.

"Lemme tell you something, this land is worth more than that," Butler said indignantly. "I'll lease it for one hundred dollars."

Butler had no idea what his land was worth at the time, but that sounded like a good, round number to ask for. "I just knew one hundred dollars would help," he explains.

The ConocoPhillips man got up to walk away and find another landowner willing to take his deal. Butler didn't flinch. About five minutes later, after it was obvious Butler wasn't going to buckle, the landman came back and told Butler he had a deal. They sat down at the desk and Butler agreed to lease two thousand acres for an up-front payment of $100 for each acre, or a total of $200,000. The contract called for Butler also to receive a cut of future revenue from the wells.

On the outside, Butler played it cool with the ConocoPhillips man. Inside, he was shaking. After the landman left, he picked up the phone to tell his wife what had happened.

"It was shocking," he says. "I thought the world was coming to an end."

Within an hour, two of Butler's neighbors came by, one of whom also had been approached by the landman. "Did you really lease your land for a hundred dollars an acre?" one neighbor asked. "You think that's a good price?"

"Well, it was good for me," Butler told him.

Later, when Butler handed Vera the $200,000 check, she laid it on their table at home and they both stared at it in disbelief. Butler eventually leased the rest of his five thousand acres to other oil companies.

It turned out he was sitting on choice pockets of oil, resulting in hefty checks when drilling began. In the first year of oil production from his property, Butler cashed checks for over $1 million, from just one well on his property. In 2012, he received about $3 million. Finally he was able to pay back the money he owed local banks, sending registered letters with checks in increments of $500,000 until his debt was clear. Soon Vera noticed her stress easing.

"I'm okay since we paid off our debts," she says.

In early 2013, Butler estimated that his acreage was worth more than $40 million, not including the checks he already had cashed and the million dollars he had in the bank.

"I don't owe a crying dime, for the first time in my life," he says. "It's too good to be true."

Unlike the Beverly Hillbillies, Butler and his family didn't use their newfound wealth to move to California or anywhere else. Butler spent some of the windfall to buy back land he had been forced to sell after he had been taken advantage of, righting an earlier wrong, in his view.

He encouraged his sons to spend at least some of the cash on their own families. "It's all right to blow your money on some things, Uncle Sam's gonna get it anyway," he told one of his sons. "But God damn, don't blow it all, it won't last forever."

When a ConocoPhillips representative told Butler that the wells on his property were proving so prolific that the oil could keep pumping for another forty years, Butler turned wistful. "The only thing wrong is it came forty years too late," he replied. "I'm too old now, I don't know what to do with the money."

Butler does have one extravagance in mind. One day, he hopes to hire a driver to take him to see Wyoming.

"That would be my treat," he says.

*   *   *

In early 2010, a young director named Josh Fox debuted a film called *Gasland,* documenting his cross-country trip to visit individuals living near drilling sites in Colorado, Wyoming, and Texas. Fox said his family had been offered $100,000 to lease their land in rural Pennsylvania and he was out to see if hydraulic fracturing was safe.

The film told horror stories of individuals and animals allegedly harmed by drilling. Fox's interviewees said methane from nearby gas drilling had invaded their water. In the most eye-catching scenes of the film, a homeowner turned on a faucet in his kitchen, held a match under the running water, and watched it catch fire, in what seemed like sure evidence of the perils of gas drilling.

Fox's film said fracking would endanger the Delaware River watershed, threatening New York City's main water source. That created nervousness among members of the city's media elite, most of whom hadn't been paying much attention to shale production.

Most scientists considered *Gasland* an effective polemic rather than an accurate depiction of the risks of fracking. Drilling for oil and gas is a messy, noisy, dirty business and it made sense that some local residents would resent the impact on communities. Some noted that methane—a colorless gas that's generally considered nontoxic unless it's in super-high quantities—occurs naturally in shallow bedrock and has been known to naturally seep into water wells and springs around the country. That potentially helped explain the flaming faucets, the scientists said.

The energy industry wasn't sure how to respond to the film. Range Resources, the company that discovered the Marcellus Shale, began sharing the names of the chemicals it was using to frack certain wells, but most drillers remained reluctant to share the details of their liquid concoctions. Some industry executives put out white papers, trying to demonstrate that the movie's arguments weren't based on science. But who wants to read a white paper from Big Oil when you can watch a film that inflames? And while the oil industry is notorious for exerting influence on politics and the media, most of the smaller companies behind the shale revolution had relatively little experience dealing with the public. In April 2010, when an oil rig in the Gulf of Mexico operated by oil giant BP began spewing oil, leading to the biggest spill in history, it sowed more suspicion of drillers. *Gasland* would gain an Academy Award nomination and help shape popular opinion about fracking.

At that point, concern was rising that the chemicals used in the fracking process were harming residents and animals near gas wells in Pennsylvania and elsewhere. The industry insisted the liquid wasn't especially dangerous, but drillers wouldn't say what exactly was in their concoctions, arguing that the details were a trade secret. That only added to the growing fears.

It still wasn't clear why the wells in Dimock were having problems. But there were enough verifiable examples of spills involving fracking chemicals, such as one by Chesapeake and drilling company Schlumberger that killed nineteen cows in Louisiana, to cause nervousness.

In late February, McClendon went to Cambridge, Massachusetts, to speak to students and faculty at Harvard University about how natural gas was "fueling America's clean energy future." As he answered questions from the crowd, a young woman screamed out, "Drilling is killing people in Pennsylvania!"

"Really?" McClendon shot back. "How many people have been killed in Pennsylvania?"

"You're really going to ask me that?" the woman said.

Another woman said arsenic, benzene, and other dangerous chemicals were found in water near gas wells, though they weren't Chesapeake's.

McClendon grew uncomfortable by the unexpected pushback, but he didn't back down. He said one million Americans—or one in three hundred of the entire country—already had signed lease contracts with Chesapeake. They were clearly in favor of hydraulic fracturing and weren't being harmed, he said.

A segment of the crowd didn't seem won over by McClendon. As he prepared to leave the stage, a student called out, "You suck!" Another yelled, "Frack you!"

★　　★　　★

C rude was surging from the Bakken in North Dakota and Montana in 2010. That year, the formation pumped 225,000 barrels a day, with Harold Hamm's company, Continental Resources, responsible for about 10 percent of the output. Five years earlier, just three thousand barrels a day were coming from the Bakken.

By 2010, Continental had figured out how to drill four wells from a single drilling "pad," an approach that allowed the company to simultaneously tap oil from the Bakken formation as well as from layers of the Three Forks formation below it, another technological innovation that pushed crude production higher.

Oil prices rose throughout 2010, moving from seventy-nine dollars a barrel to over ninety by the end of the year, helping Continental shares move from forty-three dollars at the beginning of 2010 to almost fifty-nine by the end of the year.

Hamm tried not to look at his company's stock price. He knew his shares were climbing and he was becoming a multibillionaire. But he barely mentioned the surge to colleagues. He had a feeling they were about to deliver some big news to Wall Street and didn't want them distracted.

"I wanted them to look at production," he says.

Hamm and Continental boosted their acreage in the Bakken to 856,000 net acres by the end of 2010. In September 2010, when Hamm appeared on cable business network CNBC for the first time, he spoke of how much oil was being discovered in the Bakken. The show's hosts seemed a bit skeptical, though.

"What's the time horizon to get this stuff out of the ground," asked Andrew Ross Sorkin, the CNBC personality and New York Times financial columnist. "When is this actually going to work?"

At that point, major newspapers, such as the Wall Street Journal, the New York Times, and the Washington Post had given scant details about what was going on in the region, let alone pointed out Continental's progress.

Jack Stark, Harold Hamm's senior exploration executive at Continental, knew his speech had a chance to change perceptions of the company. In February 2011, Stark traveled to Houston to give a speech at the North

America Prospect Expo, a big industry conference. He carried a new estimate for how much crude Continental's team thought was in the Bakken.

He knew Continental's newest wells were pumping one thousand barrels a day, as the company embraced multistaged fracking. The Three Forks formation was looking even more promising, and Continental was adding drilling rigs.

"Areas that weren't that good all of a sudden were commercial," Stark says.

Still, Stark was nervous. He had never spoken to an audience of over a thousand people and he knew many in the crowd would be skeptical of the figures he was about to present.

Before the crowd of industry experts, Stark said his company had a new figure for how much oil could be recovered from the Bakken: twenty-four billion barrels. He couldn't judge the reaction of most of the crowd. Lights were fixed so strongly on him that he couldn't see beyond the front row. But in that row, Stark saw heads shaking and even some looks of derision.

At that point, the entire country was said to be sitting on just about 19 billion barrels and the government was saying there were up to 4.3 billion barrels in the Bakken. And here was Stark saying several rock layers in North Dakota and Montana alone held 20 billion barrels of recoverable crude and the natural gas equivalent of 4 billion barrels of oil. That was almost twice as big as the oil reserve in Prudhoe Bay, Alaska.

"They seemed over the top, even to me," Stark acknowledges. "I had a hard time with it."

There's nothing energy veterans hate more than promoters with pie-in-the-sky estimates of how much crude they are going to find, Stark knew. "I know it's a big number," he told the audience, trying to address their doubts. "In reality, the field is that big."

The next day, CNBC's Jim Cramer featured the news on his show, helping build excitement for Continental's shares. Soon they were racing past seventy dollars.

Hamm began speaking more frequently to media outlets and government officials about how the nation was moving toward energy independence. He hoped North Dakota officials would speed up the construction of infrastructure in the Bakken region, so he could get even more oil there. Hamm wanted to head off higher taxes on oil producers and possible restrictions on fracking, another reason he embraced the public spotlight.

But he also believed the country's oil fields weren't getting enough respect, even after all the progress his company was making. "We can be the Saudi Arabia of oil and natural gas in the twenty-first century," he told several media outlets.

It burned Hamm up that few believed him. In the fall of 2011, he was invited to the White House for a dinner for Bill Gates, Warren Buffett, and others who, like him, had promised to give the bulk of their wealth to charity as part of the Giving Pledge. When it was Hamm's turn to talk to President Obama, Hamm said there was a "revolution" going on in the oil and gas industry, one that could help the nation replace the crude it was getting from OPEC.

Obama didn't seem very impressed, Hamm says. The president told Hamm that oil and gas would remain important for the next few years, but "we need to go on to green and alternative energy."

Hamm's frustration with the government grew. He began to detect growing delays for drilling permits. That year, Continental and other oil companies faced charges by the Justice Department for allegedly killing twenty-eight birds in the Bakken. In Continental's case, the charges revolved around a single bird, a Say's phoebe. The charges later were dropped, but it still irked Hamm.

"It's not even a rare bird," Hamm said, explaining his growing unhappiness. "There're jillions of them."[6]

During his time with media and politicians, Hamm avoided publicizing some less impressive figures from his company. Continental spilled at least 200,000 gallons of oil in North Dakota between 2009 and the end of 2012, according to documents reviewed by ProPublica, more than any other company. Continental also was fined for poisoning two creeks with thousands of gallons of brine and crude, a reminder of the downside to surging energy production in the country.[7] (Continental notes that its rate of oil spilled—0.1 barrels per 1,000 barrels produced—is improving.)

*   *   *

By 2010, Williston, North Dakota, was attracting ambitious Americans from all over the country, and even entrepreneurs from around the world, hoping to get rich from the emerging oil boom. The city also was a magnet for those seeking to restart their lives and recover from the nation's deep recession. Few were as eager to get to Williston as Elizabeth Irish.

Irish was the branch manager of a mortgage bank in Grants Pass, Oregon, a pretty town with about thirty-four thousand residents that's an hour's drive from both the Pacific Ocean and California's Redwood National Park.

Irish married at nineteen and never went to college. Nevertheless, her career flourished. Articulate and outgoing, she made an annual salary of nearly $120,000 in the years leading up to 2008. That year, as she turned thirty-four years old, she had seven employees reporting to her and she seemed on the fast track at work.

Liz and her husband, Matthew, a truck driver, bought a home in a nearby town for $365,000 dollars. It featured fruit and magnolia trees and was across the street from the elementary school their two daughters attended.

Then the housing crisis hit.

As the local real estate market crumbled, Irish's income evaporated. She made less than $40,000 in 2009. By the summer of 2010, she had earned less than $5,000 that entire year. Liz and her husband had neglected to build much of a nest egg and ran into trouble paying their bills. Soon they lost their home and faced thousands of dollars of additional bills.

"Everything I had done in my adult life was in real estate," Liz says. The crisis "rocked my Pollyanna world . . . it was devastating."

One day in 2010, while checking her Twitter feed, Liz clicked on a link to a *New York Times* article detailing North Dakota's economic boom. Oil was flowing from the Bakken rock formation, located near the city of Williston in the western part of the state, the article said. The state's unemployment rate was around 3 percent, and it was under 1 percent in Williston. Truck drivers were in such demand, since the average Bakken well required about two thousand truck trips to haul a million gallons of water for the fracking operation, that they were making over $100,000 a year.

Liz, who had been racking her brain for a new life plan, turned hopeful. She did a Google search with some obvious keywords, such as "North Dakota" and "trucking," and got on the phone with a trucking company in Williston.

"Things are taking off, everything you heard is true," the manager said. "Housing's tough to find, though."

Liz seized on the first part of the conversation, ignoring the last part.

"What do you think!?" she asked her husband. "It sounds crazy . . . but do you want to go?"

They enjoyed living near family members in Grants Pass, but local unemployment was about 10 percent and they didn't seem to have much of a future.

"It can't get any worse than here," Matt said.

Liz posted her husband's résumé online and it attracted immediate interest. Soon he was in his pickup truck on his way to Williston for a job interview.

Williston is at the heart of activity in the Bakken, but the city's Walmart is the region's true epicenter. Other than two busy strip clubs, and an Applebee's with a line out the door, the Walmart is pretty much the most exciting place in town. With little competition in one of the fastest-growing markets in the nation, items fly off the shelves. Staffers—who make as much as twenty-two dollars an hour—sometimes don't bother taking goods out of boxes. Instead, they just leave the boxes open for customers to dig through. Some evenings, aisles look like they've been looted.

Even the store's parking lot has more drama than some small towns. As Matt pulled in from the twenty-two-hour drive from Oregon, he saw trucks, cars, and campers, all caked with a layer of grime from nearby oil fields or from the remnants of ripped-up local roads. Some of the campers remained in the Walmart lot for weeks or even months, according to local media reports, even though there were no water or sewer hookups. The lack of heat and electricity was a particular hardship during Williston's brutally cold winters.

(A couple of years later, Walmart would expel the campers after women complained that they felt afraid to walk through the lot. There was just so much management could do to reduce the threat of danger, however. In the summer of 2013, a woman in the Walmart lot was shot in the leg.)

The parking lot was a jumble of Texans in cowboy hats, their jeans tucked into dirty boots; East Coast oil field roughnecks, some living out of their pickup trucks; and North Dakotan natives torn between resenting the newcomers and appreciating how they had transformed a city once known for little more than being the hometown of former basketball coaching legend Phil Jackson.

Williston was about the last place one could have predicted a wave of immigrants. The weather is horrible, with below-zero winters full of heavy snow and biting wind, along with hot and humid summers. There's no sports team and hardly any culture. It doesn't even have Mount Rushmore—that's the *other* Dakota.

Until the oil boom, North Dakota had been the only state to see its population shrink since the 1930s. At one point, state leaders considered dropping the "North" from the name and simply calling it "Dakota," hoping it might make the state seem a tad warmer and more attractive.

That became unnecessary when oil began to surge from nearby wells. By 2010, the city's population had more than doubled in a decade, according to most estimates, though no one was entirely sure of the accurate count since so many temporary workers were in town. The city grew so quickly that visitors relying on GPS systems got a laugh from locals; new streets were being built so quickly that navigation systems were virtually useless.

In many ways, Williston's transformation mirrored the growth of other oil boomtowns, including those that mushroomed in western Pennsylvania in the years after Edwin Drake drilled the country's first oil well in 1859. In nine months, for example, a family farm grew into a town of fifteen thousand called Pithole, though it essentially vanished by the end of the 1880s after more productive wells were found elsewhere. In *The Big Rich,* author Bryan Burrough describes how sleepy East Texas hamlets were overrun in the 1930s after oil was discovered: "When the hotels filled, the townspeople rented out rooms; when all the rooms were let, the newcomers threw up tents; when they ran out of tents, men slept in the open fields."

Among those flocking to Williston by 2010 were veterans of the Iraq and Afghanistan wars. They were particularly coveted hires because some were used to adverse conditions, were comfortable handling the military-grade explosives used in fracking operations, and had experience working as part of a team in the field. Entrepreneurs from all over the world also came, including Japanese businessmen hoping to start restaurants and others investing in new hotels. Las Vegas chiropractor Stephen Alexander had so many patients relocate to North Dakota that he too headed for the Bakken, investing $200,000 in a fifty-seven-foot recreational vehicle with a digital X-ray machine, an examination room, and other technology.[8]

Some of those flocking to the city weren't particularly picky about what jobs they might find. Marquita Wright, a thirty-five-year-old from Minnesota who had earned a college degree in interior design, came to Williston to interview for four very different positions: Walmart staffer, liquor store cashier, mail carrier, and exotic dancer—where she had the opportunity to make more than $1,000 a night.

"Williston's a little gold mine because there are no malls, so you start saving right away," Wright says.

As for Matt, he was hired on the spot in the Walmart parking lot to haul water to fracking operations. "This can't be real, right?" he asked Liz when he called from the Walmart. "This is a movie."

Most husbands head to the Bakken on their own, sending money to their families and visiting when they can. That's why there's an estimated two men for every woman in the city (and also why Las Vegas prostitutes began commuting for lucrative weekend work).

Liz and her husband didn't want to live apart, however, and she was eager for a new adventure. In the summer of 2010, she and Matt took the cash from their previous year's tax refunds, packed up their daughters, two dachshunds, and some belongings, and drove to North Dakota to begin a new life.

★   ★   ★

Charif Souki kept telling his investors that he was working on a plan. He wouldn't say what it was, though.

Shares of Cheniere Energy traded for about three dollars in late 2009 and early 2010, and most investors wrote the company off. Cheniere had spent billions on a plan to import natural gas to the United States. Now the country was swimming in gas. In 2010, 4.86 trillion cubic feet would be produced from shale formations, more than double the 2.25 trillion cubic feet in 2008.

Cheniere still was receiving revenue from Total and Chevron as part of its earlier import agreement, so Souki's company wasn't in imminent danger of bankruptcy. But it was burning through $50 million a year and still faced a billion dollars of debt due in 2012.

Board meetings turned contentious as investors became exasperated. The Blackstone Group's GSO fund and John Paulson's hedge fund, Paulson & Co., each maintained big positions in Cheniere. They weren't pleased with what was going on.

"Charif, you need a plan," Dwight Scott, a GSO executive on Cheniere's board, told Souki during one meeting.

"I'm working on a bunch of plans," Souki told him.

"Like what?" Scott asked.

Souki said he couldn't share the details just yet. That response didn't thrill the board members, including John Deutch, the former director of the CIA, who asked Souki his own set of tough questions.

At one point, Blackstone and Paulson offered to wipe out some of Cheniere's debt in exchange for control of the company, but Souki refused the deal. Like a fighter on the ropes, he didn't want to give up. He could have quit and moved on to something else. But he was rope-a-doping, hoping for a way out of his jam.

"We'll come up with something, I'll figure something out" if importing gas doesn't work out, Souki told the board. "Something good will happen."

"He didn't really have a plan besides trying to sell the company, that was our problem with Charif," says a big investor.

The Blackstone executives on Cheniere's board of directors, including Jason New, began hearing criticism of their continued support for Souki and the company. When New met with pension funds and other Blackstone clients, they relayed word that other hedge funds were calling Cheniere the best "short" in the market, or the most likely candidate for a hard fall. The hedge funds were betting big money that Cheniere would go bankrupt, they told New.

"Well, reasonable minds can differ," New responded, trying to stick up for Cheniere. He didn't share his own frustrations with what was going on at the company.

Souki's unusual background didn't help alleviate the concerns of the investors and board members, one large Cheniere investor says.

"If he was a guy from Midland, Texas, it would have been easier for people to deal" with the repeated reassurances, says a board member. "But he was Lebanese, it was a cultural thing. People would say, 'Where the hell did this guy come from—he owned Mezzaluna!'" referring to the infamous Los Angeles restaurant.

Behind the scenes, Souki and his top executives were becoming more committed to the idea of exporting gas and convinced they could retrofit the Louisiana terminal and turn it into one capable of turning natural gas into LNG so it could be exported.

In the spring, Souki and his team received an estimate from Bechtel, the global construction firm, with a cost to reconfigure their plant. Bechtel judged it would cost about $450 for each ton of LNG it wanted to export. At that price, it would cost over $8 billion to convert the terminal into one that could export natural gas using four "trains," or liquefaction and purification units. That would be enough to ship eighteen million tons of gas a year.

Most anyone would blanch at a price quote like that. With financing costs and other expenses, Souki knew he'd have to raise as much as $12 billion to make the switch, even as financial markets remained unstable in the aftermath of the global economic crisis.

Souki clearly was not like most anyone, however. If there was one thing he was confident about, it was his ability to raise boatloads of money, even billions of dollars, if necessary. *This is going to be child's play,* he figured.

"When it came to raising money, that I knew I could do," he says.

Other Cheniere executives turned almost giddy when they digested the Bechtel report. The price tag was sky high, but the staff had feared the cost would be even higher. If Bechtel was accurate, Cheniere actually might make money exporting gas, Souki and his executives determined.

Souki didn't dwell on the fact that his existing investors still were nursing losses and had had it up to here with him. He also didn't worry that Bechtel had only given an initial estimate and made it clear that the final might end up 30 percent higher.

In April 2010, Souki went to his board to pitch his new idea. "I have a big advantage here," he told the board confidently. "We already have the terminal, the storage tanks . . . everything but a liquefaction facility. . . . I think I can get the contracts and the permit." He said the company eventually could export as much as two billion cubic feet a day, or about 3 percent of U.S. domestic gas production. He looked around the room, waiting for a response.

For all the acrimony in the previous months, Souki's plan didn't elicit much reaction. Some board members already had a sense that this was going to be his new strategy. It wasn't like he had many other options. Others were so tired of Souki's maneuvering, and the roller coaster the company had been on, that they were too exhausted to give much input or put up a fight.

Cheniere had billions of dollars of debt coming due, so most investors and board members figured this was their best chance to salvage a bad situation. "The board's attitude was that it was a Hail Mary pass, fourth and twenty-five on our own five-yard line. . . . They were fatigued," says a former board member. "It was a small company trying to beat the Exxons and Chevrons to the export market, and those companies had better connections to Washington regulators."

In June 2010, Cheniere made a public announcement of its intention to begin exporting gas. Investors let out a collective yawn. Many had written off the company, so the stock didn't move very much.

When Souki called John Paulson to tell him about the idea, the hedge fund manager was friendly and wished him luck. Paulson wasn't particularly enthused, though. If he somehow could get his firm's original investment in Cheniere back, without incurring a loss, "you will have performed a miracle," Paulson recalls telling Souki.

Industry members made fun of the export idea, teasing executives such as Davis Thames, the marketing specialist in charge of selling the idea to customers. One told Thames that his company looked "desperate" by floating the export strategy.

In early 2011, Cheniere shares traded for less than seven dollars and the company's Louisiana terminal still stood idle much of the time. But the Energy Department had granted Cheniere a permit to export gas to the nation's free trade partners, and the company was close to getting a permit to export to other countries, including China. It helped that Souki had enlisted a group of former politicians and others who were well connected in Washington power circles to lobby for his plan.

Meanwhile, natural gas prices in Europe and Asia were twice as high as the $4.35 per thousand cubic feet rate in the United States, creating an opportunity to export cheaper gas from the United States. At that point, Cheniere had the market to itself—as long as it actually could revamp its facility for exports. Only one American terminal, built in Alaska thirty years earlier, could cool gas into liquid for shipment on giant cargo ships for sale abroad.

Souki was confident Cheniere could begin exporting gas as early as 2015. Already, Cheniere had a commitment from Aubrey McClendon and Chesapeake Energy to send large amounts of gas to the Louisiana facility for export.

"If you keep digging, digging, digging, you find something," an upbeat Souki said in late January 2011.

There was one big problem, though. Cheniere had to raise between $10 and $12 billion to convert the new facility and pay the cost of financing and other expenses related to the new terminal. Banks had curbed their lending in the aftermath of the global financial crisis and many investors were still hurting. It wasn't clear where Souki was going to get the money.

"This is somebody who basically enjoys being on a roller coaster," Fadel Gheit, a senior oil analyst at Oppenheimer & Company, told the *New York Times* that month. "It is more likely to see snow in New York in July than to see exports of gas from LNG terminals in the United States."[9]

Cheniere ended the year under nine dollars, as many investors remained dubious about Souki and his latest stroke of genius.

CHAPTER FIFTEEN

Aubrey McClendon stepped up his buying once again.

Throughout 2010, McClendon led Chesapeake Energy on a new land grab. There was acreage in the Niobrara Shale in Colorado and Wyoming. There also were the Woodford Shale and Granite Wash formations in Oklahoma and Texas.

It all looked too tempting. McClendon and Chesapeake didn't *have* to own it, of course. He just wanted it.

There was a limit to how much debt Chesapeake could pile up to pay for the deals. But McClendon told Ralph Eads, the senior Houston investment banker at the firm Jefferies & Co., that he had an idea where to get the billions he needed.

"If we're going to look for new money, let's go to Asia," McClendon told his old college friend, well aware of the growing coffers of companies and sovereign wealth funds on that continent. "They have the money and they're short on energy."

McClendon and Eads began traveling to Asia to sell stakes in new American fields suddenly churning out oil and gas. With remarkable speed, Chesapeake established a 600,000-net-acre position in the Eagle Ford in 2010 for $1.2 billion and then sold a one-third interest to the China National Offshore Oil Corporation for $2.2 billion.

Most of the land Chesapeake and its rivals were leasing carried "held by production" clauses. These terms required the companies to drill wells within three to five years, while also paying potential royalties to landowners, or face the loss of the leases, even after sometimes paying leaseholders bonuses of up to $20,000 per acre as part of the leasing transaction.[1] Once production began, the landowners also received a royalty on the producing wells that typically amounted to 15 to 20 percent of the wells' revenues.

So a 320-acre drilling unit could cost Chesapeake a bonus payment of $6.4 million—a fee that simply gained the company the right to drill a well or two on that unit. After paying that kind of up-front cash, there was no way the company wanted to lose that lease by not drilling the wells in short order, explaining why drilling kept apace despite weak gas prices.

On a May 2010 conference call with analysts, Chesapeake chief financial officer Marc Rowland said, "At least half and probably two-thirds or three-quarters of our gas drilling is what I would call involuntary. It's being incentivized by something other than the gas price." Chesapeake's drilling added to the nation's growing glut of natural gas, helping to push gas prices lower.

When McClendon decided to boost Chesapeake's spending on the Mississippian Lime, an oil-rich limestone formation in northern Oklahoma and southern Kansas, some members of the company's board of directors became a bit nervous. McClendon promised the board that this was going to be Chesapeake's last big push. He said that he, too, was concerned that the company's debt was more than 50 percent of its equity, or the value of the company.

"We said if it's half as good as he says, we really have to think about it," says Charles Maxwell, a board member. "It's like seeing a golden apple, you grab it because you don't know if you're going to find one again. . . . And Aubrey sounded sincere" about its being the company's last expensive grab.

In October 2010, Jeffrey Bronchick, the investor who a year earlier had called the Chesapeake proxy statement "shameful," traveled to Oklahoma City to attend an "analyst day" at Chesapeake's campus, where McClendon and other executives outlined their strategy to two hundred investors, stock analysts, and others.

After Bronchick got a look at Chesapeake's gorgeous campus, featuring a huge health club and pool, catered food, and an elaborate child-care center, he was horrified. All the money the company was spending seemed like a monumental waste. "I wish I went to college here," Bronchick joked to a colleague from their Los Angeles investment firm, Reed, Conner & Birdwell. Spotting antique maps on the walls, he said, "That's my money being spent," a reference to Chesapeake's earlier purchase of McClendon's antique map collection.

But as Bronchick listened to McClendon detail how he was going to reduce spending and begin "harvesting" Chesapeake's prime acreage, his skepticism slowly melted away. McClendon seemed to have learned his lesson and was turning over a new leaf, Bronchick decided. "Why is this stock at twenty-one?!" an excited Bronchick asked his colleague as they walked out of the meeting. "I just heard a great stock and a great story."

McClendon began to feel confident once again. He even involved his company in one of his passions: rowing. He had financed the transformation of a dry and weed-choked riverbed in Oklahoma City into an Olympic-class rowing venue. Now McClendon told colleagues the city could become the rowing capital of the Southwest and that it had unique advantages over rowing sites on the East Coast used by teams from Harvard and Princeton; there would be fewer days in Oklahoma City where inclement weather prevented teams from practicing, he said.

Chesapeake had built a $3 million boathouse, of which he contributed $1.5 million. He also financed a nearby glass-sheathed, $7 million Chesapeake Finish Line Tower. He even put ten Olympic hopefuls on Chesapeake's payroll, paying them about $35,000 a year plus benefits and assigning them to the finance and community relations departments. One Sunday in December 2010, McClendon and his son flew to San Diego with five of the rowers; he reimbursed Chesapeake for the $34,000 cost of the trip.[2]

McClendon continued to take an aggressive stance toward health concerns about fracking, as if he wasn't aware—or very concerned—that the public had turned against the activity, even as most knew very little about it. "We frack all the time. What's the big deal?" he asked *Rolling Stone* magazine in March 2012. "Where is the mushroom cloud? . . . Where are the dogs with one leg?"

McClendon might have been trying to suggest that there was nothing inherently dangerous about the fracking process, as long as it was done properly. Sometimes it wasn't being done properly, however. A year earlier, the *New York Times* had reported that wastewater containing high levels of radioactivity was sometimes being brought to Pennsylvania sewage plants not designed to treat it and then discharged into rivers supplying drinking water. The paper also said a study written by a consultant to the Environmental Protection Agency concluded that some sewage treatment plants weren't capable of removing certain drilling waste contaminants.

(After the *Times* piece, the acting secretary of Pennsylvania's Department of Environmental Protection said that studies conducted in seven of the state's rivers showed "normal" or below-average levels of radiation.[3])

Scrutiny on fracking would grow, thanks to an unlikely new interest group. In May 2012, about fifty residents of the village of Sidney, New York, met in a cramped local library to discuss how to stop a pipeline slated to run through the area carrying natural gas from the Marcellus Shale formation in nearby Pennsylvania.

The group was concerned that the pipeline might burst, causing damage. Their even bigger fear was that if a pipeline was built, gas drilling would come next. And that would mean hydraulic fracturing right in their backyards.

Sidney residents wanted no part of it. The guest speaker that evening was Vera Scroggins, the rabble-rousing former schoolteacher in nearby Pennsylvania who had once faced off against local stone cutters and now was getting media from all over the world to publicize various mishaps in Dimock and elsewhere. That evening, she brought a new ally, Craig Stevens, an intense, muscle-bound middle-aged man with close-cropped hair who wore an air purifier on his neck.

There was a lot of gray hair in the room that day; most of the attendees were men who were middle-aged or older. They were articulate, thoughtful, and well informed. Most looked like they once had taught grade school on the Upper West Side of Manhattan and probably followed the Grateful Dead on a tour or two.

At the end of an evening in which Scroggins and Stevens shared tips with the Sidney residents about how to organize resistance and court the media to stop the pipeline, a hirsute man with a ponytail and large-framed glasses approached the front of the room. Few had noticed the man quietly sitting in the last row throughout the evening, slowly stroking his beard, but all eyes were fixed on him as he began speaking to the crowd.

"I can't believe what I'm hearing," he said, according to Scroggins and Stevens. The unfamiliar man spoke about how his family had grown to love the area, and how he wanted to help publicize the antifracking cause.

After the meeting, the group found out that the man was Sean Lennon, the son of late Beatle John Lennon. Soon he and his mother, Yoko Ono, were joining other celebrities, such as the actors Mark Ruffalo and Susan Sarandon, to prevent natural gas drilling in New York.

The stars clearly were committed to the cause. But their activism also gave their careers new life, notes Scroggins, who soon began taking busloads of celebrities, media from around the world, and others on tours of the area to point out places she said had experienced environmental damage. "It's helped increase interest in Sean's music," Scroggins says. "Mark Ruffalo got an Academy Award nomination and was in the *Avengers* movie after" becoming involved in the antifracking movement.

Either way, the publicity helped pressure New York State officials to maintain the moratorium on fracking in the state and it helped focus national attention on the drilling, dealing a public relations blow to those in the business.

In late 2010, billionaire investor Carl Icahn disclosed that he had purchased nearly 6 percent of Chesapeake's shares. Icahn is a so-called activist investor, or someone who buys big chunks of a company and then levies pressure on its management to enact changes aimed at getting shares higher.

Weeks later, Icahn reached out to McClendon to let him know the company had piled on too much debt. McClendon got the message. Chesapeake quickly announced plans to sell $5 billion in assets, including a deal to sell a one-third interest in its acreage in Wyoming to China National Offshore Oil Corporation, China's largest offshore oil producer, helping Chesapeake reduce its debt.

Chesapeake shares hit $35.61 in February 2011, their highest level since before the 2008 crisis. The company announced a plan to sell its Arkansas natural gas field to Australia's BHP Billiton for $4.75 billion, just as investors wanted.

Icahn soon exited the stock, however, pocketing about $500 million in profits. On the way out, he placed a call to McClendon to thank him for the enormous windfall. It soon became clear that McClendon didn't have his heart in the debt-trimming strategy, raising concerns on the board of directors.

"You're making an assumption that natural gas prices will be strong and the company will be able to handle its debt," Louis Simpson, a new board member who once ran insurance company Geico's investment portfolio, told McClendon during one meeting in 2011. "We could have problems if it doesn't turn out the way you're saying."

"We've done our homework, we should be okay," McClendon responded.

Simpson wasn't reassured. Later, he told other board members that he was getting fed up with McClendon's leadership of the company.

"I think we need to get rid of Aubrey," Simpson said at one meeting. "The culture of the company has to change."

At one point, Simpson tried to get under McClendon's skin, asking him why shares of EOG and Continental were doing so much better than Chesapeake's.

"They got lucky in the Bakken," McClendon told Simpson.

Other board members lent McClendon support, despite Simpson's qualms. "We were dealing with a guy who took us from nineteenth to second" place in natural gas production, says one. "No one had done anything like that!"

Within months, McClendon grew more excited about the Utica Shale and was directing purchases of acreage in Pennsylvania, Ohio, and West Virginia that seemed to have both oil and gas. By then, Chesapeake had spent $2 billion to lock up 1.25 million acres in the area.

This one really was going to be a doozy, McClendon promised. In an October 2011 appearance in Columbus, Ohio, McClendon told a crowd that the Utica formation could be worth $500 billion. "I prefer to call it half a trillion, it sounds bigger," McClendon said, adding that the Utica would be the "biggest thing economically to hit Ohio, since maybe the plow."[4]

Chesapeake was the second biggest natural gas producer in the country, after ExxonMobil, and was drilling even more wells than the huge oil company. Aubrey McClendon was back on top. In the fall of 2011, he entertained a reporter in Oklahoma City's Deep Fork Grill. Over steak, fries, and $10,000 worth of wine from McClendon's collection, he expounded on the greatness of natural gas and of Chesapeake Energy.

"We have found something that can liberate us from the influence of OPEC, that can put several million Americans back to work, liberate us from four-dollar gasoline," McClendon said. "Is it too good to be true? Sometimes it seems that way."

Chesapeake was pumping three billion cubic feet a day from the 13.7 million acres it controlled, close to the size of West Virginia. Harold Hamm at Continental had barely five hundred employees, but Chesapeake counted twelve thousand people on its staff, as well as forty-five hundred land scouts searching the country for the next big play.

McClendon's company, which had paid out $9 billion in lease bonuses to landowners in the five previous years, was on track to score $2 billion of profit that year. *Forbes* estimated that McClendon's personal fortune was above $1.2 billion, more than enough for him to agree, after three years of court battles with shareholders, to repurchase the antique maps he had sold to his company.

"He is without a doubt the most admired—and feared—man in the oil patch," *Forbes* intoned, while also calling McClendon "reckless."[5]

McClendon's sunny disposition belied clouds forming on the horizon. Natural gas prices finished 2011 below three dollars per thousand cubic feet, down 30 percent in a year. At those prices, Chesapeake couldn't make money drilling some of its shale plays. The Haynesville Shale in Louisiana, for example, required gas prices of approximately $3.50, while Chesapeake's wells in the Barnett needed prices of about $4.50.

The company kept drilling away, though, to avoid losing its leases.

Chesapeake closed 2011 at twenty-two dollars a share, down by one-third in five months, a reminder that for all of Aubrey McClendon's boasts and big plans, his company's fate was tied to the price of gas.

Natural gas prices kept falling in January 2012, dropping to about $2.50 per thousand cubic feet, the lowest level in twelve years, partly due to a record warm winter. Aubrey McClendon had been among those most responsible for the remarkable amount of gas production in the country. Yet he and his staff seemed among those most surprised by the very phenomenon they helped create.

More than eight trillion cubic feet of gas would be produced in 2012, up from over two trillion in 2008. Chesapeake itself would pump 1.1 trillion cubic feet in 2012, up from 775 billion cubic feet in 2008, as all the fracking and horizontal drilling in shale paid off.

In January, in his State of the Union address, President Barack Obama declared that there was enough natural gas buried in the United States to "last one hundred years," and that a booming gas industry would lead to hundreds of thousands of American jobs.

In January, Chesapeake announced it would slash spending on gas drilling by half and shift to produce more oil from the Eagle Ford, Utica, and elsewhere. The company eventually promised to raise about $9 billion before the end of 2012 to fund drilling and reduce debt.

McClendon stayed upbeat, predicting gas prices would rally. Part of the reason prices were so low was that winter temperatures were especially mild, raising the possibility of higher prices when it got colder, he said.

"This won't last long," McClendon told a colleague one day, referring to natural gas prices. But they fell to $1.91 on April 19, 2012, a move that felt "pretty scary," a board member says.

By then, McClendon was dealing with a new headache. It emerged that entities controlled by McClendon had borrowed more than $1 billion from a private equity firm called EIG Global Energy Partners. McClendon had used the money to cover the cost of the 2.5 percent stake in wells that Chesapeake drilled. McClendon had talked about the well perk in media interviews and it was mentioned in security filings. But few investors had focused on the arrangement until Reuters broke news of the loans, and some viewed it as excessively generous.

Much more important was that McClendon had borrowed the billion dollars from EIG even as Chesapeake was negotiating to sell EIG hundreds of millions of dollars in assets. People close to McClendon and EIG insisted the firm didn't cut McClendon any breaks on the terms of the loan. At the same time, there were a limited number of firms who knew Chesapeake's acreage and could write a big check like that, another reason McClendon likely turned to EIG.

Still, it looked really bad that McClendon was borrowing so much cash from a firm looking to buy assets from his company, and the Securities and Exchange Commission opened a probe of the well purchase arrangement. The probe was still ongoing as of August 2013.

The resulting outcry over the loan was such that a week later, Chesapeake's board of directors, which long had given McClendon virtual carte blanche, said they would review his personal financial interactions with firms dealing with Chesapeake. The company also announced that the well perk would end in 2014.[6]

The spotlight remained fixed on McClendon as additional news reports provided details of the hedge fund he and Ward had helped advise years earlier to invest their money. Then the *Wall Street Journal* reported that several banks had lent McClendon money while also receiving lucrative work as underwriters or financial advisers for Chesapeake.

As Chesapeake shares fell and criticism grew, McClendon became the butt of some jokes. To defend himself against the parade of accusations, bloggers said he should try "the George Costanza defense." They were referring to the classic episode of the television comedy *Seinfeld* when Costanza is fired for having sex with his cleaning lady on his desk. Costanza pleads ignorance, asking, "Was that wrong? Should I not have done that? 'Cause I've worked in a lot of offices, and I tell you, people do that all the time."

In May 2012, with gas prices still around $2.50 and Chesapeake shares down to about seventeen dollars, shareholders began to get antsy. Investors including O. Mason Hawkins, whose $34 billion mutual fund firm, Southeastern Asset Management, was Chesapeake's largest stockholder, forced McClendon to step down as chairman, a remarkable turn of events for an executive who had wielded so much power, and commanded such respect, just months earlier.

In mid-May, McClendon, who remained Chesapeake's chief executive officer, spoke to several hundred Chesapeake employees, trying to keep the staff calm amid the firestorm.

"I encourage everybody to inhale," McClendon said. "I'm fine. You're fine. And we're in the middle of a pretty unprecedented media firestorm. . . . I would not have wished the past month on my worst enemy."

McClendon thought there might be a way to give life to Chesapeake shares. He had heard that Carl Icahn had renewed interest in Chesapeake, and he arranged to visit the billionaire in New York.

Icahn's reputation for pressuring underperforming companies preceded him; a friend, investor Wilbur Ross, once said that Icahn is "especially good at terrorizing people."[7] But Icahn had deep respect for McClendon and already had made a killing a year or so earlier in the stock. The winnings allowed Icahn to view McClendon in a slightly better light than some others. Icahn bought Chesapeake again in 2012 because he viewed the stock as being inexpensive, rather than ideal for a shakeup.

Still, as they dined in Icahn's 14,000-square-foot apartment in the Museum Tower on West 53rd Street, Icahn made it clear that he thought McClendon had to change the way he ran Chesapeake. "Look, you've got to sell assets," Icahn told McClendon, according to someone familiar with the conversation. "You're a brilliant guy and all, but you've got to sell."

McClendon promised to ramp up the company's efforts to reduce leverage. By then, Chesapeake had more than $13 billion in debt on its balance sheet and total obligations of close to $24 billion, according to rating agencies.

Soon, Icahn disclosed a 7.6 percent stake in Chesapeake, helping Chesapeake's shares rise a bit. But financial pressure grew on McClendon. In June, he pledged his personal "oil and gas memorabilia" against a loan from billionaire George Kaiser, according to securities filings.[8] News also emerged that McClendon had mortgaged his 20 percent stake in the Oklahoma City Thunder basketball team to secure other loans.[9] That debt was in addition to the $846 million that he had borrowed by the end of 2011 to pay for his slice of the drilling costs of Chesapeake wells, according to securities filings.[10]

The normally outgoing McClendon turned more pensive. "It was gradually wearing him down," says a Chesapeake executive. "He was less gregarious, walking with his head down, as if he was thinking."

In June, Chesapeake raised its bet on the Utica formation by ramping up drilling in the area, although it eventually would prove to have less oil than optimists like McClendon expected, adding to investor unhappiness.

"Assets weren't being sold fast enough," says a major investor.

Soon, Hawkins and Icahn forced Chesapeake to replace four of its nine directors with their own representatives. Together with Louis Simpson, the former Geico executive who had challenged McClendon with questions about the company's debt a year earlier, five of the company's nine board members were unhappy with McClendon.

It was clear that Aubrey McClendon's days at Chesapeake were numbered.

Natural gas prices dropped throughout 2010 and 2011. Oil prices went the other way, though, climbing from seventy-nine dollars a barrel to ninety-nine dollars.

The price moves made Tom Ward look good. In late 2008, Ward had shifted his company's focus from natural gas to oil. After buying Forest Oil's wells in West Texas's Permian Basin in late 2009, SandRidge Energy paid $1.6 billion in April 2010 to buy another oil company called Arena Resources. Rivals were enamored with drilling in dense, shale oil formations, as they tried to catch up to Continental and EOG, but Ward was buying more inexpensive, traditional U.S. oil wells, viewing them as better deals.

Ward wasn't nearly done wheeling and dealing. In 2010 and 2011, he and his team profited by selling parcels of their newly acquired land at a big gain; they also raised cash through complex joint ventures and stock sales.

SandRidge also spent $400 million to purchase two million acres in the Mississippian Lime, the formation in northern Oklahoma and southern Kansas. These were aging oil fields, but to Ward they seemed overlooked and attractive. With oil at nearly $100 a barrel, he wagered that SandRidge could profit using horizontal drilling and multistaged fracking techniques, while also employing sophisticated methods to remove water and free up oil from the wells.

Revenues and reserves were rising, and the Mississippian formation was emerging a winner, but investors remained unimpressed. SandRidge shares hit thirteen dollars in April 2011, but they fell back to about eight dollars by the end of the year. Some big holders were getting impatient, just like at Chesapeake. SandRidge had finished its first day of trading in 2007 at thirty-two dollars a share, and it hit sixty-eight dollars in 2008. But the stock finished 2011 at 90 percent below those heights.

Something else was beginning to concern shareholders: Ward was selling some of his holdings of the company. In October 2010, Ward sold six million shares, netting over $35 million. By the end of the first quarter of 2012, he held just over twenty-four million SandRidge shares, down from nearly thirty-two million in the same period in 2009, according to securities filings. The company was handing Ward shares as compensation, but he was selling even more of the stock, worrying some investors that his interests weren't fully aligned with his company's.

Ward didn't see what the big deal was. He still controlled about $190 million of SandRidge stock in early 2012, or almost 6 percent of all shares. But as the stock flatlined, investors began to grumble.

SandRidge was a relatively small company, with a market value of less than $3.5 billion at the end of 2011, and yet Ward was being paid as if he ran an oil giant. He received compensation totaling $47 million in 2010 and 2011, making Ward among the highest-paid energy executives in the country.

In Ward's view, SandRidge was set up to be a major energy company. It wasn't there yet, but it had to spend top dollar to hire and retain talent if it wanted to compete with major oil companies. Ward felt he deserved his own hefty payout because he had raised over $6 billion for the company over two years. He also insisted that he could get much more if he left and started a new venture, perhaps with the backing of a private equity firm. He had directed the Mississippian discovery and was sure he could leave the company and make more by drilling the area on his own. Besides, a number of sizable gas-focused companies had fallen by the wayside during that period, but SandRidge was still growing, thanks to Ward's shift to oil.

The arguments didn't sway some big investors. Executives of a New York hedge fund called Mount Kellett Capital Management, run by a former Goldman Sachs executive, began calling Ward, asking why Sand-Ridge was spending so much on acreage that would be costly to drill. Norman Louie, a Mount Kellett executive, told Ward that the holdings seemed "too much for you to handle." Ward responded that the push into Kansas and Oklahoma would pay off.

Mount Kellett, which grew its position in SandRidge to more than 5 percent of the company by late 2011, or almost as many shares as Ward held, was always polite with Ward, and the firm kept its complaints private. But the hedge fund executives were losing patience.

In February 2012, Ward shocked shareholders when SandRidge announced the $1.3 billion purchase of Dynamic Offshore Resources, a company producing oil in the Gulf of Mexico. SandRidge had been focused on onshore wells, and Ward had spent his career working on domestic wells. All of a sudden, the company was spending big bucks to buy a company with mature, offshore oil wells.

SandRidge shares tumbled 11 percent on news of the deal. Ninety minutes after the deal was announced, Norman Louie, the Mount Kellett executive, along with a colleague, Marcus Motroni, got on the phone with Ward.

"I don't know why you're doing this," Louie said, sounding irritated. "You've never operated in the Gulf of Mexico."

Ward calmly explained that SandRidge was paying a cheap price for assets pumping large amounts of oil and that the deal would lower the company's ratio of debt to cash flow, a key metric for the company's lenders.

The Mount Kellett executives weren't assuaged. Operating in the Gulf of Mexico is more difficult and time-consuming than drilling in the United States, they said.

"Why don't you digest this and get back to us," Ward said, ending the call.

By the summer of 2012, SandRidge shares were weakening once again, falling to just above six dollars a share. The company regularly beat quarterly earnings expectations and the Mississippian oil field in Kansas and Oklahoma was pumping out impressive amounts of crude.

But the field was turning out to be a bit less successful than Ward originally hoped. Shareholders weren't thrilled that SandRidge said it likely wouldn't be "cash-flow positive," or taking in more than it was paying out, until 2017.

In late 2011, TPG-Axon Capital, a hedge fund run by another ex–Goldman Sachs star trader, Dinakar Singh, bought SandRidge shares of its own. In the spring of 2012, as Singh and his team debated purchasing more SandRidge shares, they invited Ward to visit TPG-Axon's midtown Manhattan offices. Singh grilled Ward, asking him if he operated as aggressively as his old partner, Aubrey McClendon.

"Dinakar, Aubrey is a friend of mine," Ward told Singh and his team. "But we do things very differently."

The answer was just what Singh wanted to hear. But after Ward left, he directed his team to dig into SandRidge's operations. What they reported back was information that largely had been publicly available to investors. But the more documents they placed on Singh's desk, the more outraged he became.

Yes, Ward didn't own interests in SandRidge's wells, like McClendon, just as Ward had said. But that was only because Ward had sold the interests back to his company for $67 million during the panic of 2008. The deal had been vetted by an outside party, which said the purchase had been at a discount to the value of the wells, but the transaction still seemed to Singh like a misuse of SandRidge's precious cash.

Singh also viewed SandRidge's annual overhead of $250 million as huge for such a small company. They even spent a million dollars for someone to do accounting for Ward's personal accounts, according to the information gathered by Singh's team.

"He's incredibly greedy," he told a colleague.

By November 2012, Singh's firm owned 4.5 percent of SandRidge's shares. Singh decided that Ward and SandRidge had accumulated some impressive assets, but the only way the stock could go higher was if Ward was axed. And the only way that would happen was if Singh went public with his grievances.

Singh wrote a letter to SandRidge's board of directors calling for Ward and certain of the company's board members to be replaced. "Tom Ward's credibility is too damaged to continue in his role," Singh wrote. "SandRidge's stock performance has been nothing short of disastrous."

On November 14, Ward made a return visit to Singh's office to discuss the hedge fund's grievances. Singh and his staff appreciated that he was willing to come back, and they were struck by how courteous and gentlemanly he was in the meeting. But the tension quickly intensified. Sitting in a large conference room with a full view of Central Park's Great Lawn, Singh rattled off a litany of complaints. Sitting directly across from Ward, he went on for about forty-five minutes. SandRidge shares had performed terribly for years. The company's strategy made little sense and it had too much debt. And Ward was overpaid, Singh said.

"Tom, the facts are pretty horrible," Singh said, his voice rising, according to people at the meeting.

Ward sat quietly as Singh lectured him on all the reasons he should be fired. Just a few years earlier, Ward had been viewed as an energy visionary. He and McClendon had been lionized in Oklahoma City and elsewhere for anticipating new demand for natural gas and for opening up new gas fields in the country, helping the United States develop enough gas to last generations and laying the groundwork for the nation's energy revival. Now the fifty-three-year-old was being told by a hedge fund manager with no energy background that he was the biggest reason SandRidge was faring so poorly, and that he needed to leave the company he had founded.

When Singh finished, Ward hardly showed any emotion, according to people in the room. He spoke a bit about how the company was going to slow its expenses, but that was about it.

Ward seemed largely unmoved by the criticism. A stock analyst on a conference call said, "You look like you're following McClendon over at Chesapeake, which seems to be personally very greedy. . . . Wouldn't the company be much better off if you were more modest in your take personally?"

"Do I have other options . . . that I could get paid as much or more?" Ward responded. "I would think so. So I'm here at the discretion of the board and shareholders."

In an e-mail message to a reporter, Ward wrote: "If I chose I could make more money doing something else as very few people have done what I have done in this industry."

Soon more details emerged about SandRidge's dealings. The company had paid nearly $28 million to Ward, or to firms linked to him and his family, according to securities filings. For example, SandRidge had leased land from a company controlled by Ward's son Trent. A Ward-linked entity also purchased mineral rights in Kansas months before SandRidge leased adjacent plots, according to land records.

The payments had been fully disclosed, and Ward argued that his family had helped the company gain a quick foothold in the Mississippian formation, which proved a big winner. The questionable leases represented less than 1 percent of the acreage leased by SandRidge, Ward said, and the company's board had decided on the moves. And it made sense that his family entity bought in similar locations as SandRidge since Ward was a big believer in the area.[11]

"The reason I was paid royalties had nothing to do with" SandRidge, Ward says. "They just happened to drill where I had bought land years before."

The criticisms piled up, though, from both TPG-Axon and Mount Kellett, but also from California pension fund CalSTRS. TPG-Axon launched a campaign to replace SandRidge's board of directors, and shareholders began deliberating over a March 15 proxy vote on the matter. SandRidge's stock closed just above six dollars at the end of 2012, which didn't help Ward's chances for survival.

At a company meeting to address the situation, Ward was defiant but also fatalistic. "If I don't get the vote, then I don't deserve it," he told eight hundred employees. "I'm doing my job. . . . I think I'll still be here next year."

★   ★   ★

The investor outcry wasn't really about Ward's pay or his dealings. Ward wasn't the first oil executive to be richly compensated; Singh himself made much more money running his hedge fund. The real problem was Ward couldn't get his stock to rebound after the 2008 collapse, a failure that made his hefty salary and other activities glaring targets for investors.

Like McClendon, Ward had bet his career that an abundance of oil and natural gas could be found in the United States. By 2012, more energy was pumping than anyone ever expected, making both McClendon and Ward look prescient. For years, Wall Street investors had lent their support for the push by the wildcatters into U.S. oil and gas fields, enabling them to borrow big money to scoop up acreage at record clips. During the early stages of the nation's energy revolution, it was enough to accumulate attractive fields and add to reserves, no matter the cost.

But by 2012, it was clear that the push into new fields by McClendon, Ward, and other pioneers of the era was turning out less immediately lucrative than expected. Investors deserved a payoff for all the money they were investing, but McClendon and Ward were still building a path to profits. Just as early Internet companies saw their shares soar based on innovative ideas and strategies for future earnings, and then tumble when earnings took too long to materialize, innovators in the oil patch suffered as Wall Street changed its tune and demanded quicker results.

Had natural gas prices stayed above five or six dollars per thousand cubic feet, McClendon might have looked like a genius for grabbing so much prime shale acreage. Then again, a big reason prices fell as far as they did was due to all the production from Chesapeake and other pioneers, production that McClendon didn't seem to see coming.

McClendon wasn't the only energy mogul to miss the impact of all the drilling in the country. In 2012, Rex Tillerson, ExxonMobil's chief, said, "We're all losing our shirts today," due to tumbling natural gas prices. "We know we're making no money."

But McClendon had planted the seeds for the growth of the nation's gas supplies, not Tillerson. And ExxonMobil had a fortress balance sheet, while Chesapeake's was weighed down by all the debt accumulated from McClendon's moves, making the company less capable of dealing with a downturn in gas prices and the resulting falloff in revenues.

McClendon seemed unprepared for the growing glut of gas. It was as if Johnny Appleseed, after spreading seeds around the country, was shocked when apples actually grew beneath his feet.

Liz Irish was just as surprised as Aubrey McClendon at how the shale revolution was turning out.

After moving to Williston, North Dakota, in 2010, Liz's husband, Matt, began working as a trucker, hauling water for the fracking operations of various wells in the state. Liz looked for her own job, eager for a new start.

But housing emerged as an early problem for the family. Prices had skyrocketed throughout the region because there wasn't enough supply to handle the raging demand. By the time the Irish family arrived in Williston, three-bedroom homes were renting for nearly $3,600 a month. Prices climbed so fast that a priest in Sydney, Montana, a city about an hour's drive from Williston and near some oil drilling, gave a Sunday sermon urging neighbors not to gouge each other on rents.

Some moving to the area receive housing from their employers, or enjoy a subsidy from their job enabling them to afford reasonable housing. Others settle into "man camps," military-style complexes with communal showers and bathrooms. The rooms are clean, private, and efficient, usually featuring a bed, a small desk and refrigerator, and a flat-screen television. Some man camps even have elaborate cafeterias, gyms, laundries, and video-game rooms.

The camps aren't especially rowdy places. Most field hands are too exhausted after working twelve straight hours to do much carousing, and management forbids alcohol—though violence sometimes results from having so many cooped-up men in the same facility.

Unable to afford a home of their own, Liz and her family moved into a twenty-four-foot trailer originally used by the Federal Emergency Management Agency, or FEMA, to house displaced residents from the New Orleans area after the devastation from Hurricane Katrina in 2005. The trailer was lent by Matt's employer, but it wasn't a long-term solution.

In October 2010, the Irish family moved into a single wide mobile home parked in the Buffalo Trails campground, a particularly ugly spot where the bathrooms are shut from 9 p.m. to 9 a.m. each day and most of the residents are long-term oil workers. The monthly rent was $1,500, and the landlord liked to walk in on them at all hours, but they considered themselves fortunate to have found a home. Work seemed more promising. Matt Irish, who had been making about $20,000 a year before the move to Williston, now was making between $7,000 and $15,000 a *month*. Liz didn't really need a job, but she was bored in the trailer park. She had more trouble finding a position that suited her skill levels, though, and had to settle for a cashier job at an Albertsons supermarket.

"I was a girl and I wasn't one of them," she says, to explain her inability to find a better position.

Back in Oregon, the family had only seen a dusting of snow, so they got a kick out of their first winter in Williston, where it snowed more than 140 inches and one day registered a windchill effect of minus sixty-five degrees.

"I'm a horrible optimist and my attitude was we were going to stay there forever," Irish says.

Life got tougher as winter wore on, however. A blizzard took out the electricity for four days. Driving home one day, all Liz could see was white in front of her, and she plowed her Ford pickup into a snowdrift.

"You drive where you think the driveway is," she says.

Matt was making great money, but the work was hard. He was driving a semi truck over frozen roads that were in terrible shape, fighting through blizzards and using a blowtorch to unfreeze the water he was carrying.

Soon the family's rent was hiked to $1,700 a month, even as Matt's work hours shrank, amid growing competition from new trucking companies. Matt was still making more than $90,000 a year, but the family's expenses were higher than they had assumed. A gallon of milk cost $5.60 and they had to buy expensive, heavy coats for the girls, along with generators and snowblowers.

In the spring, Irish noticed that trucks kicked up so much dust, their trailer, car, and truck were always filthy. "Two hours after you clean, it looks the same again," she says.

Liz found a better job, making twenty-three dollars an hour as the personal assistant to an oil executive, and she tried to keep a positive attitude. But it bothered Liz and Matt that oil workers regularly propositioned their young daughters; the cursing the girls heard around town also was getting to them.

There were other troubling experiences. One day, a man wearing a black-and-white-striped suit and orange Crocs raced past her. The man seemed to be escaping from a local jail. Liz called 911 and the disinterested operator said, "He's not one of ours." Eventually, the man was found hiding behind some bushes.

"It was like a cartoon," Liz says.

In 2012, Matt lost his job, as drilling became more efficient and more frack jobs were completed reducing the need for drivers. He ended up hauling fuel for an annual salary of about $60,000.

The final straw for the family was when their landlord gave notice that a man camp would be built on their lot and that they needed to vacate the trailer. Before the Irish family even finished packing up, two Russian oil workers moved in and began doing shots of liquor on their kitchen table.

There were no homes available so the family paid $2,100 a month to rent a cramped apartment in Williston. The high rent made it almost impossible to save any money. Finally, Liz and Matt decided to bolt the Bakken and move the family back to Oregon. Back home, Liz rejoined her mortgage company. She and her husband managed to pay down some bills during their time in Williston, suggesting that their experience wasn't a total loss. They don't make as much as they did during the heyday for housing, and Matt can't find a job paying the same wages as in the Bakken, but they're thrilled to be in Oregon once again.

By 2013, there were signs that the quality of life in Williston was improving. The city routed trucks around town to reduce traffic. Investment firms such as KKR & Co. were busy building new housing, and a $70 million municipal recreational center was being constructed, featuring a water park, tennis and basketball courts, fitness areas, and more.

Liz has few regrets about her decision to leave, however. "Locals who own land or run a business are making money hand over fist," she says. "And if you're young and not attached, it's a great place to make some money . . . but you're cold and alone and you work your butt off. You can't do that forever."

EPILOGUE

After a year spent traveling the world searching for customers interested in buying natural gas from his not-yet-built terminal in Louisiana, Charif Souki finally convinced someone that his plan might work.

In late October 2011, Souki inked a deal with Martin Houston, a senior executive of BG Group, a huge global trader of LNG. BG Group agreed to pay Cheniere $410 million a year for twenty years for the right to use Cheniere's terminal to convert 3.5 million tons of LNG annually. BG made plans to sell liquefied American natural gas to consumers and businesses in Europe, Asia, and elsewhere.

The deal meant that a big foreign buyer was so sure U.S. natural gas prices—then about four dollars per thousand cubic feet—would remain below global gas prices for two decades that it was willing to pay a sizable premium to get its hands on the cheap U.S. gas.

Souki wasn't out of the woods yet. Investors still worried that he wouldn't be able to finance the cost of retrofitting his facility. Cheniere didn't even have approval from the Federal Energy Regulatory Commission to start construction at Sabine Pass.

Some investors gave up waiting. John Paulson's hedge fund sold its Cheniere interest. GSO, the Blackstone hedge fund, also decided it had had enough.

"We were tired, to be honest, and they were going to spend ten billion more," says a GSO executive. "Charif is comfortable with risk, so if you have a chance to get out you take it."

Souki kept jetting around the world, signing three more deals with customers in early 2012. He used Cheniere's tiny size to his advantage, reminding global gas players that his company didn't compete with them, unlike ExxonMobil, Chevron, and other giants examining the idea of exporting gas from the United States.

Eventually, Souki found an investor willing to help finance the construction of his revamped terminal. Blackstone's GSO fund had sold its investment, thrilled to escape with some profit after a bumpy ride with Souki. But David Foley and Sean Klimczak, executives in Blackstone's private equity group, decided to take a chance on Souki's idea. In February 2012, Blackstone agreed to invest $2 billion in Cheniere, enough for the company to begin work on two processing units, or "trains," to compress, chill, and ship gas from the United States.

Souki made additional trips, courting global investors to raise the rest of the cash he needed. He discovered that a growing number had become convinced the U.S. shale gas boom was for real, and that gas exports were inevitable. In May 2012, Temasek Holdings, the huge Singapore state investment firm, teamed up with Hong Kong private equity firm RRJ Capital to invest nearly $500 million in Cheniere.

Banks and lenders kicked in the rest of the $5 billion Souki needed to complete the first two of four gas processing units at his Louisiana facility, and the company received government approval to begin construction.

It was becoming clearer that Souki's plan was going to work. He even sent a pair of snow boots to Fadel Gheit, the stock analyst who had said it would "snow in New York in July" before any natural gas was exported from the Gulf of Mexico.

At least Cheniere "got the size right," Gheit told the *Wall Street Journal.*[1]

During the first three months of 2013, Cheniere shares climbed from eighteen dollars to twenty-five, as investors became believers. Souki managed to raise an additional $5 billion, to build the other two liquefaction units, enabling even more gas to be shipped out of his terminal. By 2015, Cheniere expects to export an average of five hundred million cubic feet of gas a day.[2]

As Souki's own shares became more valuable, he spent some of his newfound wealth on an old love. He paid over $14 million to buy a group of commercial buildings in downtown Aspen, the city he continued to visit, as he plotted a new venture.

Controversy erupted, however. While some environmental activists supported natural gas exports, pleased that gas would displace dirtier coal energy in places like China and India, some big American users of natural gas marshaled forces to pressure the government to prevent gas exports. Chemical companies that depend on natural gas, such as Dow Chemical, feared siphoning supplies from the country would send prices higher, raising costs for businesses and consumers.

A national debate began about the benefits of exporting natural gas. It didn't have much impact on Souki and Cheniere, however. The company's Sabine Pass terminal had already obtained government approvals to sell LNG, even to countries like China and India that don't have free trade agreements with the United States. It didn't seem likely that the government would withdraw the permission.

Cheniere was in the unique position of being the only company in the lower forty-eight states with both federal permits and supply contracts to ship gas overseas. In May 2013, the Department of Energy granted permission to Michael Smith's Freeport LNG to export gas, but it wasn't clear how many other companies would join Cheniere in shipping natural gas from America.

In July 2013, Cheniere shares hit thirty dollars, up 200 percent in two years. The company was on schedule to become the largest single buyer of U.S. natural gas by 2016.[3]

That summer, Souki took a much-needed vacation to Saint-Tropez, the French resort town. By then, he and his wife, the former *Sports Illustrated* swimsuit model, were separated, so he brought his three sons and their wives along on the trip.

Sipping iced coffee on a porch overlooking the city's spectacular harbor, Souki saw someone he recognized. It was a hedge fund manager from New York, one of many who two years earlier was losing money on Cheniere and was fretting about how Souki was going to keep the company alive. Now the investor approached Souki to congratulate him for moving closer to exporting gas.

At that point, Souki was worth more than $300 million, ample reward for years of hustle, perseverance, and optimism, even when it wasn't really called for.

"I feel real satisfaction," he says, looking back on his long journey.

EOG Resources' production from the Eagle Ford formation in Texas continued to grow in 2012, helping the nation produce about 2.2 million barrels of crude a day from shale and other tight rock formations, up more than 50 percent in just a year and representing about 30 percent of the nation's oil output.

The Eagle Ford gusher helped overall U.S. daily crude production hit 6.5 million barrels in 2012, up nearly 800,000 barrels in a year, the fastest growth in U.S. history.[4]

"What can I say about the Eagle Ford except that it's an 800-pound gorilla developing into a 1,000-pound gorilla?" EOG chief executive officer Mark Papa told investors.

Things got even better in 2013 for both EOG and the country. During the first three months of the year, the company completed twenty-seven "monster wells," or those with initial production of more than twenty-five hundred barrels a day. Nine of the wells even began producing at thirty-five hundred barrels a day, on par with wells in Saudi Arabian fields. EOG estimates that it can recover 2.2 billion barrels of crude from its 639,000 Eagle Ford acres.

By the middle of 2013, the country's overall oil output hit 7.5 million barrels a day.[5] A revitalization of Texas's Permian Basin, thanks to horizontal drilling and multistaged fracking, was partly responsible for the surge, along with continued growth in the Bakken and Eagle Ford regions.

The rush for land in the Eagle Ford was turning it into one of the fastest-growing regions in the country, just like the Bakken region in North Dakota. "It's a madhouse," according to J. E. Wolf III, a real estate broker in Yorktown, a city of fewer than three thousand residents in South Texas. "I've been selling real estate here for forty-three years and I've never seen it like this."[6]

By the summer of 2013, EOG stock was soaring to $150 a share, giving Mark Papa's company a market value of nearly $41 billion, or more than the combined values of Alcoa, Southwest Airlines, and Hershey.

In July, Papa, who was sixty-six years old, handed the reins of the company to Bill Thomas, his senior exploration executive and the one who had been the company's biggest proponent of trying to get oil from shale. Papa prepared to retire at the end of 2013.

"We're having a positive impact on the U.S. economy and reducing U.S. dependence on OPEC oil," Papa says. "It feels great."

By 2012, George Mitchell was a revered statesman in the energy business. His determination to extract natural gas from shale in the Barnett region had paved the way for the discovery of over a dozen shale formations in the United States. Production from those fields hit record levels in 2012 and accounted for about 25 percent of the nation's total gas production. In addition, Mitchell's techniques were partly responsible for oil pouring out of various shale formations in the country, a surge that increased the likelihood that the United States might achieve energy independence in the years ahead.

Late in life, Mitchell continued to buck convention. He gave millions to research clean energy even as he, along with his son and Joe Greenberg, invested in a new shale formation in Canada. Wearing a black velvet sports jacket during an interview in his Houston office, Mitchell said he supported taxes on fossil fuels to help alternative energy compete. He also urged stiff regulation of drillers, some of whom had committed environmental abuses.

"Fracking can be handled if they watch and patrol the wildcat guys," he said. "They don't give a damn about anything; the industry has to band together to stop the isolated incidents."

When Mitchell died in the summer of 2013 at the age of ninety-four, it was clear that few Americans could match his impact on the country and the world.

★   ★   ★

B y the summer of 2012, Harold Hamm reached his goal of achieving "ancient wealth." He also had arrived on the national scene.

At one time, Hamm had been on a quixotic quest to tap oil in overlooked U.S. fields. By this time, Continental Resources held more than 900,000 net acres in the Bakken, making Hamm's company the largest landholder in the flourishing formation. Continental was pumping more than 100,000 barrels of oil a day, double its production of 2010, making it America's ninth largest oil producer.

"The fields keep getting better," Hamm said that summer.

By late September 2012, Hamm, together with his wife and five children, owned an $11 billion stake in Continental, or more than three quarters of the company's shares. Hamm moved Continental's headquarters from Enid to Oklahoma City, one more step to becoming a true oil baron. He settled into a plush new office with a huge cowhide on the floor, a matching cowskin chair, dark wood paneling, and a rifle by his desk.

In some ways, both Hamm and his family remained genuine and unpolished, despite their enormous wealth. While taking a visitor to lunch in Continental's headquarters that summer, Hamm forgot his identification card and had to call several times for someone to let him into his own offices. At one point, he tried to treat his sister Fannie to a mansion, but she refused to budge from her tiny home in little Lexington, Oklahoma. Fannie did consent to let her brother remodel her bathroom.

Hamm seemed to hunger for a still higher profile. He testified before Congress about energy independence, arguing for "reasonable" environmental regulations, drilling on federal lands, and the continuation of tax policies beneficial to the exploration business.

"I am here as an American patriot that loves my country and as a person that is grateful for the opportunities I have been given by being an American," he told Congress, while noting that America now imported less than 45 percent of its oil, down from 60 percent a few years earlier, a figure likely to drop much further.

After months of grumbling about how President Barack Obama was holding back the U.S. energy industry, Hamm gave nearly $1 million to a "Super PAC" supporting Republican presidential candidate Mitt Romney. Soon, Hamm was enlisted as Romney's energy adviser, appearing with the candidate at a series of high-profile events.

Hamm made regular appearances on business television networks to extol the wonders of American crude and to insist that there was much more to be pumped. He and his wife, Sue Ann, smiled for the cameras at a black-tie dinner hosted by *Time* magazine when Hamm was named one of the one hundred most influential people in the world. That year, the couple hosted Romney and seven hundred Republican donors at their mansion in the exclusive Oklahoma City enclave of Nichols Hills, not far from where Aubrey McClendon lived.

It wasn't hard to see Hamm's impact on the nation. By 2012, North Dakota had passed California and Alaska to become the second-largest-producing state after Texas. By the end of the year, the Bakken pumped close to 700,000 barrels of crude a day, or about 10 percent of the entire nation's production. The Bakken has a shot at producing one million barrels a day, joining only six other fields in the world ever to reach that level of production.[7]

Andy Rihn knew his boss wanted more. Two years earlier, Rihn, a twenty-six-year-old geologist in Continental's Oklahoma City headquarters, thought he'd found a perfect place to drill. It was a spot in south-central Oklahoma, not terribly far from Continental's own offices.

Wildcatters as far back as the beginning of the twentieth century had found oil in the area, and veterans of the state, including Aubrey McClendon and Tom Ward, knew it well. But Rihn and a few colleagues became convinced that horizontal drilling could enable much more oil to flow from rock in this region.

"You really believe in it?" Jeff Hume, Continental's president, asked Rihn in a meeting one day.

"Yeah, I really do."

"Okay, but it's your neck, buddy," Hume told him, only half jokingly.

The Continental team gave the spot a new moniker: Andy's Neck.

Quietly, Continental began leasing land in the area, which included the southern section of the Woodford Shale, amassing about 200,000 acres by the fall of 2012. When the company began drilling wells, the results were better than Rihn had hoped. The rock was so full of oil it even reminded some of the Bakken.

During a speech to investors and analysts in the first week of October 2012, Hamm described the new find. It didn't seem right to call it Andy's Neck, so it was renamed the South Central Oklahoma Oil Province, or the SCOOP formation. Hamm said it could add 1.8 billion barrels to Continental's reserves. Andy's neck was safe, it seemed.

The news sent Continental's shares climbing more than 5 percent and set off a search by rivals for nearby land. At sixty-six, Hamm had come full circle. He was back in Oklahoma, out to uncover one of the last meaningful shale plays in the country.

Hamm already owned more oil in the ground than any American. It didn't seem life could get any better.

On April 8, 2013, Hamm met his old friend Mickey Thompson at an Oklahoma City bar to watch the NCAA basketball championship. It was a much-anticipated game between Louisville and the University of Michigan, and Hamm should have been in good spirits. Continental shares were on their way to $100, an all-time high, and Hamm soon would be worth more than $12 billion.

Hamm seemed subdued and preoccupied, though. Thompson had an idea why. For the previous several years, word had circulated around Oklahoma City that Hamm and his wife had a strained marriage. A few weeks earlier, Continental had announced that Sue Ann had quietly filed for divorce a year earlier, just ten days after the couple hosted the event for Romney in their home.

By the night of the game, friends had noticed a change in Hamm. He had become more serious and reserved. The public airing of the marital dispute in various media outlets, including allegations that his wife had discovered Hamm was having an affair three years earlier and a suggestion that she had been videotaping him at home, had embarrassed him, the friends said.

Thompson, who was close to both Hamm and his wife, tried to cheer his friend up. He joked that he was the last person who should give

Hamm marriage advice, since Thompson already had been married and divorced four times. There's no one "more matrimonially challenged," Thompson told Hamm, making him smile.

Because Hamm and his wife didn't have a prenuptial agreement, and because Sue Ann had spent time as a senior Continental executive, experts said she likely had claim to as much as half of Hamm's wealth. When members of the media did the math, it made some heads spin. Sue Ann seemed in line for at least $3 billion, making her wealthier than Oprah Winfrey. The divorce likely will be the costliest in history, with a bigger payout than those of Rupert Murdoch, Stephen Wynn, and Michael Jordan.

Even British tabloids began to cover the story, weighing in with details on the four homes the couple owned and the debutante ball their daughter had attended. Investors began to speculate that the divorce might hurt the value of Continental shares if Sue Ann were to receive half of Hamm's shares and sell some of them. On Hamm's victory lap, the divorce had tripped him up in a humiliating way.

When a reporter caught up to Hamm outside an Oklahoma City courtroom and asked how he was doing, he replied simply, "Not very good."[8]

By January 2013, it was clear to most everyone that Aubrey McClendon had to go. Clear to most everyone but McClendon, that is.

Chesapeake Energy ended 2012 below seventeen dollars a share, down from above twenty in the fall that year. More important, big investors, including O. Mason Hawkins of Southeastern Asset Management and billionaire Carl Icahn, had become convinced Chesapeake had an "Aubrey discount." Wall Street was being extra skeptical of the company because it had lost faith in McClendon's leadership, Hawkins and Icahn believed.

Newly installed board members had disagreed with McClendon over capital spending and expenses, with McClendon pushing for a larger budget than the board was comfortable with. During the last week of January, board members told the company's new chairman, Archie Dunham, to give McClendon the bad news. They had the votes to oust McClendon from the company he had cofounded twenty-three years earlier, and they wanted him gone.[9] It was a coup d'état, and there was nothing McClendon could do about it, Dunham told him.

McClendon, fifty-three years old, was stunned by the news. Ever the optimist, he still believed in Chesapeake and was sure he could improve the company's performance and please his new board members.

After Dunham told McClendon he was out, he started to pout, telling Dunham the board had mistreated him, according to someone close to Chesapeake's board. At one point, McClendon wanted to examine details the company had about its wells but the board wouldn't let him, the person said.

When the company announced McClendon's departure, investors immediately pushed Chesapeake shares up 9 percent, a final slap in the face for him.

Once amounting to about $3 billion, McClendon's wealth had plummeted due to the forced selling of his shares in 2008 and the reduced value of his stakes in the company's wells. By March 2013, he fell off *Forbes* magazine's billionaire list, even though he had received a severance package of about $47 million. Even his friends were unsure how much cash and assets McClendon had after subtracting mountains of debt he owed related to his ownership stake in Chesapeake's wells.

As he made plans to leave the company in April, colleagues sensed McClendon was looking forward to the next chapter of his career. He sent an upbeat letter to his employees, saying his departure was due to "philosophical differences" with the board, while promising that the "separation will be amicable and smooth."

He added, "In many respects, our accomplishments are unique and I will always remain immensely grateful for the time I have spent with you."

In the end, McClendon couldn't slow his pursuit of new drilling opportunities in America. He also had a remarkable affinity for debt and lavish pay, irking a nation still digging out of a historic downturn caused by leverage and greed.

McClendon had helped build Chesapeake into the second largest natural gas producer in the country. Though Chesapeake shares ran into difficulties during the last few years of McClendon's tenure, they remained up over 200 percent since 1993, beating almost every rival's performance.

More important, McClendon's single-minded focus on producing natural gas from once dismissed U.S. shale formations was a big reason the nation managed to produce so much oil and gas so suddenly. It also helps explain the steep drop in American energy costs, even as prices rose around the world.

At a company farewell picnic in March, McClendon walked through the crowd and chatted with employees, wearing a rumpled white dress shirt, sleeves rolled up as usual, and khaki pants. Choking back tears during a goodbye speech, he said his "greatest joy was knowing that this company played such a leading role" in changing the nation's energy outlook. He left the stage to a standing ovation.[10]

B y early 2013, the dissident investors in Tom Ward's company, Sand-Ridge Energy, weren't going anywhere. If anything, their anger was intensifying. Dinakar Singh, who ran hedge fund TPG-Axon, insisted to friends that his effort to oust Ward was about more than pushing Sand-Ridge shares higher. Singh told them that Ward had fleeced shareholders and shouldn't be allowed to get away with it.

"It's just wrong," he told one friend.

Singh's firm had proposed a vote to remove Ward and the rest of SandRidge's board of directors. It was the kind of proposal usually brushed off by companies. Shareholders generally ignore these votes or simply cast their votes with incumbent managers.

Singh's proposal tapped a vein of unhappiness that Ward hadn't fully appreciated, however. When a respected proxy advisory firm called ISS recommended that shareholders vote to replace a majority of the board, writing that "the apparent failures of stewardship on this board are legion," it was clear that the odds of Ward remaining at his company were slim.

In early March, SandRidge executives hosted an annual day for investors and analysts, hoping to spark enthusiasm for the company. Ward was well aware that almost everyone was focused on the upcoming vote and whether he would be ousted. It was the elephant in the room.

As Ward started his opening speech, he tried for a little levity. "Sorry we're running a little bit late," he said. "I was back to talking to Aubrey about a board seat . . . just kidding, just kidding."

Ward did little to win over investors or analysts that day. By then, he was resigned to stepping down at his company. Negotiations over terms of his departure became so ugly, though, that a team of bankers from Morgan Stanley was brought in to find a solution. The bankers failed miserably, however, resulting in a showdown between Ward, SandRidge's directors, and the dissident investors.

With results of the vote just two days away, Ward realized he was boxed in. He knew it was only a matter of time before he also would be kicked out.

On Wednesday, March 13, a press release announced a deal that seemed likely to lead to Ward's departure. In the release, Ward was only allowed to give a perfunctory goodbye to his president, who was resigning immediately. It was never clearer how powerless Ward had become at the company he had founded. Six weeks earlier, Aubrey McClendon had been forced out at Chesapeake. Now it was happening to Ward as well.

McClendon and Ward had helped change the country in just a few short years. Along with their fellow frackers, they helped alter the course of American economic and geopolitical history with a remarkable drive to discover new energy deposits. Businesses and consumers were enjoying falling energy costs, the oil and gas industry was thriving, and energy independence was in sight.

In the end, however, both men met condemnation and denigration. Instead of riding off into the sunset, they'd both lose their jobs.

When the press release was issued, SandRidge shares rose 6 percent; investors cheered Ward's likely departure, just as they had celebrated McClendon's.

A few months later, Ward was out at SandRidge, a precipitous fall that was cushioned by a $90 million severance package and the $150 million in shares he retained.

"It's been a hard few years," says Ward, who predicts that his moves at SandRidge will be proven wise over time. "It's easy to throw stones."

*   *   *

Almost immediately after the press release went out, Ward walked out of the office into a brisk Oklahoma evening. He got into his black BMW X6 and drove a few blocks to the Chesapeake Energy Arena, where the Oklahoma City Thunder were getting ready for a key matchup. There, he joined James, his younger son, and Frank Alberson, who had grown up in the Ward home, in front-row seats behind the basket.

From the opening tip-off, the Thunder began to pummel the Utah Jazz in a nationally televised game. Ward loosened his blue tie and seemed to enjoy himself. After a dunk by the Thunder, he gave his son a wide smile. By halftime, Ward was beaming, as if a load was off his back.

At the other side of the court, in his own prime spot under the opposing basket, Aubrey McClendon sat with his wife. McClendon wore a smart dark suit and a red tie, as if he were interviewing for a new career. Weeks earlier, McClendon had quietly inked a deal to rent office space in a six-story glass-and-concrete building about a mile from Chesapeake's headquarters. He had already begun courting investors to raise money for his next venture, a company that would explore for oil and gas. They would do it in the United States, of course.

During a break in the action, McClendon met an old friend near midcourt. McClendon was ebullient as he spoke of his plans for a comeback.

"Are you almost ready to go?" the friend asked.

McClendon flashed a big smile. "I'm almost there."

AFTERWORD

Perfect is the enemy of good.

—Voltaire

ENVIRONMENTAL TENSIONS

The wildcatters, entrepreneurs, and hopeless dreamers who took a chance on drilling in shale and other challenging U.S. rock formations have brought clear benefits to the country and possibly the world. U.S. natural gas prices, as of the summer of 2013, were about a third of those in Asia and less than half of those in Europe, thanks to surging American shale-gas production. As a result, U.S. consumers and companies are paying less to heat and cool their homes and businesses. Chemical, plastic, fertilizer, and other companies relying on natural gas to manufacture their products also are benefitting, a key reason some foreign companies are moving to the United States or building factories in the country.

Expanding energy production and related activities could lead to more than two million jobs by 2020. They also could add more than one percentage point to annual economic growth over the next ten to fifteen years, while the nation's trade deficit is expected to drop by 85 percent by 2015, helping the value of the U.S. dollar.

But how much damage have the pioneers created in their pursuit of oil and gas? Is fracking as bad as activists say, and what will its impact be as drillers continue to pursue energy from shale and other rock formations?

The short answer: Fracking has created less harm than the most vociferous critics claim, but more damage than the energy industry contends. And it may be years before the full consequences of the drilling and fracking are clear.

One of the most dramatic accusations leveled by those campaigning against fracking is that methane—the main component of natural gas—

has invaded various water systems as a result of nearby drilling, jeopardizing the health of local residents. One of the most frightening moments in Josh Fox's 2010 documentary, *Gasland,* is when a Colorado homeowner opens a kitchen faucet, strikes a match to his tap water, and sees it explode in a fireball.

The scene seems like clear-cut evidence that methane has leaked from nearby gas wells.

This accusation likely is overstated, however.

Methane—a colorless gas that's generally considered nontoxic unless it's in superhigh quantities—occurs in shallow bedrock. It's been known to naturally seep into water wells and springs and has done so throughout history in various regions, whether or not there's been any nearby gas production. In fact, three states—New York, Kentucky, and West Virginia—have towns with the name Burning Springs, a testament to this age-old phenomenon.

Joyce Siracuse, the sixty-year-old owner of Joycie's café in Montrose, Pennsylvania—six miles from the epicenter of the fracking controversy in Dimock—says she and her friends regularly lit water afire in their grade school bathroom in the late 1960s, long before fracking came to her part of the state. "We all lit the faucets growing up," she says. "It wasn't a big deal."

There are times when poorly sealed wells, rather than hydraulic fracturing, can lead to methane-filled drinking water. But that has occurred in a minority of wells, according to a 2013 study by Duke University examining drilling in the Marcellus Shale region in Pennsylvania.

Critics also contend that the toxic chemicals used for fracking can enter nearby water systems, making people and animals sick and putting public health at risk. But there have been virtually no proven cases of any fracking chemicals traveling up and into water tables, and there are reasons to doubt it will happen. Aquifers in the Barnett, Marcellus, and Haynesville regions are between four hundred and twelve hundred feet below the surface, according to a 2012 report by the U.S. Government Accountability Office, while the shale in those regions is between 4,000 and 13,500 feet from the surface. Most scientists say it will be nearly impossible for chemicals to migrate the distance between these shale formations and the aquifers, though they can't say for sure it won't happen.

"I'll take my chances on winning the lottery over the chances of frack fluid in the groundwater," environmental engineer Radisav Vidic of the University of Pittsburgh told *Scientific American* magazine in May 2013.

That's not to say water quality is great in the drilling areas. It's just hard to determine how it got that way. A 2011 Penn State study found that about 40 percent of water wells tested before drilling began failed at least one federal drinking water standard, usually for coliform bacteria, turbidity, or manganese.[1]

Hand-wringing about earthquakes caused by fracking also seems overblown. For one thing, it isn't the fracking process itself that may be lubricating preexisting fault lines; rather, it seems to be the disposal of wastewater by those reinjecting it into adjacent rock. To date, only small, locally produced tremors have been detected—much like with other mining activities—rather than huge quakes. And seismic events likely can be avoided by conducting seismic surveys or adjusting injection pressures.

Still, industry members are too quick to gloss over concerns about their activities, an insouciance that has created a public relations nightmare for their business. Leaks and surface spills of dangerous chemicals used in fracking fluids have been documented, and wells that aren't properly sealed have allowed gas and chemicals to migrate through the wellbore.

Professor Vidic has found evidence of contamination in Pennsylvania streams receiving wastewater from centralized treatment facilities in the Marcellus, and blowouts have caused fracking fluids to enter local streams. State environmental regulators have demonstrated that oil and gas development damaged water supplies for at least 161 Pennsylvania homes, farms, churches, and businesses between 2008 and the fall of 2012, though the majority of the complaints were deemed unfounded and the industry says it has improved its standards since then.[2]

Bad casing and cementing can cause problems in any drilling activity, of course, but wells that are hydraulically fractured usually are exposed to higher pressures than those that aren't fracked, raising additional concerns.

Another lingering concern: Operators aren't always required to disclose every component of the frack fluid they use, because they say such disclosure would reveal trade secrets. Since 2011, more than twenty states have begun to require the disclosure of the bulk of the components of the fluid, but the rules provide exceptions. Since the fracking cocktail can include acids, detergents, and poisonous chemicals, along with more innocuous materials, there's ample reason for unease.

Drilling is an industrial activity that brings intense noise, traffic, and disturbance to towns near active wells. Because so much water is used for fracking, this drilling requires even more truck trips than usual. And shale production impacts air quality, just as any oil-and-gas drilling does. Dust and engine exhaust from truck traffic is a health hazard, as are emissions from diesel-powered pumps. Silica sand, which is used in the fracking mixture to prop open fractures in rock, can lodge in lungs, potentially causing silicosis.

Sometimes the impact of the drilling is startling. In 2011, residents of Pinedale, Wyoming, a bucolic town of two thousand that's near gas fields, began complaining of watery eyes, shortness of breath, and bloody noses because of ozone levels exceeding those in Los Angeles and other major cities on their worst days.

"I never would have guessed in a million years you would have that kind of danger here," said Debbee Miller, a manager of a Pinedale snowmobile dealership.[3]

Despite these ills, and the awful image many have of the frackers, there's actually reason to think their activity has in some ways been good for the environment. The glut of gas has sent prices falling. This has encouraged the United States to shift much of its energy consumption from coal to gas. That's likely healthy for the environment because burning natural gas releases about half as much carbon dioxide—a greenhouse gas believed to contribute to global warming—as coal when producing the same amount of energy. And natural gas emits far less nitrogen oxide, which causes smog; it also spews no mercury or particulate pollution, unlike coal.

In fact, U.S. energy-related emissions of carbon dioxide fell 12 percent between 2005 and 2012 and now stand at the lowest levels since 1994, according to the U.S. Energy Department's Energy Information Administration. That unexpected improvement is due to more electricity coming from gas instead of coal, as well as the lingering effects of the economic downturn, which has slowed energy demand.

Before toasting the frackers, however, there's one big caveat: Natural gas wells and pipelines frequently leak methane. And methane is a more potent contributor to climate change than carbon dioxide. If too much methane is being leaked into the environment by the gas wells and pipelines, it could be more than offsetting the reductions in carbon dioxide coming from the shift from coal to gas. Some states require natural gas producers to use technology to capture gas during the construction and drilling of wells, but many have resisted the requirements.

So how much leaking is going on? It's hard to tell, scientists and environmental activists say. The Environmental Protection Agency doesn't do a good enough job tracking methane leakage, many argue. Some private parties, like the Environmental Defense Fund, are working on studies of the issue, but it's too early to get a sense of their results. Just as troublesome, the extent of most risks from shale development is unknown because most studies don't track the long-term, cumulative effect of this drilling.

In some ways, the debate about fracking is a lot of misdirected energy (no pun intended). It's just not realistic to expect a nation still suffering from the deepest economic downturn since the Great Depression to ignore some of the largest energy fields in the world, forgo the substantial savings that result from drilling in those domestic spots, and keep funneling money to Russia, Iran, Qatar, and other energy powers.

A better tack is to put pressure on oil and gas producers to improve their behavior. There's been tremendous progress already. Drillers are recycling more frack water, for example. Meanwhile, reports of contamination cases dropped by two-thirds between 2011 and 2012, according to Pennsylvania's Department of Environmental Protection.

But more regulation of fracking activity is needed, such as rules ensuring that well casings are set at proper depths and have tight seals, to be sure chemicals never leak into aquifers. Ohio and Texas have agreed to extensive rules regulating and testing well casings, but other rules are inadequate, even according to some major drillers. More federal rules to control air pollution from drilling are also needed.

"It's all totally fixable, but just because the problems are manageable doesn't mean they will be managed," says Fred Krupp, president of the Environmental Defense Fund. "It's going to take action by state regulators, industry, and citizens to make it happen."

It's clear that oil and natural gas will be needed for the foreseeable future, another reason to focus on improving the activity of operators rather than try to push them out of the oil and gas business. A world with an eventual population of ten billion people, most of whom are more concerned about putting food on their tables than the impact of fossil fuels on the environment, will need a variety of energy sources.

Few believe renewable energy sources can provide the bulk of the world's needs anytime soon, but it would be a shame if work on wind, solar, and other cleaner sources slowed amid the burst of oil and gas production. For one thing, it's not yet clear how long this period of abundant fossil fuels will last. Production from wells from shale and other tight rock is notorious for declining rapidly. The recent growth has been impressive but the output may not last as long as optimists expect.

Hopefully, natural gas will buy time so that renewable energy sources can grab a larger share of the market. In 2012, U.S. usage of solar panels rose 76 percent to the equivalent of enough electricity generation to power 1.2 million homes. Growth is expected to be about as strong in 2013, amid record solar installations.

There are other reasons for optimism, such as a continued drop in energy demand in the United States, a shift that in some ways is as surprising as the nation's growth in energy production. As recently as 2006, American drivers pledged their allegiance to huge automobiles. Our cars, like our homes, couldn't be too big. It seemed a bit suspicious—even a tad European—to fret about the environment or gas prices.

By 2007, however, when President George Bush signed legislation reducing gasoline consumption in vehicles, energy demand was on the wane. It's continued to drop over the past few years as four-dollar-a-gallon gasoline pushed gas guzzlers off the road, as did young people embarrassed at what their parents were driving.

A drop in demand in the United States and around the globe could continue. Companies and homes are becoming more energy efficient. And the average American new car and truck will get nearly fifty-five miles per gallon by 2025, under fuel efficiency standards enacted by President Barack Obama, up from an average of almost twenty-four miles per gallon in 2012. Edward Morse and other analysts at Citigroup also note slowing oil demand in China. Any slowing of global energy demand will bring benefits to the environment and put pressure on prices.

Alternative-fuel vehicles, such as all-electric cars and hybrids, are also gaining popularity. Brokerage firm Raymond James says electric vehicles could claim 1 percent of the market in 2013, a share likely to keep rising. Any widespread embrace of self-driving cars could cripple oil demand.

George Mitchell's son, Todd, says his father's work will have had a negative impact on the world if it forestalls progress on renewable energy, instead of giving innovators time to improve wind, solar, and other cleaner energy sources.

"I think that it will be clear in decades or more that extracting hydrocarbons from tight shale formations blew up all previous assumptions about the availability and economics of oil and gas development," Mitchell says. "What's harder to understand now is how good a thing that is."

THE GLOBAL VIEW—THE REVOLUTION CONTINUES

Never make predictions, especially about the future.

—Casey Stengel

Jack Stark, Continental Resources' senior production executive, sat on a dais at the Hilton Americas hotel in Houston in early March 2013 before a standing-room-only crowd of hundreds of energy executives.

Just a few years earlier, industry leaders had little interest in what Stark or anyone else at Continental had to say. Now Stark was on a first-day panel at CERAWeek, one of the energy world's most important annual conferences.

Alongside Stark were senior executives of several big oil companies, including Schlumberger, Apache Corp., and Statoil ASA. Their topic—the latest developments in oil production from "tight" rock, including North Dakota's Bakken formation—was among the hottest on the day's agenda. It was a sign of how much had changed for Stark, Continental, and the energy business.

Stark and his fellow panelists received loud applause when they finished, leaving Stark in great spirits. Descending the podium, he walked straight to the conference's main hall, where hundreds of industry members had gathered, looking for someone to chat with. He didn't recognize a soul, however. There were executives from China, the Middle East, and South America. But Stark couldn't find a familiar face in the entire room. He walked the length of the hall once, but didn't see anyone he knew. He searched again—nothing.

Eventually, Stark found someone, though she was no oil executive, engineer, or geologist. She was a reporter for a trade magazine.

Stark had received a glimpse into the future of the shale revolution. For more than a decade, headstrong wildcatters worked to discover techniques to extract oil and gas from shale and other challenging bedrock in the United States. Now the focus is shifting to whether energy can be pumped from shale deposits around the world.

Some foreign locales are packed with even more oil and gas than American formations, raising the possibility that the shale gale will have a tailwind affecting global economics and geopolitics. But there are good reasons to be skeptical about whether many foreign formations will produce significant amounts of energy anytime soon. For now, the new age of energy plenty appears to be a distinctly American phenomenon, creating the potential for a new era of U.S. economic outperformance.

Europeans have been worried for years about where they might find future energy supplies. Almost every country on the continent is a net gas importer, and many resent a reliance on countries like Russia and Algeria, which in the past have turned off their spigots amid pricing disputes.

Now they have reason to be excited about ending that dependency.

An estimated 470 trillion cubic feet of gas is buried in various European shale formations, or the equivalent of about thirty years of gas use on the continent, even excluding Russia or the Ukraine, which have nearly as much gas.

The continent has far greater population density than the United States, however, and more powerful environmental lobbies. France, Bulgaria, and a few other nations have already banned hydraulic fracturing due to concerns about environmental damage. A lack of experienced drilling crews and limited pipeline and other infrastructure also handicaps Europe. A sign of how little is going on is that the continent had about seventy rigs operating as of 2013, according to brokerage firm Raymond James, compared with seventeen hundred in the United States.

Poland, a fiercely independent nation that's been under Russia's thumb for centuries, has been among the most eager nations to tap its own shale gas resources. Poland's natural gas prices were about 50 percent higher than those in the United States as of early 2013, another reason the country has pinned its hopes on shale drilling.

Early predictions pointed to an estimated 187 trillion cubic feet of proven, recoverable shale reserves in Poland, enough to satisfy the nation's needs for three hundred years at current levels of demand. The reserves, which lie under Warsaw and Lublin and stretch all the way to the country's Baltic Sea coast, were considered among the largest in Europe outside Russia.

Major players like ExxonMobil, ConocoPhillips, Marathon Oil, and others raced to Poland to buy up acreage in 2010, and the Polish government subsidized the development and handed out exploratory concessions covering a third of the country. The first flare from a shale gas well, dubbed the "Flame of Hope," was religiously symbolic to many in the predominantly Catholic nation. The Polish foreign minister spoke about the country becoming "a second Norway," referring to Poland's wealthy, oil-rich neighbor to the north.

Air quickly came out of the Polish bubble. In early 2012, estimates for the country's shale reserves were cut by 90 percent. The richest parts of Poland's shale formations are so deep, at over sixteen thousand feet, and the shale has such high silicon content that extraction has proved difficult.

ExxonMobil, ConocoPhillips, and others beat a hasty retreat from the country, though at least part of the decision likely stemmed from a desire by some to avoid offending Russia, one of the largest energy players in the world. Chevron and others still seem committed to Poland's shale formations, but by the spring of 2013 fewer than fifty wells had been drilled there.

The United Kingdom is a likely participant in the shale energy revolution, at least at some point, for a number of good reasons. The nation is seeing a marked slowdown in production from its prized North Sea oil beds. The country has been importing gas since 2004, relying on Norway and the Netherlands, and is in an increasingly precarious position when it comes to key energy supplies. In March 2013, Britain came within six hours of running out of natural gas entirely, the *Financial Times* reported, as wholesale gas prices surged to record levels.

Two tremors in the spring of 2011 around the town of Blackpool caused deep unease when they were linked to fracking efforts, leading to a ban on fracking. The government ended the ban in late 2012, however, and has announced tax breaks to kick-start shale drilling. Writing in the *Daily Telegraph* in August 2013, Prime Minister David Cameron argued that "if we don't back" fracking technology, "we will miss a massive opportunity to help families with their bills and make our country more competitive. Without it, we could lose ground in the tough global race."

The county of Lancashire is seen as a potential hotbed of shale activity, as is the area between Liverpool and Manchester, which was a major locus of the Industrial Revolution. That's led to visions of an industrial revival in the area. Acreage about thirty miles south of London has also sparked excitement.

It likely will be years before a true gusher of oil or gas results from this dense rock, however. For one thing, it's still unclear how much gas is trapped in the nation's shale. Some estimates have placed the country's recoverable gas reserves at a puny 26 trillion cubic feet; others judge there to be more than 130 or even 200 trillion cubic feet, enough to power Britain's homes for over a hundred years.[4]

So far, few major, international energy producers have drilled in the country. Cuadrilla Resources, which shifted its sites to the UK after the cost of acreage in Pennsylvania proved too expensive, was one of the nation's only shale gas driller as of the summer of 2013. The company, now chaired by Lord Browne, the controversial former head of oil giant BP, remains among the most bullish on the nation's shale prospects.

The hurdles are high, though. In Britain, mineral rights don't belong to landowners, as they do in the United States, but to the Crown. That makes licensing agreements to drill the rock very complex. Cuadrilla had drilled just six wells in the UK as of 2013. Britain is one of the world's most densely populated countries, and unease among the populace remains high. When an energy executive visited Leeds University in late 2012, a group of protestors disrupted the lecture and mooned the speaker through the window, with "FRACK-OFF" written on their bums, a sign of the popular unease that likely will slow drilling. In August 2013, Cuadrilla suspended oil-drilling activity in Balcombe, a village south of London, after local police warned of threats against the exploration site, though drilling resumed within days.

It didn't help that in 2013 Greenpeace taped someone it claimed was a Cuadrilla public relations executive saying, "I know that everything I say sounds like utter, fucking bullshit," as he tried to argue that his company could drill safely in the densely populated region.

If environmental concerns can be addressed, the UK likely will pursue shale drilling, if only to help the nation shift from its growing dependence on coal, a dirtier energy source. The country relies on coal for more than 40 percent of its electricity needs, up from 30 percent in 2011.[5]

In fact, use of coal has grown even as the country's production has slipped. The United Kingdom had to import nearly forty-five million tons of coal in 2012, up almost 40 percent in a year, according to the UK Department of Energy and Climate Change. Coal use is expected to fall, but for now the UK indeed is carrying—and importing—coal to Newcastle.

There's more excitement over Mexico's potential to become a serious shale producer, partly because the country shares similar geology with its northern neighbor. Mexico boasts 545 trillion cubic feet of recoverable shale gas, the sixth largest such reserve in the world, according to the U.S. Energy Information Administration. There's also a serious amount of shale oil buried in Mexican rocks, helped by the fact that the Eagle Ford formation that's proved so prolific in South Texas extends well into northern Mexico. Mexican officials have pushed to exploit the country's shale as the nation's net energy imports grow and a nation that once was an oil power frets about its shrinking stature in the energy world.

But there's real concern that Mexico, which suffered a severe drought as recently as 2012, doesn't have sufficient water supplies to expend on fracking. Just as important, Mexico's state-owned energy power, Petróleos Mexicanos (PEMEX), doesn't yet have the resources to drill in shale and has resisted working with U.S. oil and gas companies.

Foreign energy companies have been reluctant to make meaningful commitments to Mexico because they haven't been allowed to gain any ownership of the company's shale wells, though the government says it is committed to making that possible. There's also limited independent data to verify how much gas is in the country.

Cheap gas might easily be imported from nearby fields in the United States, also reducing the urgency for Mexico to develop its own shale resources. As a result, fewer than a dozen wells have been drilled on the Mexican side of the border so far, with just a handful more expected by 2016.

When an American geologist named Charles Edwin Weaver visited Argentina's western Neuquén province in 1931, he got a good look at a unique outcropping of shale that was black in color, but speckled with white. The formation, in the arid Patagonian province, resembled the hide of a cow, so Weaver named it the Vaca Muerta, or dead cow.

Today, the Vaca Muerta and other Argentinian shale basins have spurred more excitement than those of almost any other country. Argentina has an estimated 802 trillion cubic feet of recoverable shale gas, the second most in the world and enough to satisfy the nation's consumption for a remarkable five hundred years, experts say. The country's shale deposits are believed to hold twenty-seven billion barrels of oil, the fourth most in the world. Not only that, but Argentina's formations resemble some of the most prolific areas in the United States. "The rock looks quite similar to the Eagle Ford," says Mark Papa of EOG Resources.

Foreign investment was chilled when President Cristina Kirchner expropriated a 51 percent stake in Argentina's state-run energy company, YPF SA, from Spanish energy giant Repsol in 2012. But in the summer of 2013, Chevron agreed to fund most of a $1.5 billion joint venture with YPF to develop the country's shale deposits.[6]

Argentina's ongoing dispute with Repsol could continue to dissuade foreign companies from investing in the country, and the cost of fracking and drilling Argentinian shale can be twice that of U.S. rock. "It's too early to form a judgment about whether it will be commercial," Papa says. "It's the first inning."

Russia is sitting on an estimated 285 trillion cubic feet of gas, as well as a humongous oil formation, the Bazhenov in western Siberia. It may be eighty times larger than the Bakken and stretches over an area larger than Alaska and California combined. The Bazhenov could supply hundreds of billions of barrels of oil, explaining why Exxon and others have been trying to gain a foothold in the area.

It may be a while before Russia gets around to tapping its shale reserves, however. Its reserves of conventional energy are so vast that the country hasn't seen a need to focus on shale.

Even the most ardent environmentalist hoping to contain climate change through renewable energy might feel tempted to root for shale development in China. The nation, the world's second largest economy, burns coal like it's a national pastime. Coal accounts for as much as 80 percent of Chinese electricity production, resulting in persistent air pollution and reduced life expectancy.

China's greenhouse gas emissions are almost twice those of the United States, and they're growing at more than 8 percent a year. There's really no way to make headway against global warming unless China can be weaned off coal, but the country is expected to boost coal-fired power by twice the total generating capacity of India by 2020. At that point, China will emit greenhouse gases at four times the rate of the United States. Even if American emissions somehow were to disappear, global emissions would be back to current levels in four years, due to China's growth alone, according to Elizabeth Muller, cofounder of Berkeley Earth, a nonprofit organization focused on climate change.

Muller and others argue that the United States should take a more aggressive stance on helping China tap its shale. China has the largest measured shale gas reserves in the world at 1,115 trillion cubic feet, according to the Energy Information Administration, as well as the third largest reserves of shale oil in the world. In 2012, China's state council opened the gates to limited foreign investment in local shale plays, and the government has been a fan of shale drilling. Chinese companies have invested billions in U.S. shale plays, hoping to learn the tricks of the trade.[7]

It won't be easy to shift to shale, however. China's more challenging rock formations require more water to frack each well compared with the United States. That's a problem for a country with a water system that is already strained. China also doesn't have the pipelines, service companies, or other infrastructure necessary to ramp up shale production. It doesn't help that most of China's shale is in difficult, mountainous terrain that's prone to earthquakes.

Foreign nations lack perhaps the key element behind the U.S. energy revolution: an entrepreneurial culture and ample incentives for the years of trial and error necessary for shale breakthroughs. George Mitchell, Harold Hamm, Mark Papa, and other headstrong wildcatters persevered because they knew they could gain both fame and remarkable fortune finding economic ways to tap shale. Comparable prizes don't always exist in other countries, where governments can play a larger role in society.

The United States also boasts an extensive energy infrastructure, such as pipelines and elaborate databases of underground geology, deep capital markets to finance newfangled drilling, more rigs than anyone else, collection and storage facilities, and an experienced labor force.[8]

The U.S. legal system gives individuals ownership of mineral rights under their land and the ability to lease the rights to others. That has accelerated drilling in comparison with foreign nations, where mineral rights are controlled by slow-moving governments. And the United States benefits from light population density in places like North Dakota and Texas, where much of the richer shale beds are located.

There are a few things the United States seems to do better than anyone else, such as create computer apps, drones, and rap stars. Fracking, so far, has been another area where there's a distinct American advantage.

It may be a while before oil and gas production surges around the world, but the shale revolution already is impacting global geopolitics.

Once, the fate of the United States, like dozens of other nations, was dictated by having full access to Middle Eastern oil. Starting in the 1970s, embargoes by members of the Organization of the Petroleum Exporting Countries (OPEC)—and even the threat of such action—restricted U.S. foreign policy, forcing leaders to commit precious resources to keep the region secure.

Oil never was the sole reason why America entered into conflicts in the Middle East, of course, but a series of American leaders has acknowledged the imperative of keeping crude from Saudi Arabia and other Gulf states flowing.

"An attempt by any outside force to gain control of the Persian Gulf region will be regarded as an assault on the vital interests of the United States of America," President Jimmy Carter declared in his 1980 State of the Union address. "Such an assault," Carter said, "will be repelled by any means necessary, including military force."

The first U.S.-led invasion of Iraq came after the nation took control of Kuwaiti oil fields. It's no coincidence that it also came at a time when a consensus was building around the peak oil theory and the view that oil production was in terminal decline.

"Kuwait was about Iraq gaining too much control of oil," says Dennis C. Blair, the retired U.S. Navy admiral who served as the country's third director of national intelligence under President Obama. "American military involvement in that part of the world really all goes back to oil, and it's all because the Saudis and Gulf states were swing producers."

As recently as 2006, fears abounded that America's global influence was waning and that the country was beholden to various Middle Eastern states, few of whom shared many values with the Western democracies.

Now that's all changing, as is the perspective of many U.S. political and military leaders. Exploding oil and gas production from U.S. fields has reduced American oil imports from OPEC members by about 25 percent since 2009. Today, the United States gets about 8 percent of its oil from the Middle East; these imports could be reduced to a mere trickle in the next few years. The United States has already become a net exporter of refined petroleum products, including gasoline, for the first time in decades. Growing crude output from Canadian oil-sands fields means the United States has another friendly supplier with which to replace oil from the Middle East and other countries.

It all means the United States' economic future is less dependent on energy-producing countries such as Saudi Arabia, Russia, Iran, Turkmenistan, and Venezuela, all of which have at times had strained relations with various Western nations.

America's new bounty likely gives it more flexibility to conduct policy as it sees fit. Already, new U.S. oil supplies are a key reason the country was able to engineer a boycott of Iranian oil and ratchet sanctions on the country to try to discourage it from developing nuclear weapons.

In 2012, Secretary of State Hillary Clinton created a dedicated energy bureau within the State Department and said energy would play a key role in future diplomacy. The United States could push its allies to develop their own shale resources, so they, too, can be weaned away from energy supplies from unstable and unfriendly nations.

If U.S. oil production continues to grow, the country could further reduce its involvement in the Middle East, helping it avoid costly conflicts, likely saving the United States money and lives and perhaps improving relations in that volatile part of the world. Heavy involvement in the Middle East succeeded in "keeping oil flowing, but we built huge resentment in the region that played a big role in the September 11 attacks and its cost us enormously," Admiral Blair says.

To be sure, America likely will always have extensive involvement in the region. Unlike natural gas, oil prices are set by global buyers and sellers, so additional American production won't be enough to keep American prices low or to allow the nation to ignore the next flare-up in the Middle East. As long as we care about oil, we're going to care about the Middle East. Just as important, America's allies have huge stakes in keeping the region as calm as possible, with oil pumping, necessitating continued American involvement.

But in an age of rising oil and gas supplies and growing budget limitations, America likely will be less willing to get bogged down in the dangerous region or pay deference to the interests of Middle Eastern energy producers. The United States has already reduced its aircraft carrier fleet operating around the Strait of Hormuz, which connects the Persian Gulf to international oil markets and has been a flash point with Iran since the late 1970s.

Wild cards abound in the new era. Nations that are more dependent on Middle Eastern oil, such as China, may be asked to assume more of the burden of keeping the region stable. It's not yet clear how China might handle an increased role on the international stage. And conflicts could arise as traditional energy powers, such as Russia, Iran, and Venezuela, lose some of their influence and feel threatened.

Still, the shifts brought by growing oil and gas production in U.S. shale formations likely will make the country more secure and could spark a shift in power away from OPEC and Russia.

The unexpected and extraordinary shale revolution, which is affecting just about anyone who heats a home, flips a light switch, or drives a car, is a reminder and reaffirmation of America's enduring greatness. But the antagonism and animosity generated by the ongoing fracking debate raise disturbing questions about the nation's future.

The great leap forward should have involved alternative energy, not oil and gas. The U.S. government allocated over $150 billion to green initiatives between 2009 and 2014, according to the Brookings Institution, including money for wind farms, solar panels, and other renewable energy sources. Investors from Silicon Valley and Wall Street poured billions of their own into alternatives. Overall, the world has made more than $2 trillion of investments in renewable-energy projects over the past twenty years, according to the International Energy Agency.

There's too little to show from the investments, however. Cars don't run on waste, and wind and solar aren't yet ready to power the world.

Instead, a group of frackers, relying on market cues rather than government direction, achieved dramatic advances by focusing on fossil fuels, of all things. It's a stark reminder that breakthroughs in the business world usually are achieved through incremental advances, often in the face of deep skepticism, rather than government-inspired eureka moments.

George Mitchell's team spent seventeen frustrating years trying to get meaningful amounts of gas from shale, Harold Hamm's men failed to pump much oil out of the Bakken until 2007, while Charif Souki's company was on its last breath before he seized on the idea of exporting gas. Their achievements are a reminder of the role of perseverance and obstinance in history's advances. Breakthroughs in alternative energy will come, but they, too, will take trial and effort.

The successes of the architects of the shale era are attributable to creativity, bravado, and a strong desire to get really wealthy. It doesn't get more American than that. Indeed, while the huge rewards promised in the market-driven American economy have led to an unfortunate income divide, they also provide incentive for remarkable achievement.

Two of the largest energy deposits of the past decade were discovered by sons of Greek immigrants—George Mitchell and Michael Johnson. Each met frustrations earlier in life and won acclaim after their seventy-fifth birthdays, becoming true embodiments of the American dream.

For all the criticism the country has fielded for losing its edge in innovation, surging American energy production is a reminder of the deep pools of ingenuity, risk taking, and entrepreneurship that remain in the country. Author Niall Ferguson and others argue that Western civilization has entered a period of decline, but many smaller American towns are experiencing a rebirth, with some young people in the energy business enjoying six-figure salaries, suggesting an underlying resilience in a country still recovering from the deep economic downturn.

Sadly, the story of the nation's energy rejuvenation also is a reminder of how divided the nation has become on almost every topic of importance. When it comes to energy production from shale formations, one camp argues that fracking poisons and should be abolished, while the other snickers at legitimate health concerns while clinging to the mantra of "drill, baby, drill."

As with other raging political, social, and economic debates, the nation would be best suited edging back toward the forsaken middle ground, finding ways to work together for the greater good.

ACKNOWLEDGMENTS

I had the privilege of spending over one hundred hours with George Mitchell, Harold Hamm, Mark Papa, and other architects of the shale revolution, and for that I'm appreciative. Nicholas Steinsberger, Kent Bowker, Dan Steward, Ken Bowdon, Terry Engelder, Michael Johnson, James Kochick, James Henry and others who laid the groundwork for the remarkable changes to the country were patient and helpful.

Critics of the industry, who play a crucial watchdog role, also were generous with their time. I'm just as thankful to the countless individuals I met as I crisscrossed the nation—those who work in the energy industry and others impacted by it. They took the time to lend their perspective, explain the difference between shale oil and oil shale, or just point a lost reporter to a local motel.

My publisher, Adrian Zackheim, had boundless enthusiasm for this book, while my editor, Maria Gagliano, lent expert insight and judgment. Bruce Giffords, Roland Ottewell, and Ingrid Sterner provided ace editorial production, copyediting, and fact checking assistance, respectively.

Deep appreciation and thanks go to Moshe Glick, who had the original idea for this book, well before most had heard of fracking, and encouraged me to stay with it. I could not have completed this project without Scobe and Hal Lux, who patiently answered questions and scrutinized sections of the manuscript at all hours of the day. Industry experts, including Ed Morse and Scott Anderson, lent crucial perspective.

I'm indebted to Doni Bloomfield, a truly remarkable and indefatigable research assistant, as well as Rachel Louise Ensign, who gave crucial and appreciated research assistance. I'm also grateful for the invaluable counsel and critiques of colleagues, former colleagues, friends, and family members, including Ezra Zuckerman Sivan, Vanessa O'Connell, Brad

Reagan, Ron Pollack, Erik Mielke, Karen Richardson, Liam Pleven, Craig Karmin, Doni and Eric Landy, Josh Marcus, Susie Nussbaum, Harold Simansky, Adam Brauer, Robin Sidel, and John Phillips.

I'd also like to thank the *Wall Street Journal*'s managing editor, Gerard Baker, and Rebecca Blumenstein and Matt Murray, deputy editors in chief of the paper, for giving their blessings for this project. Heartfelt thanks to Francesco Guerrera, the editor of the paper's Money and Investing section, who was especially supportive.

Thanks go to Miles Davis, Paul Kelley, Liam Finn, Neil Young, Kathleen Edwards, Simon and Garfunkel, and Yaz for keeping me company very, very late at night.

My mother, Roberta Zuckerman, and late father, Alan Zuckerman, were the best parents a young man could hope for. Their lessons and love guide and propel me.

Last but not least, I thank my wife, Michelle, for showing me so much patience, understanding, and extraordinary support during the course of this project. Readers also owe her some gratitude for somehow convincing me to keep the book under five hundred pages. My love and appreciation go to the best fracking boys in the world, Gabriel Benjamin and Elijah Shane. You kept me company as I wrote, played catch with me when I needed a break, and rejuvenated me when I lagged. You'll never know how much joy and happiness you bring me, every single day.

NOTES

INTRODUCTION

1. Stacey Vanek Smith, "North Dakota, Land of Jobs," Marketplace.org, October 18, 2011; Craig Karmin and Gregory Zuckerman, "A Boomtown Is Born in North Dakota," *Wall Street Journal*, November 14, 2012.

CHAPTER ONE

1. Laura Elder, "Billionaire Looks to Conservation," *Galveston County Daily News*, January 2, 2005.
2. Joseph W. Kutchin, "How Mitchell Energy & Development Got Its Start and How It Grew," self-published, January 1, 1998.
3. Ibid.
4. Diana J. Kleiner, "Smith, Robert Everett," Texas State Historical Association, *Handbook of Texas Online*.
5. Kutchin, "How Mitchell Energy & Development Got Its Start and How It Grew."
6. Elder, "Billionaire Looks to Conservation."
7. Ibid.
8. Kleiner, "Smith, Robert Everett."
9. Barbara Shook, "Cracking the Code on the Barnett Shale: How It Happened," *World Gas Intelligence*, July 8, 2009.
10. Ibid.
11. Ann Zimmerman, "A Lot of Gas," *Dallas Observer*, May 15, 1997.
12. Dan Steward, *The Barnett Shale Play: The Phoenix of the Fort Worth Basin; A History* (Fort Worth and North Texas Geological Societies, 2007).
13. Andrew Maykuth, "Tapping Shale, Seeking Sustainability: A Rare Oilman," *Philadelphia Inquirer*, August 29, 2010.

CHAPTER TWO

1. Robert P. Hauptfuhrer, "The Story of Oryx Energy Company: Continuity and Change," Oryx Energy Company, 1994.
2. Andrew Cassel, "Retirement Spurs Shift at the Helm of Sun Co.," *Philadelphia Inquirer*, November 7, 1986.

3. Daniel Yergin, *The Prize: The Epic Quest for Oil, Money and Power* (New York: Touchstone/Simon & Schuster, 1991).

4. G. B. Morey, *The Search for Oil and Gas in Minnesota*, Minnesota Geological Survey 6 (St. Paul: University of Minnesota Press, 1984).

5. "Horizontal Drilling Method Could Wring Lots of Oil Out of Old Fields," Reuters, October 7, 1989.

6. "Austin Chalk Getting Another Look," *Explorer*, July 2012, http://www.aapg.org/explorer/2012/07jul/austin_chalk0712.cfm.

7. "Oryx's Hauptfuhrer: Big Increase Due in U.S. Horizontal Drilling," *Oil and Gas Journal*, January 15, 1990.

8. Jack Willoughby, "Putting on Heirs," *Institutional Investor*, March 1, 1997.

9. "Can Oryx Energy Cap Its Gusher of Red Ink?," *BusinessWeek*, November 6, 1994.

10. Gregg Jones, "Gamble in the Gulf," *Dallas Morning News*, December 26, 1994.

11. Ibid.

CHAPTER THREE

1. Caleb Solomon and Robert Johnson, "Lone Star Legend: One Tycoon in Texas Still Is Dreaming Big, Even If It's Out of Style," *Wall Street Journal*, July 28, 1993.

2. Leonardo Maugeri, "Oil: The Next Revolution," Belfer Center for Science and International Affairs, Harvard Kennedy School, June 2012.

3. Daniel Yergin, *The Prize: The Epic Quest for Oil, Money and Power* (New York: Touchstone/Simon & Schuster, 1991).

4. Solomon and Johnson, "Lone Star Legend."

5. Gary Peach, "Estonia Eager to Teach World About Oil Shale," Associated Press, June 1, 2013.

6. Robert Johnson and Allanna Sullivan, "Mitchell Energy Picks William Stevens to Be Its President, Operations Chief," *Wall Street Journal*, December 13, 1993.

CHAPTER FOUR

1. Ann Zimmerman, "A Lot of Gas," *Dallas Observer*, May 15, 1997.

2. Russell Gold, "The Man Who Pioneered the Shale-Gas Revolution," *Wall Street Journal*, October 23, 2012.

3. Dan Steward, *The Barnett Shale Play: Phoenix of the Fort Worth Basin; A History* (Fort Worth, TX: Fort Worth Geological Society/North Texas Geological Society, 2007).

4. Daniel Yergin, *The Quest: Energy, Security, and the Remaking of the Modern World* (New York: Penguin Press, 2011).

5. Daniel Yergin, "Stepping on the Gas," *Wall Street Journal*, April 2, 2011.

6. Jim Fuquay, "Q&A, George Mitchell, Founder of Mitchell Energy," *Fort Worth Star-Telegram*, April 2, 2008.

CHAPTER FIVE

1. Jerry Shottenkirk, "OKC-Based SandRidge Energy CEO Tom Ward Says He Has Much to Be Thankful For," *Oklahoma Journal Record*, January 8, 2007.
2. Michael W. Sasser, "His Own Terms," *Oklahoma Magazine*, December 2012.
3. Jerry Shottenkirk, "Hard Work, Luck Make Billions for Oklahoma Executive," *Journal Record*, August 13, 2007.
4. Mary A. Fischer, "The FBI's Junk Science," *GQ*, January 2001.
5. David Geffen, "Jewish Black Gold?" *Jerusalem Post*, July 3, 2008.
6. Dan Piller, "A Tale of Two Fields: Giddings Offers a Lesson in How Quickly Things Can Change," *Fort Worth Star-Telegram*, March 7, 2006.
7. Gretchen Morgenson, "Pie in the Sky," *Forbes*, December 16, 1996.
8. Rich Robinson, "Oil Company Grows to Maturity," *Oklahoman*, July 21 2002.

CHAPTER SIX

1. Keith A. Eaton, "Harold Hamm: All Cattle and No Hat," *Distinctly Oklahoma*, November 2010.
2. Ibid.
3. Harold Hamm, "Birth of a Wildcatter," *Forbes*, December 24, 2012.
4. Steven Mufson, "For Oil Driller Harold Hamm, Bakken Boom Brings More Billions and a Chance to Dabble in Politics," *Washington Post*, August 12, 2012.
5. David Zizzo, "Unusual Feature Lies Under Ames," *Oklahoman*, May 4, 2010.
6. Joshua Schneyer, Brian Grow, and Jeanine Prezioso, "Lack of a Prenup Imperils Oil Billionaire's Fortune," Reuters, June 14, 2013.
7. Eileen and Gary Lash, "Kicking Down the Well," American Association of Petroleum Geologists Web site, September 2011.
8. Bryan Burrough, *The Big Rich: The Rise and Fall of the Greatest Texas Oil Fortunes* (New York: Penguin, 2009); "H. L. Hunt Leaves $5 Billion—and Two Sets of Heirs to Claim It," *People*, December 16, 1974.
9. Ruth Sheldon Knowles, *The Greatest Gamblers: The Epic of American Oil Exploration* (Norman: University of Oklahoma Press, 1980).

CHAPTER SEVEN

1. James Norman, "Petrodollars," *Platts Oilgram News*, July 6, 1998.

CHAPTER EIGHT

1. Brian Grow and Anna Driver, "Chesapeake Board Member Lent Money to CEO McClendon," Reuters, April 27, 2012.
2. Shashana Pearson-Hormillosa, "Facetime with XTO Energy's Bob Simpson," *Dallas Business Journal*, June 28, 2009.

3. Bryan Gruley, "Wildcatter Finds $10 Billion Drilling in North Dakota," Bloomberg News, January 19, 2012.

CHAPTER NINE

1. Sean Murphy, "Volatile Gas Industry Takes Chesapeake on Roller Coaster Ride," Associated Press, November 20, 2005.
2. Asjylyn Loder, "McClendon Eating Healthy No Help in Bet Undermining Chesapeake," Bloomberg News, June 27, 2012.
3. Russell Gold and Daniel Gilbert, "The Many Hats of Aubrey McClendon," Wall Street Journal, May 7, 2012.
4. Ibid.
5. John Shiffman, Anna Driver, and Brian Grow, "Special Report: The Lavish and Leveraged Life of Aubrey McClendon," Reuters, June 7, 2012.
6. Ken Kolker, "Reserved for the Rich, or Preserved for All?," Grand Rapids Press, December 16, 2007.
7. Shiffman, Driver, and Grow, "The Lavish and Leveraged Life of Aubrey McClendon."
8. David Graham, "McClendon Gift Raises Eyebrows," Chronicle (Duke University), March 8, 2007.
9. Jerry Shottenkirk, "OKC-Based SandRidge Energy CEO Tom Ward Says He Has Much to Be Thankful For," Journal Record, January 8, 2007.
10. Ibid.
11. Eric Konigsberg, "Kuwait on the Prairie: Can North Dakota Solve the Energy Problem?," New Yorker, April 25, 2011.
12. Ibid.
13. John J. Fialka, "Wildcat Producer Sparks Oil Boom on Montana Plains," Wall Street Journal, April 5, 2006.
14. Luisa Kroll, "Harold Hamm on Diabetes," Forbes, October 11, 2010.
15. Joshua Schneyer, Brian Grow, and Jeanine Prezioso, "Special Report: Lack of a Prenup Imperils Oil Billionaire's Fortune," Reuters, June 14, 2013.
16. Colin Campbell, ed., Peak Oil Personalities: A Unique Insight into a Major Crisis Facing Mankind (Skibbereen, Ireland: Inspire Books, 2012).

CHAPTER TEN

1. Asjylyn Loder, "McClendon Eating Healthy No Help in Bet Undermining Chesapeake," Bloomberg News, June 27, 2012.
2. Colin Campbell, ed. Peak Oil Personalities: A Unique Insight into a Major Crisis Facing Mankind (Skibbereen, Ireland: Inspire Books, 2012).
3. Steve Toon, "The Dash for Cash," A&D Watch, February 2008.
4. Jerry Shottenkirk, "Hard Work, Luck Make Billions for Oklahoma Executive," Journal Record, August 13, 2007.

5. Sean Murphy, "Volatile Gas Industry Takes Chesapeake on Roller Coaster Ride," Associated Press, November 20, 2005.

6. Christopher Helman, "The Sordid Deal That Created the Okla. City Thunder," *Forbes*, June 13, 2012.

7. John Shiffman, Anna Driver, and Brian Grow, "Special Report: The Lavish and Leveraged Life of Aubrey McClendon," Reuters, June 7, 2012.

8. Ana Campoy, "Natural-Gas Producers Cut Output," *Wall Street Journal*, September 6, 2007.

9. John J. Fialka, "Wildcat Producer Sparks Oil Boom on Montana Plains," *Wall Street Journal*, April 5, 2006.

10. Joe Carroll, "Peak Oil Scare Fades as Shale, Deepwater Wells Gush Crude," Bloomberg News, February 6, 2012; Daniel Yergin, *The Quest: Energy, Security, and the Remaking of the Modern World* (New York: Penguin Press, 2011).

CHAPTER ELEVEN

1. Christopher Helman, "In His Own Words: Chesapeake's Aubrey McClendon Answers Our 25 Questions," *Forbes*, October 5, 2011.

2. Joshua Schneyer, Jeanine Prezioso, and David Sheppard, "Inside Chesapeake, CEO Ran $200 Million Hedge Fund," Reuters, May 2, 2012.

3. Ryan Dezember, "Texas Oil Man Finds a New Groove," *Wall Street Journal*, June 15, 2012.

4. Margaret Cronin Fisk, "Chesapeake Loses Bid to Void Texas Oil, Gas Rights Award," Bloomberg News, September 12, 2012.

5. Clifford Krauss and Eric Lipton, "After the Boom in Natural Gas," *New York Times*, October 20, 2012.

6. Joe Carroll, "Exxon Quits 2nd-Biggest U.S. Gas Area Amid Price Drop," Bloomberg News, October 2, 2008.

7. Fareed Zakaria, "Why We Can't Quit," *Newsweek*, March 15, 2008.

CHAPTER TWELVE

1. Christopher Helman, "In His Own Words: Chesapeake's Aubrey McClendon Answers Our 25 Questions," *Forbes*, October 5, 2011.

2. William D. Cohan, "The Man Who Walked Away from Goldman Sachs," *Fortune*, January 26, 2010.

3. Helman, "In His Own Words: Chesapeake's Aubrey McClendon Answers Our 25 Questions."

4. Ibid.

5. Russell Gold, "Margin Calls Spark New Wave of Sales," *Wall Street Journal*, October 14, 2008.

6. Joe Carroll, Jef Feeley, and Laurel Brubaker Calkins, "Chesapeake CEO Disavowed Role in 2008 Plunge, Sold Shares," Bloomberg News, June 27, 2012.

7. Jef Feeley, Laurel Brubaker Calkins, and Joe Carroll, "Chespeake's McClendon Said Company Overpaid for Leases," *Bloomberg News*, June 26, 2012.
8. Ben Casselman, "Credit Crunch and Sinking Prices Threaten Chesapeake Energy's Growth," *Wall Street Journal*, October 10, 2008.
9. Russell Gold, "Bad Call," *Wall Street Journal*, February 8, 2009.

CHAPTER THIRTEEN

1. Ben Casselman, "Chesapeake Energy CEO Defends Stewardship," *Wall Street Journal*, June 13, 2009.
2. Ben Casselman, "Chesapeake Under Fire from Shareholders over CEO Pay," *Wall Street Journal*, April 28, 2009.
3. Casselman, "Chesapeake Energy CEO Defends Stewardship."
4. Laura Legere, "Nearly a Year After a Water Well Explosion, Dimock Twp. Residents Thirst for Gas-Well Fix," *Scranton Times-Tribune*, October 26, 2009.
5. Nathan Vardi, "The Last American Wildcatter," *Forbes*, February 2, 2009.
6. Ibid.

CHAPTER FOURTEEN

1. Clifford Krauss and Eric Lipton, "After the Boom in Natural Gas," *New York Times*, October 20, 2012.
2. Peter Lattman, "K.K.R.'s Energy Billionaires Club," *New York Times*, November 25, 2011.
3. Shashana Pearson-Hormillosa, "Facetime with XTO Energy's Bob Simpson," *Dallas Business Journal*, June 28, 2009.
4. Russell Gold, "Oil and Gas Boom Lifts U.S. Economy," *Wall Street Journal*, February 8, 2012.
5. Selam Gebrekidan, "100 Years After Boom, Shale Makes Texas Oil Hot Again," *Reuters*, May 3, 2011.
6. Stephen Moore, "How North Dakota Became Saudi Arabia," *Wall Street Journal*, October 1, 2011.
7. Josh Harkinson, "Who Fracked Mitt Romney," *Mother Jones*, November/December 2012.
8. "Chiropractor Follows Patients to Oil Patch," *Talkin' the Bakken* magazine, June 2013.
9. Clifford Krauss, "U.S. Company, in Reversal, Wants to Export Natural Gas," *New York Times*, January 27, 2011.

CHAPTER FIFTEEN

1. Clifford Krauss and Eric Lipton, "After the Boom in Natural Gas," *New York Times*, October 20, 2012.

2. John Shiffman, Anna Drive, and Brian Grow, "Special Report: The Lavish and Leveraged Life of Aubrey McClendon," Reuters, June 7, 2012.

3. Scott Detrow, "More on DEP Tests: Hanger Downplays Radiation Threat," WITF.org, March 7, 2011.

4. "McClendon Values Utica Shale at Half a Trillion Dollars, NGI Reports," BusinessWire, September 21, 2011.

5. Christopher Helman, "Billionaire Wildcatter, Risk Addict Aubrey McClendon Has Bet It All on Shale," *Forbes,* October 6, 2011.

6. Russell Gold, "Board Turns on Chesapeake's CEO," *Wall Street Journal,* April 26, 2012.

7. Max Abelson, "Icahn, Icahn! Son's $2.9 M. PH, Lawyer's $4.2 M. Condo," *New York Observer,* October 13, 2009.

8. Miles Weiss and Zachary Mider, "Chesapeake CEO Pledges Mementos for Billionaire Kaiser's Debt," Bloomberg News, June 13, 2012.

9. Christopher Helman, "The Sordid Deal That Created the Okla. City Thunder," *Forbes,* June 13, 2012.

10. Asjylyn Loder, "McClendon Eating Healthy No Help in Bet Undermining Chesapeake," Bloomberg News, June 27, 2012.

11. Michael Erman, Anna Driver, and Brian Grow, "Insight: How SandRidge Energy's CEO Adapted the Chesapeake Playbook," Reuters, January 14, 2013.

EPILOGUE

1. Ben Lefebvre, "Gas Glut Favors Would-Be Exporter," *Wall Street Journal,* January 23, 2012.

2. Steven Mufson, "The Natural Gas Revolution Reversing LNG Tanker Trade," *Washington Post,* December 7, 2012.

3. Ben Lefebvre, "Cheniere CEO: Traders Needed as Natural Gas Appetite Grows," *Wall Street Journal,* March 7, 2013.

4. Tom Fowler, "U.S. Oil-Production Rise Is Fastest Ever," *Wall Street Journal,* January 18, 2013.

5. Jennifer Hiller, "EOG Resources: Eagle Ford Shale Is 'Steaming Ahead,'" *FuelFix,* May 16, 2013.

6. Frank Bass, "Eagle Ford Shale Boom Fuels 'Madhouse' in South Texas Counties," Bloomberg News, March 15, 2013.

7. John Kemp, "Is Bakken Set to Rival Ghawar?," Reuters, November 9, 2012.

8. Joshua Schneyer, Brian Grow, and Jeanine Prezioso, "Special Report: Lack of a Prenup Imperils Oil Billionaire's Fortune," Reuters, June 14, 2013.

9. Daniel Gilbert and Tom Fowler, "Chesapeake Investors Tired of the 'Aubrey Discount,'" *Wall Street Journal,* January 30, 2013.

10. Brianna Bailey, "Outgoing CEO McClendon Bids Chesapeake's Oklahoma City Workers Farewell," *Oklahoman,* March 28, 2013.

AFTERWORD

1. Laura Legere, "Sunday Times Review of DEP Drilling Records Reveals Water Damage, Murky Testing Methods," *Scranton Times-Tribune*, May 19, 2013.
2. Ibid.
3. Mead Gruver, "Wyoming Air Pollution Worse Than Los Angeles Due to Gas Drilling," Associated Press, March 8, 2011.
4. Cassie Werber and Sarah Kent, "U.K. Increases Estimates of Shale-Gas Reserves," *Wall Street Journal*, June 27, 2013.
5. Mark Seddon, "The Long, Slow Death of the UK Coal Industry," *Guardian*, April 10, 2013; Andrew Bounds, "Coal No Longer King but Still Important," *Financial Times*, April 21, 2013.
6. Taos Turner and Daniel Gilbert, "Chevron, YPF Sign $1.5 Billion Shale-Oil Deal," *Wall Street Journal*, July 16, 2013.
7. Elizabeth Muller, "China Must Exploit Its Shale Gas," *New York Times*, April 12, 2013; Edward Wong, "Pollution Leads to Drop in Life Span in Northern China, Research Finds," *New York Times*, July 8, 2013.
8. The Boston Company, *End of an Era: The Death of Peak Oil*, February 2013.